DYNAMIC PROCESSES IN THE CHEMISTRY OF THE UPPER OCEAN

NATO CONFERENCE SERIES

I   Ecology
II   Systems Science
III   Human Factors
IV   Marine Sciences
V   Air–Sea Interactions
VI   Materials Science

IV MARINE SCIENCES

Recent volumes in this series

DYNAMIC PROCESSES IN THE CHEMISTRY OF THE UPPER OCEAN

Edited by

J.D. Burton
University of Southampton
Southampton, United Kingdom

P. G. Brewer
Woods Hole Oceanographic Institution
Woods Hole, Massachusetts

and

R. Chesselet
Centre des Faibles Radioactivités
Laboratoire Mixte CNRS–CEA
Gif-sur-Yvette, France

Published in cooperation with NATO Scientific Affairs Division

PLENUM PRESS · NEW YORK AND LONDON

Library of Congress Cataloging in Publication Data

NATO Advanced Research Institute on Dynamic Processes in the Chemistry of the
   Upper Ocean (1983: Jouy-en-Josas, France)
   Dynamic processes in the chemistry of the upper ocean.

   (NATO conference series. IV, Marine sciences; v. 17)
   "Proceedings of a NATO Advanced Research Institute on Dynamic Processes in the
Chemistry of the Upper Ocean, held July 6–12, 1983, at Jouy-en-Josas, France"—
Verso t.p.
   "Published in cooperation with NATO Scientific Affairs Division."
   Includes bibliographic references and index.
   1. Chemical oceanography—Congresses. 2. Ocean-atmosphere interaction—
Congresses. 3. Oceanic mixing—Congresses. I. Burton, J. D. (James Dennis), 1931–
   . II. Chesselet, R. III. Brewer, P. G. IV. North Atlantic Treaty Organization. Scientific
Affairs Division. V. Title. VI. Series.
   GC110.N37 1983   551.46′01   86-22520
   ISBN-13: 978-1-4684-5217-4   e-ISBN-13: 978-1-4684-5215-0
   DOI: 10.1007/978-1-4684-5215-0

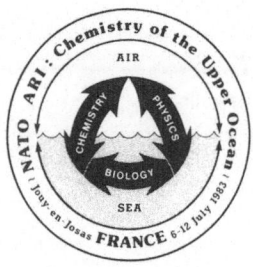

Proceedings of a NATO Advanced Research Institute on Dynamic Processes
in the Chemistry of the Upper Ocean, held July 6–12, 1983,
at Jouy-en-Josas, France

© 1986 Plenum Press, New York
Softcover reprint of the hardcover 1st edition 1986

A Division of Plenum Publishing Corporation
233 Spring Street, New York, N.Y. 10013

PREFACE

The Advanced Research Institute (ARI) on Dynamic Processes in the
Chemistry of the Upper Ocean had its origins in discussions by the NATO
Special Programme Panel on Marine Sciences during 1978 when a wide range of
topics for future ARIs was being considered. What was then envisaged was a
workshop on chemical aspects of the oceanic mixed layer, at which consideration would be given to the inputs, cycling and removal of material, and
the problems involved in the quantitative assessment of fluxes. It was
realised that any attempt to model chemical processes would need the active
collaboration of workers from other fields, especially physical oceanographers concerned with air-sea interaction and turbulence, and biological
oceanographers with expertise in primary productivity and the cycling of
particulate and dissolved organic material.

As plans for the ARI developed further a somewhat different emphasis
emerged, focused on the question as to how chemists should set about
observing an environment as variable and dynamic as the upper ocean and
selecting the appropriate scales for the framework of measurements to study
a particular process, especially in the light of current knowledge of
physical processes of transport and mixing. It was plain that the capability of physical oceanographic methods to resolve differences on small
spatial and temporal scales is considerably ahead of the capabilities of
biologists and chemists who rely upon discrete sampling and complex laboratory manipulations in order to obtain most of their data. The objective of the ARI thus became one of examining critically the status of
observations and experiments on the upper ocean and evaluating the possibilities for future development in chemical studies of this important
oceanic regime.

The Panel was fortunate in persuading Dr R. Chesselet to lead the
Steering Committee and to provide the excellent facilities for the workshop, which was held at Jouy-en-Josas, 6-12 July 1983, in warm summer
sunshine. The other members of the Steering Committee were Prof. R.T.
Barber, who could not attend the workshop due to seagoing commitments but
who contributed importantly to its development, Dr P.G. Brewer, Dr J.D.
Burton and Prof. C.G.H. Rooth.

The material presented in the workshop sessions was not grouped on a
traditional disciplinary basis but those papers submitted for publication
have been arranged rather more closely along those lines for easier access
by the reader. Rapporteurs presented summaries of syntheses of each session, and working groups in the areas of gas fluxes, particle fluxes and
dissolved species tackled the problem of identifying key points for future
research. Revised versions of the reports of the working groups are given
at the beginning of this volume.

The members of the Steering Committee are grateful to the Scientific Affairs Division of NATO for financial support of this ARI. They thank the Marine Science Division of the U.S. National Science Foundation and the Programme Interdisciplinaire de Recherches en Océanographie (PIROCEAN) (Centre National de la Recherche Scientifique et Direction de la Recherche/ Ministère de l'Education Nationale) for support of some participants. The excellent organizational support provided by the Centre des Faibles Radioactivités/Centre National de la Recherche Scientifique - Commissariat à l'Energie Atomique and the staff of the Centre d'Enseignement Supérieur des Affaires, Jouy-en-Josas, is also gratefully acknowledged.

The editors are grateful to the authors of the papers and the chairmen and rapporteurs of the working groups for their cooperation and patience during the preparation of this volume. They wish particularly to thank Miss S. Driver, Mrs D. Flood and Mrs J. Watson for their help in manuscript production, and the editorial staff of Plenum Press for their advice and assistance. The ARI logo was devised by Professor J. McN. Sieburth.

J.D.Burton

CONTENTS

REPORT OF WORKING GROUP I: GASES

INTRODUCTION

Numerous volatile components are involved in dynamic photochemical and biological processes in the upper layers of the oceans. As a consequence, the ocean surface can act as a significant source and, on occasion, as a sink for trace gases which play an important role in the chemistry of the atmosphere (e.g., CO, NO, N_2O). In some instances (e.g., $(CH_3)_2S$, CH_3I) the fluxes from the ocean to the atmosphere make significant contributions to the global cycling of the elements, and in others (e.g., CO_2) the flux into the ocean exerts an important control on atmospheric composition.

Over the past five years, advances in analytical techniques and sampling procedures have expanded the range of gases that can be analysed to encompass a growing number of biologically important trace gases. The transfer velocity for the exchange of non-reactive gases at the air-sea surface has also been measured at a number of sites and over a range of sea states and wind speeds. Corresponding laboratory studies have provided the basis for useful models for the exchange of non-reactive gases at the air-sea interface. A rationalization of the differences arising between the two sets of measurements is urgently required.

Recent photochemical and microbiological studies suggest a wide range of potential sources and sinks for trace gases and indicate that many gases might be involved in dynamic processes within the oceans. It is important to assess the influence of such reactivity on gas transfer theories and to use this information together with more realistic physical models, to gain a clearer picture of the processes controlling both the oceanic profiles of the trace gases and their transport across the air-sea interface. The significant amount of information now available from satellite studies on the distribution of sea-state and wind velocity will facilitate the application of such studies to global climate models and oceanic circulation models.

THE PHYSICAL FRAMEWORK FOR DISSOLVED GAS DISTRIBUTION

Physical Scales

The conservation of a chemical constituent can be modelled in a linear system using the equation

$$\frac{DC}{DT} = - \nabla.\vec{M}_c + S_{in} - S_{out} \tag{1}$$

where C is the concentration, $\vec{M}_c$ the molecular diffusive flux of C and S_{in} and S_{out} represent production and utilisation processes for C. In practice, values of C must be averaged over time and space scales appropriate to the process under consideration to provide mean $\bar{C}$ values. To this end, the concentration and velocity fields are separated into their mean and fluctuating parts $\bar{C}$, C', $\vec{U}$ and U'), to give

$$\frac{\partial \bar{C}}{\partial t} + \vec{U}.\nabla\bar{C} + \nabla.(\overrightarrow{U'C'}) = - \nabla.\vec{M}_c + \bar{S}_{in} - \bar{S}_{out} \tag{2}$$

This equation shows that the distribution observed for $\bar{C}$ develops in time and space in response to five factors:

(i) $U.\nabla\bar{C}$ *Advection* across the gradient of the near field by the mean velocity.

(ii) $\nabla.(\overrightarrow{U'C'})$ The divergence of the mean turbulent advection of fluctuations in C ("*turbulent diffusion*").

(iii) $- \overrightarrow{\nabla M_c}$ *Molecular diffusion* which can be important in non-turbulent boundary layers and at scales smaller than the turbulent velocity fluctuations.

(iv) $\bar{S}_{in}$ ⎱   These terms may be distributed throughout space and may
(v) $\bar{S}_{out}$ ⎰   vary in time.

The processes determining the mean gas distribution will depend on the relationships established between the time and space scales, physical dispersion processes, and the rates and location of the chemical and biological source and sink functions.

To focus attention on the salient features of the processes that influence the chemistry of the gases, we have regarded the upper ocean and the overlying gas phase as divided into a number of layers and have considered the likely time and space scales associated with each layer (Table 1).

A customary distinction is to recognise an oceanic mixed layer – a nearly homogeneous, actively mixing layer with rapid vertical exchange and typically extending to depths of 20 to 200 m. When it exists, a mixed layer may be either deeper or shallower than the euphotic layer. However, the working group felt that the concept of an oceanic mixed layer has been too much abused, and that it cannot be assumed that an actively mixed and mixing layer is ubiquitous. Observations show both small scale stratification within nearly homogeneous profiles and homogeneous layers with no significant microscale velocity, indicating no significant mixing. We feel that identification of the mixed layer is problematical and so we omit this layer in the classification scheme (Table 1).

Physical Models of Air/Water Gas Transfer

Comparison of laboratory and field studies. The main resistance to gas transfer is concentrated in a thin layer (thickness ∂) near the air/water interface (levels 3 and 4, Table 1). The resistance of this layer is given for unreactive gases by

$$r = \int_{o}^{\partial} dZ/(D + D_{eddy}(Z)) \tag{3}$$

where   Z   = vertical distance from the interface,
        D_{eddy} = eddy diffusivity due to turbulent motion,
        D   = molecular diffusion constant of the gas ($\sim$0.2 cm^2 s^{-1} in air and $\sim$10^{-5} cm^2 s^{-1} in water).

For exceedingly soluble gases or for gases that are involved in reactions in the aqueous phase the exchange is usually gas phase controlled (level 3, Table 1, ∂ = 3–5 μm) and a transport film model can be used. However, when

2

Table 1. Physical length – time scales relevant to upper ocean chemistry

Level	Vertical scale	Horizontal scale	Time scale	Comments
1 Free troposphere	10 km	10,000 km	1 – 10 day	
2 Planetary boundary layer	1 km	1,000 km	1 day	Ekman Layer
3 Atmospheric microlayer	3 mm	?	1 sec	
4 Oceanic microlayer	10 – 100 μm	?	1 – 10 sec	
5 Photic layer	10 – 100 m	3 – 300 km	1 – 100 day	a
5A Bubble penetration zone	3 – 10 m	3 – 100 m	3 – 10 sec	b
		100 km	1 hour	c
6 Seasonal thermocline	40 – 200 m	100 – 300 km	30 – 300 day	d
7 Lower thermocline	1,000 m	300 – 10,000 km	1 – 30 year	e
			1 – 300 year	f
			30 day	g

a Length scales of horizontal variation range from widths of upper layer fronts to widths of major current features. Timescales correspond to exchange times under conditions of moderate stratification and turbulence. For weak stratification and moderate wind speeds, Langmuir circulations can become established giving exchange times in the photic zone of the order of 10 minutes.

b Extent of bubble plunge, surface wavelengths/periods.

c Wind field fetch/variations.

d This timescale is associated with the fluctuation due to transient eddies and fronts, and establishment/destruction of seasonal thermocline. During periods of free convection a mixing layer may overturn in a few hours, deepening to as much as 200 m. If convection is not present, strong wind mixing in the upper layer can give exchange times of the order of 10 days.

e Period of gyre rotation.

f Interaction with surface (period of rotation x 'recirculation index').

g Fluctuation due to transients.

3

the transfer is liquid–phase controlled (e.g., sparingly soluble gases, such as He, N_2, O_2) (level 4, Table 1, ∂ = 10–100 μm) a solid wall model ($D_{eddy} \sim Z^3$) can be used at low wind speeds (0–3 m s^{-1}) where the water and air move in unison and a fluid wall model ($D_{eddy} \sim Z^2$) at higher wind speeds (3–10 m s^{-1}) where there is significant slip between the two phases and wave formation becomes important. In the wind regime up to 10 m s^{-1} there appears to be a linear relationship between transfer velocity ($U_x = r^{-1}$) and wind velocity (U) with a pronounced increase in slope when the fluid wall model becomes effective. Experiments with artificially generated wave patterns indicate the importance of surface roughness and show that fetch as well as wind speed must be considered. The occurrence of waves *per se* has no clear influence on gas exchange rates. No simple functional relationships have been deduced for the combined effects of wind speed and fetch on transfer velocity, even for laboratory studies, although a qualitatively similar behaviour may be deduced from field measurements. The complex effects of fetch on the open ocean, where inhomogeneities occur in the patterns of wind velocity and in the timescale of variability, are not adequately represented by steady–state wind tunnel experiments. No quantitative explanations are available for the persistently lower transfer velocities observed in the field in comparison with wind tunnel experiments at equivalent wind velocities.

At wind velocities greater than 10 m s^{-1} wave breaking, spray formation and bubble transport become important. Since much of the deep water is formed at high latitudes in the winter, processes occurring in this velocity domain might be of considerable importance. In the shallow water conditions simulated by wind tunnel experiments (<50 cm depth) only bubbles <500 μm in diameter are important. In the deep ocean such bubbles are infrequent but it is difficult to assess the influence of high wind velocities on gas transfer in the absence of field observations and in the face of sparse laboratory data. Experiments designed to compare the behaviour of highly and sparingly soluble gases (e.g., CO_2/O_2, N_2O/Ar) indicate that the gas transfer velocities show a similar behaviour for wind speeds <10 m s^{-1}. However once wave breaking and bubble formation become significant the transfer rates of the sparingly soluble gases show a much greater dependence on wind speed than those of the highly soluble gases.

Suggestions for laboratory studies. Further consideration should be given to studies of atmospherically important trace gases (e.g., CO, N_2O, O_3) and measurements of appropriate reactivity terms should be undertaken under environmentally realistic conditions. Studies of the combined effects of fetch and wind velocity for the range 3–10 m s^{-1} should be undertaken and consideration should be given to the design and interpretation of experiments involving non–steady state wind stress. The role of capillary waves in the transition from solid–wall to fluid–wall behaviour needs to be clarified and attention should be paid to the parameterization of surface roughness. Water depth (over the range 30-60 cm) does not appear to influence the results of experiments for wind velocities <10 m s^{-1}. This will certainly not be true for higher wind speeds. Consideration should be given to the design of suitable experiments for studying air/water transfer velocities at wind speeds >10 m s^{-1} with adequate water depth. Further studies are required to determine the influence of surface active components on gas transfer across the air/water interface.

Suggestions for field studies. Consideration should be given to the more accurate determination of transfer velocities. Field data are missing, particularly for wind speeds both <4 m s^{-1} and, most importantly, >10 m s^{-1}. The effect of bubbles and spray at high wind speeds should be studied by measuring the transfer for gases of distinctly different solubility (e.g., O_2 *versus* Rn). Presently available techniques include

4

the radon deficit method and the oxygen balance method. However, further consideration should be given to alternative gases for gas balance methods (e.g., ^{3}He, chlorofluoromethanes), and to determining gas fluxes on the atmospheric side of the interface. Moreover, the use of deliberately added tracers might be considered, although quite a large–scale signal (scale 1–10 km) would have to be employed. In such an experiment a volatile tracer (e.g., sulphur hexafluoride) and a non–volatile tracer (e.g., a rhodamine dye) would be injected into the water in a known ratio at a fixed depth and the ratio of the concentration of the tracers would be followed as a function of time. Lastly, the use of enclosures might prove a cost–effective means for gas exchange measurements if accompanied by complementary studies of the disturbing effects of the enclosure. More use should be made of satellite imagery where facilities are available to relate transfer velocity measurements to sea state (wave spectrum, wind velocity spectrum) and to the immediate history of the sea state.

For wind speeds >10 m s^{-1} further measurements are required of the bubble spectra both in the field and in the laboratory. Such comparisons have been made to a limited extent for water depths <50 cm although one set of measurements indicates the possibility of bubble penetration down to 40 m. Acoustic spectrometry has proved difficult to apply in the field but laser scattering and Fresnel imaging techniques (with computer–aided data analysis) might provide alternative approaches. The formation of bubbles of >500 μm diameter is also important for the transfer of non–volatile (particularly surface–active) components out of the mixed layer into the atmosphere leading to the formation of marine aerosols. Here too, a study of bubble population density and size spectrum as a function of sea state would be useful.

Little is known about the influence of surface–active films on transfer velocities although some wind tunnel experiments indicate that the effect is potentially significant. The zonation of surface films by Langmuir circulation cells and their elongation into windrows indicates the spatial complexity of this phenomenon. More information is required on the composition of the surface layer, its temporal variation and its influence on the transfer of both reactive and non–reactive gases.

CHEMICAL AND BIOLOGICAL SOURCES AND SINKS FOR GASEOUS COMPOUNDS

Introduction

In addition to their involvement in air/water transfer reactions, the dissolved gases also participate in a wide range of chemical and biological processes in the water column and in the atmosphere. For most of the trace gases so far studied, the observed profiles represent a temporary balance between rapid fluxes associated with these addition and removal processes. Unfortunately in most instances the reaction sites have not yet been identified (Table 2) and in all instances our knowledge of the mechanisms and rates of reaction is rudimentary.

Photochemical Processes

Photic zone. A number of processes may occur in the photic zone (level 5, Table 1) which result in the simultaneous production and destruction of gaseous species. These processes are chiefly microbiological and photochemical/thermochemical. For example, methane is often supersaturated in surface waters. This condition probably results from excess anaerobic production (from organic matter decomposition and transformation at reducing microsites) over aerobic methane oxidation in and below the photic zone, and, possibly, photochemical methane oxidation in the photic zone.

Table 2. Probable sources and sinks of selected trace gases[a]

Gas	Source		Sink	
	Non-biological	Biological[b]	Non-biological	Biological[b]
^{222}Rn	5 – 8 (5)[c]	—	1, 2, 5–8 (5)	—
N_2O	—	5–7, 8 (7)	0, 1 (10)	5–7, 8 (7)
CH_4	—	6, 7 8 (7)	0, 1, 2 (9)	5, 6 (7)
O_3	0 – 3 (6)	—	0 – 4 (5)	—
$(CH_3)_2S$	—	5 (5)	2, 3, 4 (3)	5, 8 (4)
CO	4, 5, 6 (2)	5–8 (?)	0, 1, 2, 3 (7)	5, 8 (2)
C_2–C_5 hydrocarbons	—	4, 5–8 ? (3)	2, 3, 4 (4)	(5, 6) (?)
CH_3I	—	5, 6, 7 (3?)	2, 3, 4 (3)	5, 8 (?)
NO[d]	4, 5 (10)	5–8 (7)	2, 3, 4 (0)	5–8 (7)

[a] The numbers refer to levels of ocean-atmosphere system identified in Table 1. 0 corresponds to the stratosphere and 8 to levels below the thermocline in the ocean.

[b] Site 4 (oceanic microlayer) is a potential area of activity, but only during the dark period in the absence of photo-inhibition.

[c] Value of n given in brackets corresponds to timescale of process (= 10^n sec).

[d] Not considering recycling of NO_2 at site 3 (atmospheric microlayer).

6

The concentration of CO in the photic zone undergoes large diurnal variations with maximal concentrations at mid-day and minimal concentrations, approaching the equilibrium value, in the early morning. These variations occur simultaneously through the entire photic zone, but the highest concentrations are found near the top of this zone. Laboratory and field studies indicate that the instantaneous concentration of CO in the water column is determined mainly by biological uptake and photochemical production from the oxidation of dissolved organic matter. In fact, short term variations in the concentration of CO have been directly correlated with short-term variations in solar radiation. Thus, the time-scales for production and consumption processes are quite short, of the order of minutes to hours, as confirmed by laboratory experiments.

Interestingly, the few measurements available for formaldehyde (and other low molecular weight carbonyls) also show diurnal variations in surface seawater and, therefore, the chemistry of formaldehyde and carbon monoxide may be closely linked, as has been previously shown in the atmosphere. Other gaseous species, in particular COS, are also produced in the photic zone as a result of photochemical processes.

Upper photic zone and surface microlayer. Photochemical reactions in the uppermost layer of the ocean (level 4, Table 1) are enhanced due to the higher light intensity and higher concentration of chromophores (light absorbers). One of the most important species produced and released to the atmosphere is the NO radical, formed by photolysis of NO_2^-. Zafiriou calculated that $0.2 - 60 \times 10^{-3}$ mol m^{-2} year^{-1} are produced in the upper 10 m of the ocean but, due to its high chemical reactivity in seawater (residence times of the order of seconds), only that NO which is formed in the surface microlayer (level 4, Table 1) may actually diffuse into the atmosphere. A similar situation probably exists for other radicals formed in seawater (NO_2, HO_2, H_2O_2, and CH_3O_2). Reactions involving ozone (O_3) at the sea surface may also result in the formation of free radicals. For example, it has been suggested that this process may result in a production of iodine which is then released to the atmosphere. The released iodine may in turn react with atmospheric ozone. The existence of free radicals in surface waters may be of considerable importance because they may react with gaseous species diffusing from below, thereby reducing their flux to the atmosphere.

Lower marine troposphere. The oceans may play a dominant role in the photochemistry of the lower marine troposphere (levels 2 & 3, Table 1) because most reactive species produced over land are not transported long distances in the atmosphere. The oceanic release of the reactive species, NO, is particularly noteworthy. This species initiates a series of reactions in the atmosphere. The course of these reactions is completely different at NO concentrations higher or lower than about 10 pptv (based on atmospheric model simulation calculations). As a result, the concentration and distribution of other reactive compounds are significantly affected. One of these reactive species is the OH radical, which has a half life of less than one second in the atmosphere, and which reacts rapidly with a number of gases released from the ocean. Thus, the flux of these gases may in turn be greatly affected by even small variations in the concentration of NO in marine air. Since it is still unclear whether the oceans are a net source or sink for NO, the *global* importance of oceanic release of NO cannot be assessed at this time.

In addition to oceanic release of reactive species, the ocean, especially the surface microlayer, may in itself be a sink for free radicals and reactive gases (e.g., ozone) produced in the lower troposphere. The importance of this sink is unknown.

Key Problems and Experiments Related to Chemical Processes

The above examples illustrate the complexity of interactions controlling the cycling of reactive trace gases within and between the lower marine troposphere and the upper ocean. The instantaneous concentration of any particular species is the result of a number of competing production and loss processes occurring in these systems and at the interface. We attempt below to point out the key problems and measurements needed to obtain better estimates of fluxes and to model the complex interactions.

Characterization and analysis of gaseous species. To date, only a few of the possible photochemically important species formed in the photic zone have been identified and studied. These include NO, NO_2, H_2O_2, H_2CO, CH_3I, COS and CO. Work is needed to identify other important species in order to obtain a fuller understanding of the types and rates of photochemical reactions occurring near the surface of the ocean. This information may be obtained through laboratory studies based on controlled irradiation of natural seawater samples. In order to carry out these studies, as well as complementary field studies, techniques need to be developed which allow for rapid sampling and analysis and which give numbers which accurately reflect the steady state concentration in the sample. Analysis methods for trace amounts of short-lived species (e.g., half-life about 1 sec or less) are especially needed.

Characterization of photochemical processes. Basic information about the photochemical processes which give rise to reactive trace gases in the sea is lacking. For example, what are the major chromophores in seawater? To what extent is the formation of primary and secondary photoproducts influenced by redox-labile metals, surface catalysis on suspended particles, photosensitizers, and the concentration of photolabile dissolved organic matter? Again, many of these questions should be initially addressed by laboratory studies with natural seawater samples.

Factors affecting flux estimations. Cycling models require knowledge of net global air-sea fluxes. At present, only very crude estimates are available, and these for only a few species. To obtain better flux estimates, concentrations of gases should be measured not only in the marine photic zone, but simultaneously in the overlying atmosphere on a temporal and geographical basis. Also, depth profiles as close to the boundary layers as possible should be taken in order to evaluate net production or consumption of species within these layers and in the surface microlayer.

The stagnant-film models, on which the available estimates were often based, assume a thin surface film to exist acting as a passive boundary for the diffusion of gases. Reactive species, such as free radicals, may be formed at high rates in the surface microlayer. Reaction of these species with gases diffusing into the microlayer from above and below may result in large discrepancies between calculated and observed fluxes through this boundary. Furthermore, the microlayer itself may be an important source for some gaseous species such as H_2CO, CO, NO and I_2. Thus, the stagnant-film model needs to be significantly modified to take these processes into account, to replace the apparent gas-transfer enhancement factors that have commonly been used. Lack of good values for gas transfer velocities (or gas/liquid resistances), as well as of seawater solubilities (or Henry's constants) for many trace gases, has further hampered the estimation of fluxes. Only after the shortcomings mentioned have been reduced by appropriate laboratory and field work, will it be possible to calculate air-sea gas exchange fluxes which can be quantitatively compared to other fluxes transporting gases to the sea surface, namely rainout and atmospheric particle fluxes.

   Micro-organisms as sources and sinks for trace gases. Although the
micro-organisms themselves are just being identified and cultured, evidence
suggests that microbial processes can provide major or potential sources
and sinks for many of the gases considered (Table 2) as well as the methylated amines and formaldehyde. Indeed the flux of sulphur from the oceans
as dimethyl sulphide constitutes a substantial fraction of the global
transport of the element. A range of micro-organisms including methyl-
ogenic phototrophs, aerobic and anaerobic methylotrophs are involved.
Techniques are being developed for the isolation and characterisation of
the relevant micro-organisms. The culturing of natural populations in
media enriched with the appropriate trace gases can provide a means for
isolating the micro-organisms that act as sinks. The selection of suitable
methanogenic substrates for the isolation of the micro-organisms responsible for gas production also appears to be straightforward. The isolates
that have been obtained have been used to produce fluorescent antibodies
which enable the relevant micro-organisms to be detected and enumerated in
natural systems by means of epifluorescence microscopy. The activity and
ultrastructure of the isolates will provide useful criteria for assessing
their significance as sources and sinks of trace gases in the natural
systems. Specific fluorescent antibodies for key methylotrophic enzymes
are at hand and are being used to assess the distribution of the micro-
organisms involved in the cycling of trace gases in the upper ocean.

   Simple reliable methods for estimating the rates of gas transformation
at sea do not exist. The isolation of the micro-organisms involved will
assist the laboratory development of such procedures. An example is
provided by the use of recently isolated methane oxidizing bacteria to
develop a method for estimating the rate of methane oxidation based on the
rate of formation of formaldehyde. Once the microscopic and rate procedures are developed ashore, then they must be taken to sea to study the
upper ocean ecosystem.

   Proposals for field studies: Importance of the methane cycle. There
is a need to elucidate the apparent paradox of the *in situ* production of
methane by very strict obligate anaerobes in the upper layer which is oxic.
It is possible that highly oxidative bacteria such as methane oxidizers and
oxygen insensitive anaerobes such as sulphate reducers that maintain low
redox potentials could produce microenvironments which in effect would
create a "false benthos". Again electron and epifluorescence microscopy
would be invaluable for detecting and characterizing microbial aggregations
that might function in this way within the upper layer of the sea. The
formation of methane from methylamine produced by plankton may represent
another source.

   Proposals for field studies: Comparison of photochemical and microbiological cycling. Due to the fortuitous occurrence of "light potentiated
inhibition", that brings heterotrophic and methylotrophic bacterial
processes to a standstill during the light period, it may be possible to
separate the photochemical and bacterial processes of formation and
consumption of specific compounds. Studies of diel variations combined
with isotopic dilution experiments should enable us to understand how
interactions between biotic and abiotic sources and sinks control the
instantaneous concentrations of reactive trace gases, such as CH_4, C_2H_6,
H_2CO, CH_3I and COS in the upper ocean. For example, for CO and H_2CO a
probable source is photochemical oxidation of dissolved organic matter
(DOM) but it is not known which fraction of the DOM pool is the dominant
precursor, humic substances or biologically released organics (e.g.,
methylated amines and sulphur compounds in algal exudates). If it is the
latter, then CO and H_2CO production should roughly correlate with biolog-

9

ical productivity in the water column. Thus, in order to gain an understanding of the interactions of sources and sinks in any one water body,
variations in the production of reactive gases should be studied over
diurnal cycles, as well as on a seasonal and geographical basis. Appropriate physical and biological parameters must be measured simultaneously.
This information will reveal in which regions of the oceans and at what
times during the year substantial gas production and consumption occur and,
thereby, facilitate the construction of global budgets and cycling models
for these gases.

CONCLUSIONS

For the transport of sparingly soluble gases across the air-sea
interface, wind tunnel experiments have provided a reasonable theoretical
framework for windspeeds up to 3 m s^{-1}. Between 3 and 10 m s^{-1}, wave
formation introduces an ill-defined coupling between wind speed and fetch
which raises severe problems in non-steady state conditions. Above 10
m s^{-1}, wave breaking and bubble formation contribute significantly to the
transfer rates for sparingly soluble gases. For highly soluble gases these
effects can be largely neglected.

Chemical and biological processes can produce significant fluxes of
some gases within the water boundary layer (level 4, Table 1) which comp-
licates the application of the gas exchange equations. If these processes
in turn induce changes in the physical properties of the boundary layer
(e.g., surface tension) then non-linear effects will be introduced.

The profiles of gases within the water column appear, in general, to
represent a temporary balance between the fluxes generated by competing
formation and removal processes. The most important chemical processes are
photolytically induced and it is only in the past four years that mechanisms and rates have been established for a small number of processes. In
general the sites of reaction, the intermediates produced and the rates of
the reactions are poorly characterised. The degree of coupling of the
various reaction sequences is also poorly known so that modelling of
photochemically induced reactions is largely confined to the atmospheric
side of the air-water interface. More experimental work is needed here and
new techniques must be developed to enable sampling and analysis to be
carried out on timescales appropriate to the processes under consideration.

The situation with respect to biological formation and removal processes is even less well characterised. New sampling, staining and microscopic procedures are at last providing the first components of a picture
of structure, formation and trophic organisation in the nanoplankton and
picoplankton. It is apparent that micro-environments such as micro-
aggregates are very important and it is clear that in many instances the
rates of the biological processes are very rapid. However, the picture is
far from complete and studies of the rates of gas production or decomposition will need to be undertaken in the laboratory once the associated
organisms have been identified and isolated.

Clearly much laboratory work needs to be done to relate the time and
space scales of the addition and removal processes (Table 2) with those
established by the physical mixing processes (Table 1). In the meantime
careful studies of trace gas profiles under different meteorological
conditions and over the diurnal and seasonal cycles will provide valuable
clues to the processes controlling gas distribution. Analytical and
sampling techniques must be developed to facilitate such measurements.

REPORT OF WORKING GROUP II: PARTICLES

FRAME OF REFERENCE

The nature and fluxes of particles occurring in the open ocean surface layer are depicted schematically in Fig. 1. The net flux of particles from the atmosphere to the surface ocean is comprised of both an external input of continental aerosols and a recycled component of ocean derived aerosols. These atmospheric aerosols can be transported to the surface layer either as wet or dry fallout and enter the surface waters through the photochemically active surface microlayer. During these transfer steps the aerosols can be substantially modified and may enter the surface water in a form ranging from totally dissolved to totally particulate.

Particle production within the upper ocean is controlled primarily by the photosynthetic phytoplankton and bacterioplankton within the sunlit region (euphotic zone). This material covers the size range from pico-plankton (e.g., photosynthesizing bacteria) to nanoplankton and micro-plankton (Table 1). Phytoplankters utilize dissolved inorganic species such as bicarbonate, nitrate, phosphate, silicon and required trace elements to synthesize their organic matter and to secrete their tests.

Data from the U.S. GEOSECS program show that biological particles dominate the detrital phases throughout the oceans. In surface waters (<100 m), organic matter comprises 30-70% of total particulate matter. The hard skeletal phases of the marine organisms, $CaCO_3$ and SiO_2, together make up about 25–50% of total particulate matter.

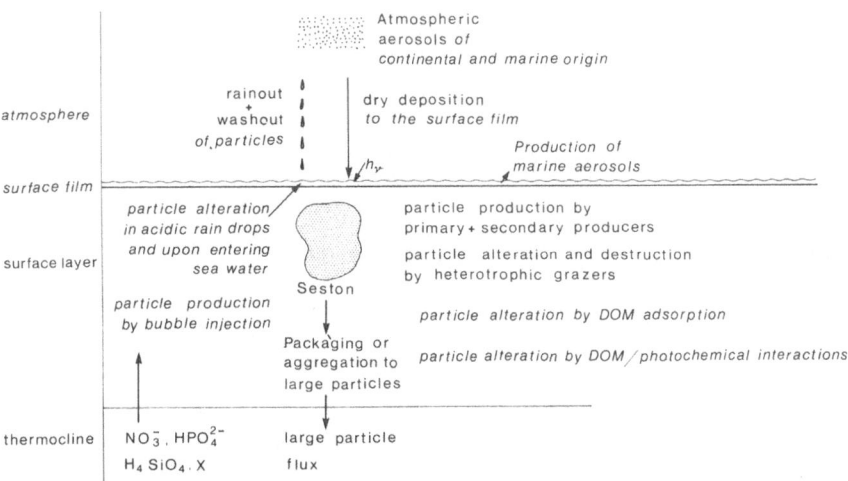

Fig. 1. Schematic diagram of the nature and fluxes of particulate material in the surface layer of the ocean.

A substantial part of the phytoplankton is grazed upon by heterotrophic zooplankters leading to secondary particle production by micro-
plankton, mesoplankton and macroplankton, in the form of organisms and
fecal material, as well as to residues (i.e., detritus) of the original
primary production. The size distribution of particles in the upper ocean
is highly variable. In general it is characterized by high numbers of very
small particles per unit volume in the size range of a few μm and a decreasing number of particles with larger diameters. These large particles
include aggregates which are inorganic particles bonded by intermolecular,
intramolecular, or atomic forces, aggregates of mucilaginous phytoplankton
cells (e.g., diatoms and coccolithophores), agglomerates which are composed
of organic and inorganic matter held together by surface tension and organic cohesion due to mucus and biological activity, floccules which are
amorphous inorganic particles held together by electrostatic forces, as
well as remnants of larger organisms like crustacea, salps, pteropods and
fish. Another source of large particles is fecal material of animals, such
as fecal pellets of copepods and fecal strings of swarm-forming macrozooplankters like salps and euphausiids.

As particles sink through the water column, some elements are added to
seawater in dissolved form, whereas some are removed from solution. The *in
situ* addition and removal of trace elements - termed the J flux and the J
efflux, respectively - modify the composition of seawater appreciably in
the case of several elements.

The biogenic particulate material can remove dissolved organic and
inorganic species from the surface waters by both metabolic uptake, e.g.,
by bacteria, and by passive adsorption mechanisms. Such dissolved species
as ^{234}Th and ^{210}Pb appear to adsorb with varying rapidity onto surface
functional groups of the particulate organic matter. The mechanisms of
such adsorption processes are not well understood, although attempts have
been recently made to model the adsorptive properties of particulate material in the deep ocean. The apparent rates of such scavenging processes
can be determined using natural U and Th series radionuclides such as
^{234}Th, ^{228}Th, ^{210}Pb and ^{210}Po. The data available at this time demonstrate

Table 1. Size classes of planktonic organisms

Plankton size class	Length	Particles
Picoplankton	<2 μm	bacteria, cyanobacteria, smallest eucaryotic organisms
Nanoplankton	2 – 20 μm	μ-flagellates, dinoflagellates, coccolithophores, small diatoms
Microplankton	20 – 200 μm	diatoms, dinoflagellates ciliate protozoa, nauplii
Mesoplankton	200 μm – 20 mm	large diatoms and dinoplagellates, copepods
Macroplankton	20 mm – 20 cm	euphausiids, pteropods, appendicularians, ctenophores
Megaplankton	>20 cm	colonial appendicularians, medusae

a strong correlation between scavenging rates and the rates of particle production.

The extent to which the various particulate forms are transported out of the surface waters depends to a large degree on the relative amount of recycling within the upper ocean. Recycling processes range from phytoplankton exudation of dissolved organic matter and excretion by zooplankton, to oxidation of the particulate organic matter by bacteria. Trace species associated with the particulate material can also be regenerated back to the dissolved form.

In general, the large, rapidly sinking particles caught by sediment traps are the chief carriers of material to depth. These sinking aggregates, agglomerates and floccules do entrap fine particles and adsorb (scavenge) dissolved species and carry them to depth.

Residence times of organic particles in the surface layer depend largely on environmental factors like turbulence and mixing as well as on the availability of nutrients in the euphotic zone. Residence times range from about 2 to about 100 days based upon ^{234}Th and "new" production estimates.

Several time and space scales are important in particle dynamics and associated chemical processes. The transformation and recycling processes occur at reaction sites which range in size from the molecular (e.g., photochemically driven) to the scale-sizes of biological particles (including marine aggregates). Collectively, however, these reaction sites can be considered "small-scale" relative to dominant meso-scale or large-scale oceanic phenomena. Observations of *in situ* chemical distributions and remotely-sensed (e.g., Coastal Zone Color Scanner (CZCS)) structures often show large-scale coherent patterns. A set of processes must occur within the ocean which transmit and/or transform the local structure produced at the reaction sites into the observed large-scale "coherent" patterns. It is likely that meso-scale dynamical processes fulfil this role.

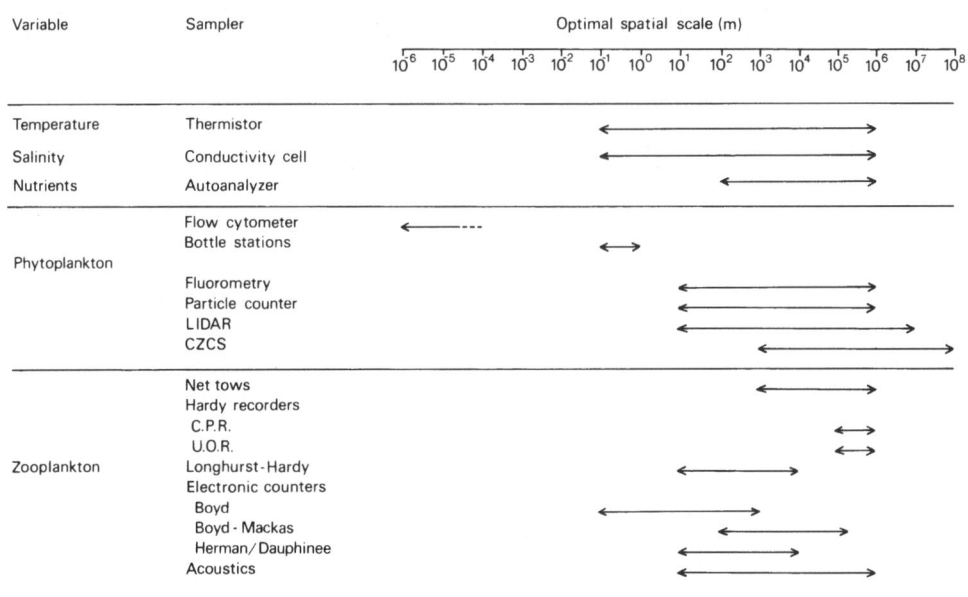

Fig. 2. Normal spatial scales which can be sampled appropriately with various oceanographic instruments (from Yentsch & Yentsch, 1984).

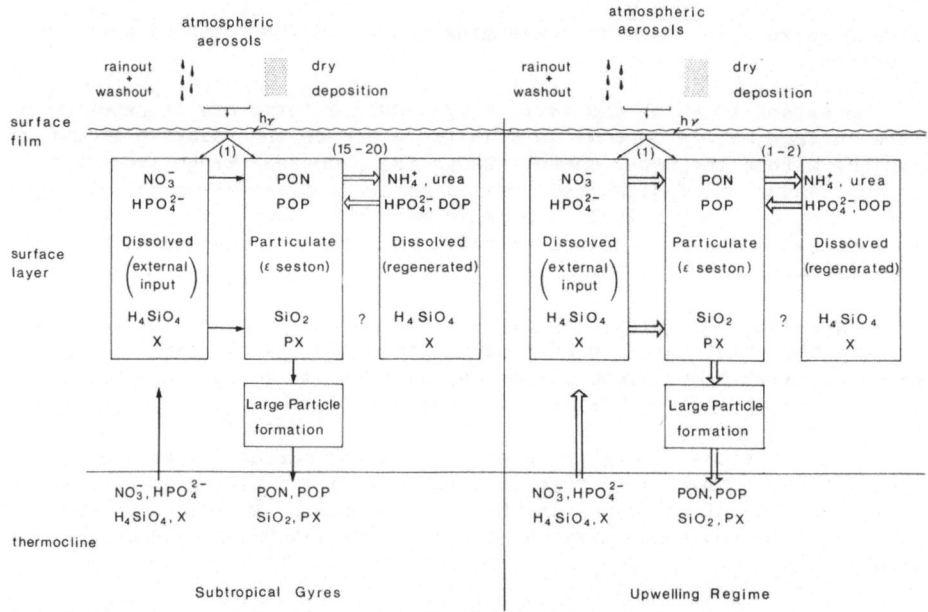

Fig. 3. Schematic diagram of particle formation in two different oceanic
systems. The arrows indicate different intensities of processes.
The figures in brackets indicate different modes of recycling.

Figure 2 gives the normal spatial scales which can be sampled appropriately with various instruments.

The formation of organic particulate matter *via* primary and secondary production as well as the extent of recycling of organic and inorganic material in the surface ocean vary drastically in space and time. Differences relative to transformations of some biogenic elements between subtropical gyres and coastal upwelling regimes are depicted in Fig. 3. The most striking difference between the two systems is that in the oligotrophic subtropical gyres, particle formation is dependent on rapid and intensive recycling of inorganic nutrients in the forms of ammonia and urea, whereas in upwelling regimes particle production is driven by ample external supply, for example, of inorganic nitrate, phosphate and silicon. The corresponding average vertical particle fluxes leaving the euphotic zone, in, for example, the forms of particulate organic nitrogen, phosphorus and silicon, are low (<50 mg C m^{-2} day^{-1}) in open ocean systems and high (>150 mg C m^{-2} day^{-1}) in upwelling areas.

The seasonal effect in temperate waters on primary production and on the loss of particles from the euphotic zone is depicted in Fig. 4. It is postulated that the structure and functioning of the pelagic system changes drastically with time, regulated by the supply of nutrients, leading to "new" and "regenerated" production. This concept has also an impact on sinking strategies of phytoplankton. It has been shown, for example, that diatoms have physiological control over buoyancy. Declining growth is accompanied by increasing sinking rates, where the frustule acts as ballast. Increased mucus secretion in conjunction with the cell protuberances characteristic of bloom diatoms, leads to entanglement and aggregate formation during sinking; the mucilaginous aggregates scavenge mineral and other particles during descent, accelerating sinking rates further. It has been assumed that the accelerated sinking of diatoms is part of their life strategies to survive unfavorable periods. Conditions are very different in

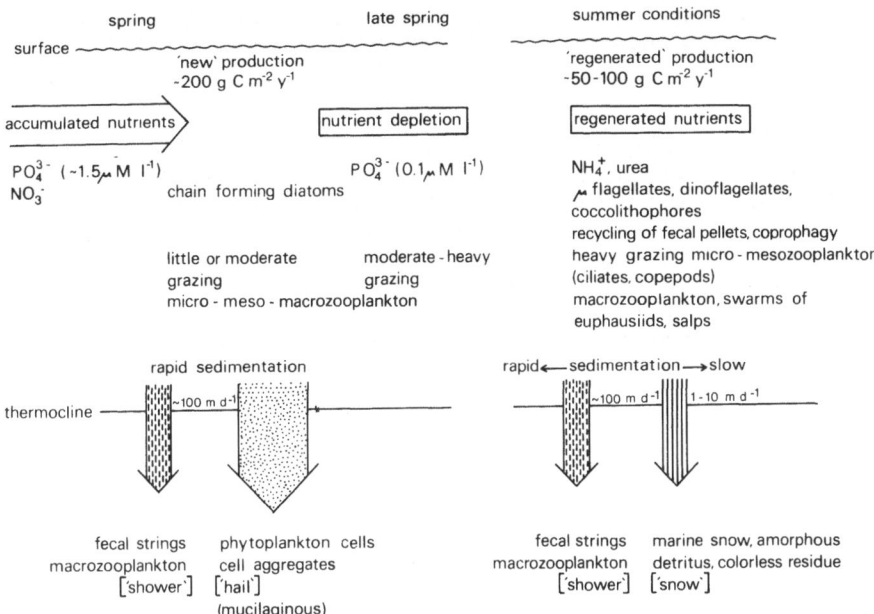

Fig. 4. Seasonal effect on primary production and loss of particles
from the euphotic zone in temperate waters (from Zeitzschel,
1984).

the regenerated nutrient mode during summer. The phytoplankton composition
has changed from diatoms to picoplankton and nanoplankton (μ-flagellates,
dinoflagellates and coccolithophores). The ecological role of the zooplankton is also very different in the described seasons. The ratio of
phytoplankton to zooplankton biomass is 5—10 in spring and 0.4—1 in summer
reflecting a very different grazing pressure on the phytoplankton. There is
evidence from sediment trap work in the Atlantic and Antarctic oceans that
the particle flux in spring is comprised mainly of cells and cell-aggregates
of phytoplankton whereas the flux during the summer period is restricted
to fast sinking fecal strings of macrozooplankters (e.g., euphausiids and
salps) and a slow sinking amorphous colourless residue. It is postulated
that most of the fecal pellets of microzooplankton and mesozooplankton
(e.g., ciliates and copepods) are retained and recycled in the mixed layer.

A very different transport mechanism of organic material to greater
depths *via* zooplankton was suggested more than 20 years ago. This mechanism is based on the observations of diurnal vertical migration of a
large number of zooplankton species. The migration pattern of different
species is confined to certain depth ranges, giving rise to the concept of
a ladder of migration. This implies that fecal material is recycled several
times in the deep water column, before it reaches the sea floor. The
validity of this concept and the implications for the particle flux in the
ocean has yet to be shown.

Sediment trap experiments in the upper zone of the ocean in recent
years have shown that flux rates of particles do vary in time intervals of
days. The classical view of Alexander Agassiz, who proposed already in 1888
that deep sea organisms are nourished by a "rain" of organic detritus from
the overlying surface waters, has to be revised. It is usually not a
"constant rain" which transports particles to depth, but a sequence of
showers, hail, long calm periods and snow, due to changing meteorological,

physical and biological conditions in the surface layer of the ocean, which produce an up to now unpredictable signal at the bottom of the sea.

PROBLEM AREAS

There are a number of special chemical problems related to particulate transfer at the air—sea interface for which attention is necessary beyond the level expected in current U.S. projects (mainly VERTEX and SEAREX). We refer here to liquid phase chemistry associated with rainout and sub-cloud-layer particle washout by rain, and to processes associated with sea spray droplets and with bubble scavenging of the uppermost water layers. Questions include: What are the reactions between dry fallout and surface films? Are some of these reactions photochemically activated? Bubbles can scavenge surface active materials and picoplankton. What is their role in the formation of organic particles in the water and their ejection into the atmosphere?

To date we lack conceptual models and multi-disciplinary laboratory work on the nature of the variety of suspended particulate material in the sea, e.g., of marine snow and of fecal material of zooplankton with its native microbial flora. This very much restricts the laboratory study of chemical processes associated with such particles. Can mathematical models be developed or must we restrict our studies to those we can carry out on shipboard?

Particles provide a mechanism for transporting materials from one part of the ocean to another that can completely bypass the physical processes of mixing and advection (thus leading to non-conservative distributions). With the exception of some major ions (Na^+, K^+, Mg^{2+}, SO_4^{2-}) in seawater there are few chemical substances that are not specifically affected by particle transport in one form or another. In addition, particles are not only important sites for thriving biological communities, they are continuously produced and consumed by biota. A full understanding of particle product- ion, transport and destruction processes, influencing the chemical recycl- ing within the ocean and biological processes, thus requires us to under- stand how particles are produced and at what rate, as well as what their ultimate fate is. More extensive study of particle formation and decomp- osition is demanded, including times and rates of recycling within the surface layer. Programs such as VERTEX have given us a tantalizing glimpse of the richness of chemical, biochemical and biological interactions. But these can only be viewed as a beginning.

An important aim of process—orientated studies should be to develop the ability to model the observed processes. This is essential for planning new experiments and for the development of models of chemical cycling within the interior of the oceans. Until recently, a model of chemical transport by particles would have used an uptake or dissolution term which would have been assumed to be constant, or exponentially decreasing with depth. Recent work with sediment traps suggests that many processes can be represented by having a non-sinking small particle population and a second, rapidly sinking, large particle population. The small particles are likely to be affected by advection and diffusion. Indeed, deep ocean observations are consistent with this. New levels of complexity will almost certainly arise as our understanding develops. Such process models should then be incorporated into models of ocean circulation and mixing in order to examine the importance of advection and diffusion of small particles in determining the ultimate fate of chemical substances subject to particle transport.

A very challenging way to improve our knowledge of the complex, often non-linear, interaction between physical, chemical and biological processes in the ocean is the adoption of the Lagrangian approach in addition to the commonly applied Eulerian mode of measurements. Most measurements at sea are conventionally carried out either in the time domain (e.g., with moored instruments like current meters or bottom-tethered sediment trap arrays) or in the space domain (sampling with water bottles, pumps, plankton nets or towed instruments like the batfish, on transects or grids). Comparison and matching of data sets from these different efforts is extremely difficult.

As a promising alternative the Lagrangian approach combines the time and space domain for a specific body of water, e.g., a surface water column of 100 m depth. Several drift experiments in open ocean water in the North Atlantic and Antarctic have shown the benefit of this technique. Moreover, the availability of fast computers has opened the way to Lagrangian modeling; it is for example possible to follow an ensemble of particles in an assumed water column over appropriate time intervals. Models now available have followed the outburst, development and fate of a phytoplankton spring bloom in an elegant and convincing way.

Several kinds of new instrumentation are needed for observing particles *in situ* and for studying particle dynamics at sea. These include optical devices for observing marine snow and its inclusions and for measuring *in situ* sinking rates of a broad spectrum of particles. Remote sensing from aircraft and satellites may be usefully extended to provide areally integrated particle distributions using, for example, the natural or induced fluorescence of chlorophyll. Table 2 gives an overview of limitations and advantages of various optical instruments. We also need new devices to measure continuous (over time or space) profiles of primary productivity and nutrients and we need techniques for studying properties within microenvironments, such as the redox potential within marine snow aggregates. We need self-contained buoys to obtain uncontaminated samples of wet and dry depostion from atmospheric fallout at sea. Extensions and improvements are needed in isotope techniques to measure the formation and decay of particles.

Multi-disciplinary thinking in marine science seems to be still in its infancy. Scientists who work with sediment traps either look from the surface down (planktonologists) or from the bottom up (benthologists, sedimentologists). The former group is interested in the production in and loss of organic material from the euphotic zone or mixed layer, whereas the latter wants to quantify the flux of particulate material which reaches the bottom and is eventually buried in the sediment. Chemical oceanographers are mainly interested in the adsorption and dissolution processes of suspended matter in the water column in between. Unfortunately the handling of the obtained samples varies fundamentally between these disciplines. The sedimentologists are used to sieving their valuable accumulated material from sediment traps in the same way as material from sediment cores, in a standard but crude way which destroys the wealth of the potential information and which is not acceptable to any biological oceanographer who can neither understand nor tolerate the magic figure 63 μm! The excuse "we sieved very gently" does not help. Sediment trapping, although a controversial tool, like any rain gauge, might provide a field where chemists, biologists and sedimentologists concentrate their experience and intelligence in improving our knowledge and understanding of the very complex dynamics of the particle flux in the ocean.

LARGE-SCALE EXPERIMENTS

The working group was somewhat reticent to propose new large-scale experiments, in part because several programs, principally in the U.S.A.,

Table 2. Limitations and advantages of various optical instruments for the analysis of particulate matter in the seas (from Yentsch & Yentsch, 1984)

Instrument	Variable measured	Platforms	Major limitations	Major advantages
Particle counters plus in vivo fluorometry	Particle volume size spectrum	Ships	Correspondence versus simultaneous measurement; depth resolution/m	Measurement on per cell basis; rapid 1000/sec
Flow cytometry; cytofluorometry (FCM)	Chlorophyll a per cell; accessory pigments per cell; Coulter	Seaside laboratories to date; ship potential	Sample size very small 0-150 μm size range	Can yield characterization and concentration of pigment groups present in phytoplankton; rapid 1000/sec
Fluorescence-activated cell-sorting	Validation of FCM	Seaside laboratories to date; shipboard	Low yield of particles at ambient concentrations; 0-150 μm size range	Fluorescence-activated; rapid 1000/sec; overcomes fractionation; based on size exclusively
In vivo fluorometry with pumping system	Chlorophyll a with temperature, salinity, nutrients	Ships: small boats to research vessels	Pumping causes mixing which limits fine scale resolution; depth resolution 1 m	Inexpensive; can use on any vessel with AC/DC current; rapid profile 1 m/min
In situ fluorometry	Chlorophyll a with temperature, salinity	Ships with winch; buoys	Fine scale resolution; depth resolution 1 cm	Loose nutrient profile; rapid profile 1 m/min

Method	Measures	Platform	Advantages	Disadvantages
In situ multichannel fluorometry	Bioluminescence potential via fluorescence of luciferin, chlorophyll a and accessory pigments	Ships with winch; 7 conductor wire; buoys	Fine scale resolution plus accessory pigmentation; characterize phytoplankton; depth resolution 1 cm	Rapid profile 1 m/min
Bioluminescence photometer	Stimulated bioluminescence	Ships: buoys	Adequate measure of bioluminescence potential; rapid 1 m/min	Stimulated bioluminescence versus in situ levels; no corresponding data; cannot resolve distance versus intensity; rapid 1 m/min
Beam attenuation: transmissometry	Light transmitted through given water volume, therefore total less absorbed and scattered	Ships; small boats; seaside laboratory	Best when corresponding in situ particle and chlorophyll data are available.	By-passes apparent light characteristics; inexpensive; rapid 1 m/min; continuous
AOI, LIDAR	Chlorophyll a and accessory pigments	Aircraft	Spectral signatures; accessory pigmentation; gross characterization of phytoplankton; little interference from detritus and yellow substances; 1 m resolution.	Limited resolution with depth; heavy power requirements
CZCS	Chlorophyll a and accessory pigments; yellow substances	Satellites	Interference from detritus, seston and yellow substances on chlorophyll a signal; data reduction costly; data and sensor accessibility.	Broad scale; 1000+ km resolution

include particle flux measurements and in part because we are not totally familiar with the present work under these programs. Some include, for example, studies of the nature of the particles while others are more closely confined to specific particle-related flux measurements.

Some of our general observations may be useful, however. To whit: an understanding of chemical cycles on an ocean basin scale is not only of intrinsic interest but may well be very relevant to proposed studies of global physical dynamics. Some knowledge of upper—layer processes may prove essential if, say, nitrogen cycles, which involve cross-pycnal fluxes, are to be used as critical tests of physical models based on iso—pycnal mixing and advection determined from T, S and $^3H/^3He$ data.

Integration of surface and deep water studies can be achieved partic—ularly, but not solely, through the use of sediment trap methods. These process studies could be combined with important generalizations of the relationship between primary production and chlorophyll concentrations at the ocean surface, determined by satellites, as proposed by Eppley and others, which could permit extrapolations over large open ocean and coastal areas. These concepts could be implemented using the satellite Ocean Color Imager to provide one significant parameter related to primary production at a wide range of space and time scales.

This combination of techniques, in the context of emerging physical theories of the "ventilated thermocline", presents major challenges to the chemical oceanographer.

Specific Experiments

First, the group supports the idea of a "North Atlantic Sediment Trap Experiment" as discussed during the meeting. This experiment would make use of the ongoing studies employing satellite-tracked buoys as vehicles for the traps. The traps would be released in January/February at ten positions between the mid-Atlantic Ridge and the American Continent (off Florida to South of Greenland). They would be collected off the European Coast around October/November. Each trap would be fitted with a sequencing sampling mechanism, in order to provide fortnightly samples. To obtain a gradient of settled material the depths of the floating traps are proposed to be 100, 200, and 500 m. The intention is to coordinate such an experiment inter—nationally, especially concerning the logistics of trap deployment and the distribution of material for chemical analysis. In addition, bottom—tethered sediment traps in conjunction with current meter moorings in the same area would yield valuable information on particle fluxes to the deep ocean floor.

This experiment would provide a large-scale view, over several months, of particle flux in the North Atlantic. The results could be compared with new production estimates from historical sources as well as with particle fluxes being measured independently off Bermuda and historically in warm core Gulf Stream rings and in the PARLUX program. Satellite chlorophyll data will be a valuable adjunct to the program to test notions of relationships between phytoplankton biomass, primary production, and the sinking flux. A pilot project to study drifting sediment traps in the proposed area, by a group of Kiel scientists, will start in spring 1986.

The vistas and insight provided by the "North Atlantic Sediment Trap Experiment" will be essential if we are to design critical experiments for frontal regions and other areas where non—linear interactions are expected.

Secondly, an experiment on particle flux interactions is recommended. One of the interesting observations in SEAREX was the large pulses of atmospheric dust arriving at Enewetak from the Asian landmass. The phosphate input to the surface ocean resulting from the dustfall could be significant for particle formation and we hypothesize that it is, according to the following scenario. Phytoplankton growth in the subtropical gyre of the North Pacific is limited by the rate of fixed nitrogen input; at the same time ambient phosphate levels are essentially zero. An extra input of phosphate, *via* dust, could stimulate the growth only of nitrogen fixing algae. The resulting primary production would be "new" production, since the nitrogen is ultimately from the air – a source external to the euphotic zone. This new production would be added to the existing sinking flux of biogenic organic particles out of the euphotic zone and this increased flux could interact with the fluxes of inorganic materials such as silicon and perhaps of trace metals scavenged from the surface waters.

A time series of sediment trap collections, phosphate measurements and population assessments of nitrogen-fixing organisms would be a start at testing the hypothesis. Advice from the North Pacific dust collection network would be essential for timely sampling.

Thirdly, a number of important physical, chemical and biological phenomena show an implicit diurnal pattern. Differences between day and night in the temperature structure in the upper water column, due to solar heating during the day and cooling during the night, create convection cells resulting in a considerable diel change of the depth of the mixed layer. This mechanism may influence the nutrient supply in the euphotic zone and particle retention or rejection in the mixed layer. Photosynthesis and respiration of phytoplankton as well as vertical migration are governed by the light input at the surface. There is growing evidence that a number of chemical reactions vary diurnally. The field of photochemistry bears a lot of relevance to the particle dynamics in the sea. The study of trace organic compounds can yield a wealth of information on fundamental, dynamic biological, physical and geochemical processes in the ocean, especially in the mixed layer.

The reality of the understanding of important diurnal processes in the surface ocean is that insight comes from all the signals – the physics, chemistry and biology.

REFERENCES

Yentsch, C.M., and Yentsch, C.S., 1984, Emergence of optical instrumentation for measuring biological properties, Oceanogr. Mar. Biol. Ann. Rev., 22:55.

Zeitzschel, B., 1985, The dynamics of organic production in the Rockall Channel area, Proc. Roy. Soc. Edinburgh, in press.

REPORT OF WORKING GROUP III: SOLUTES

GENERAL CONSIDERATIONS

The majority of analyses made up to the present distinguish the dissolved and particulate phases by means of filtration using a pore size of
about 0.4 μm. This may not be appropriate, particularly if colloidal or
bacterial particles are of importance.

For the great majority of even highly particle reactive substances,
the larger fraction of the substance is likely to be found in the dissolved
phase. This means that attempts to determine fluxes using measurements of
dissolved material demand very high precision, in order to resolve the
small changes involved. On the other hand, attempts to measure fluxes
directly (e.g., of carbon dioxide in gas exchange, or thorium in particles)
are generally complicated by high variability because of the large variations in space and time, and the difficulty and expense of the measurements.

In the past, most chemical measurements have been made at widely
spaced stations. The vertical variations have received somewhat more attention than may be warranted, compared with horizontal variations, except
in the case of vertical measurements very close to the sea surface. This
has led on occasion to inappropriate interpretations in terms of one-
dimensional vertical models, for example. It is now important to give
appropriate weight to both horizontal and vertical distributions, in both
measurements and models.

There have been many measurements of chemical substances in the past:
the NODC files cover some 20,000 stations having both nitrate and silicon
data, for example. These data are however not very easily accessible, many
are of unspecified but probably dubious quality, and recent data only
become available slowly. More attention needs to be paid to making sure
that data are quickly and easily available, and that data quality is
assessed and controlled.

The description of the speciation of substances in seawater is complex, and full information is rarely available. The details (e.g., of
the exact form of the complexation with organic compounds and inorganic
ligands) are, with the exception of the extent of hydrolysis, rarely of
major importance and interconversion between species is generally rapid.
The most important aspect is the partitioning between dissolved and
particulate phases. This (and other aspects of speciation) are often
primarily controlled by the oxidation state, which is therefore of central
importance. The simple assessment of oxidation state using an overall
redox potential is however rarely useful because few species are to be
found at thermodynamic equilibrium; moreover great variations may occur

between bulk solutions and micro-environments (e.g., those produced by organisms).

The representation of organic compounds by means of a portmanteau measurement of dissolved organic carbon is inadequate for an understanding of chemical processes. The identification of key marker compounds which can be used to indicate particular processes is a general priority in this area.

PROBLEM AREAS

The principal problem areas, identified by the working group, for dissolved material in the ocean surface layer are as follows:

To what extent does nutrient transport into the mixed layer (almost exclusively by physical transport) determine the primary biological production, and thus the carbonate chemistry of the oceans, and possibly global climate (including glacial/interglacial transitions)? Biologists usually consider primary production on a small scale, whereas geochemists generally consider it on a much larger scale. These approaches need to be married, and greater account needs to be taken of the full range of possible physical processes, notably upwelling, vertical mixing, and mixing along sloping (and possibly outcropping) isopycnal surfaces.

Do correlations between the concentrations of nutrients and other trace substances (e.g., trace metals) have any common mechanistic cause? If so, they may be capable of considerable generalisation (e.g., *via* the computation of ratios analogous to Redfield ratios). If not, such generalisations are likely to be highly misleading, and potentially dangerous. The extent to which, for example, intracellular and surface scavenging effects are involved needs to be assessed.

What controls the depth and turnover time of the mixed layer? What is the relative importance of wind stress, internal sheer, and convective overturning? This is of great importance in determining the rates of biological processes, and also the extent of disequilibrium of important substances (such as CO_2). Are frontal processes important? The working group noted that the maximum concentrations of photochemically generated substances occur where there is a conjunction of necessary conditions (light, reagent supply, etc.). It should not be assumed that the maximum concentration even of photochemicals need be found at the surface.

What are the rates of dissolution of, or desorption from, particles injected into the surface layers whether they be of atmospheric, littoral, or oceanic origin? To what extent are photochemical processes involved in changes of oxidation state of Fe and Mn and hence in their conversion from particulate to soluble forms?

The distributions of most chemical substances are determined by the interaction of physical, chemical and biological processes. Their correct interpretation may require the use of non-trivial models, and the fitting of such models may be susceptible to the application of inverse methods. Such methods are not just a tool for determining rates of *physical* processes: in principle, information relevant to all three major disciplines may be obtained. It is very likely that the existing oceanic chemical data base contains more information than has yet been extracted. It should be more fully exploited using modern methods, but bearing in mind the need to evaluate critically the quality of data. The full utilisation of the ocean surface layer experiments proposed below would benefit from prior attention to the modelling required and the possible application of inverse methods for the "deconvolution" of the data.

POSSIBLE EXPERIMENTS

Notes on a Surface Ocean Solute Experiment

In the case of dissolved gases such as CO_2, seawater isolated from external perturbations would soon reach a chemical equilibrium state which could be predicted with considerable accuracy. In practice seawater is subjected to both physical and biological forces which greatly perturb the chemistry, and there is the potential for learning much about these forces from the observations of chemical species. Classical examples are in the use of chemical tracers for physical mixing studies, and nutrient elements in biological studies. Many chemical species respond to both physical and biological changes, and a separation of the processes is critical for further understanding.

One such case involves the oceanic distribution of CO_2. The partial pressure of CO_2 in surface waters is lowered by photosynthesis, and raised by warming. Each perturbation is damped by gas exchange with the atmosphere which attempts to restore equilibrium. The biological or meteorological effects will assume greater or lesser dominance in different oceanic regions. Since CO_2 is important for climatic reasons it is important to resolve the processes.

An oceanic experiment involving mass balance of CO_2 and O_2 concentrations, reconciled with nutrient and particle fluxes out of the experimental boundary, and correlated with gas fluxes and meteorological forcing would be most important. The oxygen productivity method, long neglected, could be highly useful here and might enable transfer velocities to be determined. The flux of many biologically or photochemically produced species could be constrained in this way. Particle removal rates would clearly be important and sediment traps would need to be deployed at the base of the mixed layer.

The scale of such an experiment needs careful thought, however. An area should be chosen of sufficiently large scale so as to avoid much short-term physical noise and yet of a size that is tractable for detailed chemical and biological sampling. In the long term it would be desirable to repeat the experiment in several areas representing different oceanic regimes.

In order to balance budgets of oxygen, nutrients, and dissolved inorganic carbon species it would be necessary to sample them extensively (preferably continuously) down to depths of a few hundred metres, possibly using a pumped system or a towed sensor package. A time of rapid biological and seasonal change should be selected to maximize the signal to be detected: springtime would probably be best. Assessment of biological productivity using chlorophyll analysis and primary production estimates would be required.

The details of such an experiment remain to be worked out, but the principal goal of balancing the budgets of CO_2, O_2, and nutrients seems to be feasible, and well worthwhile.

Field Experiment to Determine the Factors Controlling the Vertical Distribution of Dissolved Substances and Microorganisms in Near-Surface Seawater

The ocean surface is a unique environment in which microorganisms, organic material and light interact to form new chemical species. Many reactions have been postulated to occur but it is only very recently that a few of these species have actually been detected. Rather better known is the role of organisms in the processing of substances to produce both dissolved and gaseous forms. The whole of this highly-reactive near-surface system is acted on by the physical processes of wind mixing and convection. For this and other reasons, it is not at all clear what the depth distribution of biologically and photochemically produced substances should look like. The aim of this experiment would be to establish the form of such depth profiles by direct measurement and to attempt to explain the distributions in terms of physical, biological and photochemical processes.

All measurements and sampling would be made in a Lagrangian frame, possibly using a vessel such as FLIP. Physical measurements would include water stability (*via* temperature and salinity), with a repetition rate fast enough to resolve the passage of internal waves, light penetration as a function of wavelength, current shear, sea-surface temperature *via* radio-meter and film probe, as well as standard meteorological parameters. The state of coverage of the sea surface with organic films would be assessed by film pressure and optical techniques. Concurrent with these measure-ments, water samples would be collected with a vertical spacing increasing exponentially down from the surface. They would be analysed for biologic-ally and photochemically produced gases (such as O_2, CO_2, H_2, CO, $(CH_3)_2S$, NO) as well as Mn^{2+}, Fe^{2+}, nutrients, DOC and POC. Biological analysis of the samples would include species identification, physiological markers of mixing, faecal pellets, and particulate size distributions. Floating sediment traps would be deployed. The whole exercise should be repeated at different seasons and locations, to encompass a range of light regime, vertical density structure, and biological water type.

This experiment is largely an extension of work carried out during the ODEX programme, and the experience gained there should be taken into account.

Investigation of Relationships Between Trace Metals and Nutrients and of Mobilization of Metals from Particles

Improved methods of sampling and analysis for dissolved trace metals in ocean waters have been applied to give accurate data on the distribut-ions of several of these elements. The vertical profiles for some elements are now quite well defined and it is apparent that there are high positive correlations between the micronutrient elements and certain metals. For example, cadmium concentrations correlate closely with those of reactive phosphate, with distributions which vary between major ocean regions in a way which is consistent with the general circulation. For zinc, the correlation is closer with silicon than with phosphate, suggesting that there is a deeper cycle of release from particles for this element.

Concentrations of these elements in the upper ocean are thus strongly controlled by scavenging on particles sinking below the euphotic zone. While these observed correlations are sufficiently well established in a few cases to allow the inclusion of the metals in models developed for nutrient transport, the mechanisms determining the relationships are not clear. It seems likely that scavenging on the surfaces of cell membranes and skeletal debris occurs, but associations with particular biochemical components through metabolic processes are not excluded.

For several metals, manganese being the most-studied, distributions show maxima close to the sea surface and a decrease in concentration with depth as a consequence of particle scavenging. This behaviour is related to the oxidation-reduction chemistry of the element, the higher concentra-tions being attributable to Mn(II). An input of manganese in this reduced state is indicated and recent work suggests that the mechanism may be photoreduction of Mn(IV) entering with particles. In the ocean environment it is difficult to distinguish the photochemical and microbiological processes. Laboratory experiments are perhaps the only way to evaluate these processes separately. In the first instance such experiments should concentrate on photochemical aspects. Microbiological experiments may be required at a later stage. For some other elements, atmospherically transported particles may provide a significant source to the surface ocean by dissolution without change in oxidation state.

Closer spaced data sets on the distributions of these elements in the upper ocean, laterally and vertically, will provide inferences as to possible mechanisms. The main requirement, however, for both groups of metals, is for laboratory experiments to test hypotheses concerning mechanisms and to establish rate constants for the reactions involved. Some work has already been undertaken on photoreduction of manganese, but there is a need to extend the approaches, using natural particulate phases and having a more critical definition of photochemical parameters and organic composition of the solution. Measurements of dissolution rates of airborne particles are needed with consideration of size-fraction effects and these should be obtained under conditions representative of natural systems. The investigation of modes of association of metals with biogenous material demands different approaches. These may include the uptake of metals by microorganisms in different size fractions in culture, and the incorporation of radioactive tracers into populations with subsequent examination of the distribution in biochemical fractions.

A Possible Synthesis

The evolution and properties (particularly the turnover time) of the mixed layer are of central importance to studies of chemical processes in the surface ocean. It might therefore be desirable to combine experiments suggested above to form a larger and more ambitious programme, with the aim of following the development of the seasonal thermocline, and the associated evolution of biologically important chemicals, over a substantial area of an ocean during the production season. This would not only address the second and third problem areas identified, but also the more general first problem area. Currents in the area would need to be determined over a long period, and the best method for doing so would need to be determined. The sampling could be designed to interface easily with an appropriate mathematical model for the interpretation of the data.

UPPER OCEAN CHEMISTRY: SPACE AND TIME SCALES

John H. Steele

Woods Hole Oceanographic Institution
Woods Hole, Massachusetts 02543 U.S.A.

ABSTRACT

The variability at all space and time scales observed in the upper ocean is in marked contrast with the relative uniformity of distributions in the deep ocean. The complexity and variability of the upper layers of the ocean at a wide range of space and time scales has been demonstrated by the dense sampling techniques available for temperature, salinity and chlorophyll. It is highly probable that for many chemical constituents there will be similar structure which is not accessible for study by present methods. Three partial resolutions of this problem are suggested: (1) Development of methods to give comparable density of observation in the three dimensions x, z & t, (2) Relations between selected organic compounds which may have large coherence scales as a result of their specific biological functions, and (3) Integration of surface and deep water studies, particularly, but not solely, through the use of sediment trap methods.

INTRODUCTION

Upper ocean chemistry is necessarily dependent on physical and biological processes and so will be influenced by the great fluctuations in these processes in the near-surface layers. Both the physics and the biology have been shown to have variability at a wide range of space and time scales. Especially, the horizontal variance at meso-scales and smaller (100 m - 100 km) is significant, not only in terms of sampling design, but in understanding energetic exchanges (whether physical or biological). Similarly, in the vertical, fine structure at centimetre scales is significant.

Physical, and some biological, instrumentation permits sampling across this range of scales. At present, however, there is relatively little chemical technology available for comparable sampling intensity, horizontally and vertically.

Conversely, in the deep ocean, chemical tracers, particularly radionuclides, have provided a unique method of study for longer term and larger

Woods Hole Oceanographic Institution Contribution No. 6093

scale investigations of both physical and biological processes. In this area of study chemists no longer provide a service to the other disciplines, but lead in the development of concepts and in the testing of hypotheses about both physical circulation and biological cycling through the deep ocean. The apparent lack of such a role in the upper ocean raises questions about shortcomings in analytical methods and, especially, sampling regimes for the chemical processes in the upper layers.

This discussion of these problems will focus on the questions of scales derived from biological dynamics in relation to chemical processes. As a start, however, it is necessary to consider the general classes of transformations which affect the chemistry.

RELEVANT PROCESSES

At its simplest (and naivest) we are concerned with three types of processes: (1) changes in chemical composition, (2) changes in distribution due to physical processes, and (3) changes due to biological activity. For many chemical elements, or compounds, all three factors affect the distribution. As end members of this spectrum one can take salinity (or NaCl) and nitrogen. Sodium and chloride ions are essentially conservative in the interior of the ocean. Changes in concentration are either at the surface boundary (evaporation/precipitation) or by internal redistribution. Nitrogen is subject to physical redistribution (mainly as NO_3) and to biological exchanges between dissolved and particulate phases; and necessarily is present in a large number of chemical forms. Many of the other chemical elements or compounds of interest participate in all these types of process. We need to know the type of reactions involved and the rates for these processes. In the context of upper ocean dynamics (and so on time scales very much less than geological) any component which behaves in exactly the same way as chlorinity, is of little extrinsic chemical interest. A sub-class of such compounds are those which change only at the boundaries. Any compound for which the exchange was in equilibrium would also be of little interest since there would be little discrimination with other elements conservative in the interior. There are two ways in which information about the interior of the fluid can be obtained. First, by major changes in the time function of the input at the boundary - as with tritium derived from testing of nuclear weapons. Secondly, if there is a chemical change in the *interior* with a *defined* rate. In practice, this has only resulted from radioactive decays where parent and daughter product can be measured. The best example is the work with ^3H - ^3He (Jenkins, 1980).

These techniques have been perhaps the main reason for the strength of chemistry in deep ocean studies. They have two drawbacks. One, inherent in the complicated and rigorous analytical techniques, is the relatively small number of results that can be obtained. A second problem arises when it may not be possible to consider the elements to be completely soluble and conservative (apart from isotopic decay). This is the case with series such as ^{226}Ra - ^{210}Pb, where interior and boundary adsorption on particles and consequent particulate fluxes are necessary to explain deep Atlantic distributions (Spencer et al., 1981). At the same time, lack of fundamental knowledge about the rates of adsorption and release permits ambiguities in the interpretation of the boundary conditions. Thus, the inclusion of particle flux rates makes the questions more interesting, but less precisely soluble. The use of sediment traps to measure vertical flux rates, may help to eliminate this ambiguity, but would not take account of the quasi-horizontal (or isopycnal) dispersion of fine particles with very slow sinking rates. Thus, a knowledge of particle size distribution is also needed.

The success of these methods in deep water results from highly sensitive measurement techniques that estimate rate processes. The numbers of observations that can be made is relatively small and so the interpretation depends on the assumption that the data are not affected significantly by fine structure in space or time. In the upper layers of the ocean this assumption is inherently unlikely to hold.

The use of radon (^{222}Rn) for rate measurements illustrates this problem. The great variability in vertical profiles near the sea surface (Broecker and Peng, 1974) appears to preclude its use for estimation of vertical eddy diffusivity, although it has been used to obtain such estimates near the sea floor (Broecker et al., 1968). It is an interesting question whether high intensity data acquisition for this measurement would permit estimates of vertical mixing rates since these values are possibly the most significant unknown for the upper layers. Further discussion of the application of ^{222}Rn to the study of near-surface processes is given in this volume by Roether.

A similar type of problem arises with trace metal concentrations. With improved methods there are now clear relations between deep water profiles of metals such as cadmium, nickel and zinc, with micronutrients such as phosphate, nitrate, and silicon. It is possible to categorize metals by their similarity in distribution to particular nutrients, e.g., the cadmium/phosphate relation. However, when data from near-surface horizontal transects are examined, there are no simple relations. Boyle et al. (1981) have reviewed the near-surface distributions of various trace metals usually with sampling at stations 10-100 km apart. They find no general relation. They conclude for cadmium, that "these data prove that there are significant differences between the surface distributions of cadmium and phosphate. Determining the mechanisms producing these differences will contribute significantly to understanding of the biological uptake and recycling of trace metals". The question arising here is whether the relations between, say, one trace metal, one nutrient, and possibly some measure of particulates, can be stated in a simple and general form. In the same paper, Boyle et al. propose a simple general model of the relation between a trace metal (X) and dissolved phosphate (P) in surface waters:

$$\left[\frac{X}{X_0}\right] = \left[\frac{P}{P_0}\right]^\gamma \tag{1}$$

where X_0 and P_0 are the winter or upwelled concentrations; γ represents a fractionation factor dependent on uptake by plants and grazing by animals. The results presented do not suggest that such a generalization is immediately apparent. Is it likely that such relatively precise patterns could be expected?

The fractionation factor γ is assumed to be of the form

$$\gamma = \prod_{i=1}^{n} \gamma_i$$

where the γ_i are the individual factors for each process such as photosynthetically induced uptake, animal excretion and bacterial decomposition. The equations governing transformation of a nutrient through each component

$$\frac{dP_j}{dt} = f(P_j) \tag{2}$$

(for a set, j, where j represents plants, animals, bacteria and the dissolved materials) are all highly non-linear, and the corresponding system

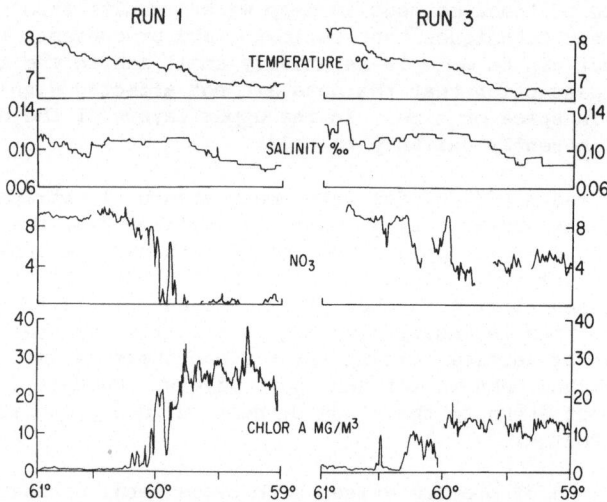

Fig. 1. Temperature, salinity, nitrate and
chlorophyll data collected at 3 m
depth on N–S transects at 0°30'E on
6 and 10 April 1974 (runs 1 and 3,
respectively) (from Steele and
Henderson, 1977).

for another chemical element

$$\frac{dP_i}{dt} = f(X_j)$$

(3)

can only exhibit simple relations if $\gamma_i = 1$ for all i.

Since this equivalence does not hold even for isotopes of the same
element, it is extremely unlikely that a relation of the form (1) will be
derived from sets of equations such as (2) and (3).

For these reasons, studies of many chemical properties, both organic
and inorganic, are likely to be faced with the same problems as those
involving say, nitrate and chlorophyll which can be measured semi-
continuously. An example of the varying inter–relations is given by a set
of measurements in the North Sea (Fig. 1), which demonstrates that: (1)
simple (negative) relations can exist between the dissolved and particulate
phases, but these are likely to be short lived and the usual patterns in
horizontal distributions do not show obvious relations, and (2) these
patterns need have no apparent relation to simultaneously measured near–
surface physical phenomena, demonstrating a need for two or three–
dimensional presentations.

At the same time these dense data sets have led to a "spectral" app–
roach to the data analysis (Platt and Denman, 1975) and to the modeling of
the results (Steele and Henderson, 1977; Fasham, 1978), in the biology as
well as in the physics. It is necessary to expect that a similar approach
requiring dense data sets will be needed for the chemistry of this upper
layer of the ocean. Thus, the problem concerns instrumentation and
sampling patterns.

CRITICAL SCALES FOR CHEMICAL PROCESSES

Much of the conceptual argument and data analysis relating to biological processes has focused on the expected space and time scales at
which the biological rates can modify distributions of biological parameters in the presence of physical mixing and advection. The simplest
representation, in terms of diffusion for a biological parameter, B, is

$$\frac{DB}{Dt} = K_y \frac{\partial^2 B}{\partial y^2} + gB \tag{4}$$

where y can be a horizontal (x) or vertical (z) dimension, K_y is the
diffusivity or mixing coefficient ($cm^2 s^{-1}$) and g represents the rate of
change (s^{-1}) of the population due to biological processes. In a very
simple dimensional analysis the physical and biological processes are of
the same order represented by equation (5), when the length scale L (cm)

$$L = O(\sqrt{K_y/g}) \tag{5}$$

For vertical processes where there is no vertical eddy diffusivity then
$K_z = O(10^{-4} cm^2 s^{-1})$. For phytoplankton with a doubling time of one day,
growth rate is $O(10^{-5} s^{-1})$ or less. Thus, length scale $L_z = O(3 cm)$ or
slightly greater. Essentially this has been the basis for the rebuttal by
Jackson (1980) and others, of arguments by McCarthy and Goldman (1979),
that nutrient and other chemical exchanges could occur at scales of
$O(100\mu)$. There is some evidence using *in situ* fluorometry (Fig. 2) in
support of 1-10 cm scales and, more recently, Goldman (1984) has proposed
that in oligotrophic water, "marine snow" may provide the sites for
chemical cycling comparable to the macro scale events of the classical
spring bloom. These scales could be relevant to behavior of individual
grazers - the traditional copepods or the newer micro-zooplankton.

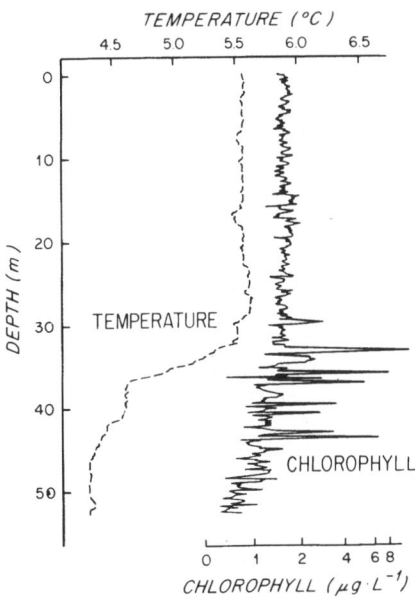

Fig. 2. Vertical profiles in the
Baltic Sea of temperature
and of chlorophyll a from
an *in situ* fluorometer
(from Derenbach et al.,
1979).

The need for such high vertical resolution in the physics has been demonstrated through the study of processes such as salt fingering (Schmitt, 1979). We need confirmation of such fine structure in biological variance in the upper layers in stable thermoclines. For a full understanding of the dynamics, a knowledge of chemical distributions and processes is required at these small scales.

It is difficult, however, in the physics or the biology, to maintain such resolution when the interactions of horizontal and vertical processes are considered. These small vertical features have to be smoothed or ignored. For horizontal dispersion in the upper layers, neglecting special features such as fronts, the same basic relation (4) applies. Using horizontal diffusive coefficients $O(10^5 - 10^6$ cm^2 s^{-1}) gives $L_x = O(km)$ as the critical scale where physical and biological processes are comparable. There is considerable literature on the near-surface variability (Steele, 1978) but the interactions between horizontal and vertical distributions are of particular interest. Data obtained with the Batfish (Denman and Mackas, 1978) (Fig. 3) show how structure at less than kilometre scales can be removed from fluorometer data when it is plotted against density rather than depth. Once more the comparison is of biology with physics, but the linking nutrient processes and other chemical parameters cannot be measured in the same way.

Thus, measurements widely separated in space and time will tend to be aliased in relation to the physical and biological processes so that only the grossest averaging will be possible. But, if the complete cyclical

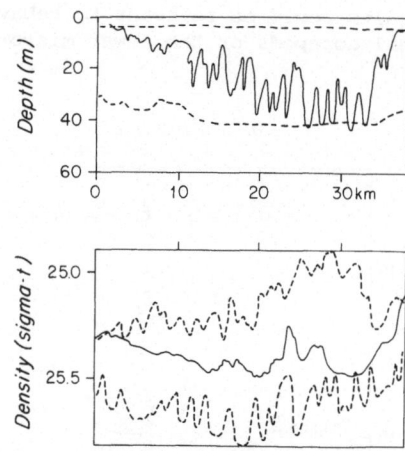

Fig. 3. Chlorophyll variability on a Batfish transect in the Gulf of Maine, plotted against depth and sigma-t. The dashed lines show the limits of undulation by the Batfish which cycles approximately once every 0.5 km. The solid line is the 1.0 mg m^{-3} chlorophyll contour within a vertical gradient decreasing with depth (after Denman and Mackas, 1978).

processes are to be understood for, say, metals such as cadmium, copper or zinc, then the dynamics in the upper layer will need to be investigated in the same detail given to the physics and to certain parts of the biology.

ORGANIC CYCLES

Although a simple stoichiometry between elements seems unlikely in terms of biomass, there is increasing interest in the relation between organic compounds and particular species or groups. This, in turn, is related to questions of the space and time scales for coherence in species composition or in age structure of animals. It has long been recognized that presentations of data as percentage phytoplankton species composition or dominant size groups (Fig. 4), or copepod stage classes, appear less erratic than the biomass data (Fig. 5). Recently, this has been quantified in studies by Mackas (1984) who showed that whereas zooplankton or dry weight data had coherent lengths O(10 km) or less, the community structure was coherent on scales O(100 km). This is explicable if one assumes that biomass patchiness is generated by processes with daily scales such as phytoplankton growth and zooplankton diurnal migration (Evans, 1978). But, the species or age structure is connected to the overall life cycle of the herbivores O(30 - 50 days) and so could have comparably larger spatial scales.

Relationships between organic compounds which are dependent on these species structures may, in turn, demonstrate similar larger scale coherence. It has been proposed that proportions of accessory pigments in phytoplankton can exhibit these features (Yentsch, 1980). The arguments about Redfield ratios of C:N:P in relation to phytoplankton production rates (Goldman, 1984) illustrate the use of relative composition in the field and in the laboratory to elucidate questions about population dynamics. More detailed analyses of particular compounds such as specific sterols (Gagosian and Nigrelli, 1979) could provide information about the growth and removal of species or groups such as diatoms. The relative abundance and type of lipids, particularly wax esters, in zooplankton may have patterns dependent not only on species (Sargent, 1976), but also on the previous food supply (Lee et al., 1971; Lee and Barnes, 1975). Such field data combined with experimental results could provide the equivalent of demographic data utilizing the nutritional status from sub-samples to indicate the time amd space integrated inter-relations of a population with its food supply. Thus, not only is there a basic need to acquire more knowledge of significant biochemical processes rather than to rely on bulk estimates such as organic carbon, but also the relative distributions of these particular compounds may integrate the effects of smaller scale variability decreasing the sampling problems and increasing their usefulness as measures of overall cycles in the upper layers.

VERTICAL EXCHANGES

There is a marked contrast between the variability at all space and time scales observed in the upper ocean and the relative uniformity in patterns of distribution, of nutrients and metals for example, in the deep ocean. Thus one can characterize different ocean basins by their vertical profiles of certain elements which show regular large scale gradients and systematic relations between elements. How are the large short-term fluctuations smoothed or integrated in transfer to deeper water? It was relatively easy to ignore or minimize this problem so long as vertical transport was considered to be a slow or negligible process. In particular, deep water concentrations of, say, silicon or phosphate could be considered as nearly conservative and so used directly as physical tracers.

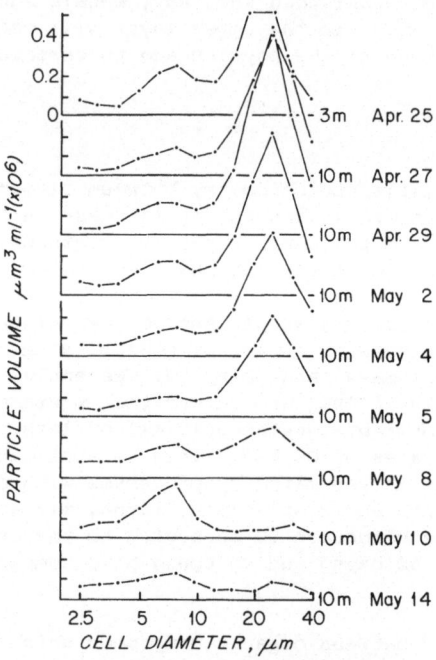

Fig. 4. Particle size distribution at 10 m depth on the
Fladen Ground, North Sea,
in 1976 (from Gamble,
1978).

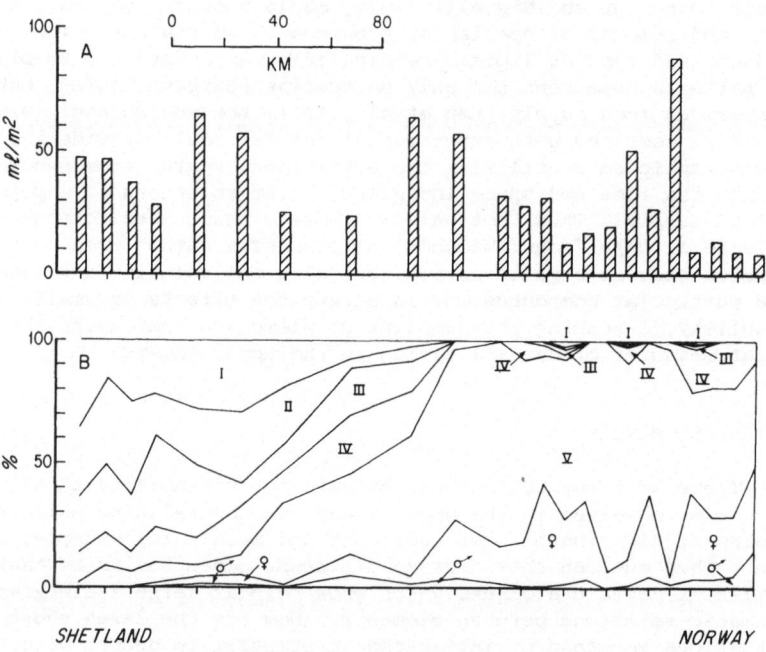

Fig. 5. (A) Plankton volumes and (B) stage distribution of
the dominant copepod *Calanus finmarchicus* along a
transect across the northern North Sea (from
Geophysical Institute, Bergen, 1976).

Data on fast flux rates, from sediment traps and from other sources, make this less acceptable. The development of sediment trap technology is a major element in both surface and deep water studies in spite of the problems about their performance and replication. Shallow floating traps are used to estimate the residual component of the rapid cycles in the surface layer and relate to the questions about "new" versus "recycled" production (see Eppley, this volume). In deep waters, traps provide an estimate of cross-pycnal fluxes and rates of decomposition processes which transfer constituents from the particulate to a dissolved phase, and possibly in the other direction.

The two uses of this technique should be complementary and yet studies from these two aspects appear to be pursued separately. Certainly the questions are different and this is mainly because of the different time and space scales. Upper ocean dynamics are now treated in terms of hours or shorter, whereas deep ocean processes are viewed as seasonal or longer. It is of interest to consider the time and space scales appropriate for a trap at, say, 200 m. If sinking rates are of the order of 10 - 100 m day^{-1} then there will be some integration over periods of 2 - 20 days. Similarly, horizontal diffusion (ignoring advection) will integrate over scales from say, 2 - 20 km (using a 1 km day^{-1} relation). At greater depths the range and the scales will be proportionately larger. Thus, the traps not only provide an estimate of the much slower rates of "new" nutrient input (assuming steady state), but will smooth out, in some unknown way, the fine structure in the upper layers. This is one of their major potential advantages whether one regards the deep ocean as a sewer for the feces from the upper layer, or alternatively views the upper layer as a black box releasing interesting chemicals for study as they progress to the sediments.

For the former, it has become apparent through the oxygen studies of Schulenberg and Reid (1981) and Jenkins (1982) that an understanding of deeper processes is essential to resolve important or apparent contradictions in the upper layers. For the deep ocean dynamics, some knowledge of the upper layer processes will prove essential if, say, nitrogen cycles which involve cross-pycnal fluxes are to be used as critical tests of physical and chemical models based on isopycnal mixing and advection determined from data on temperature, salinity and $^3H/^3He$.

CONCLUSIONS

The complexity and variability of the upper layers of the ocean at a wide range of space and time scales has been demonstrated by dense sampling techniques available for temperature, salinity and chlorophyll. It must be assumed that for many chemical constituents there will be similar structure which is not accessible for study by present methods. There are three partial resolutions of this problem:

(1) Development of methods to give comparable density of observation in the three dimensions x, z, & t.

(2) Relations between selected organic compounds which may have larger coherence scales as a result of their specific biological functions.

(3) Integration of surface and deep water studies, particularly, but not solely, through the use of sediment trap methods.

I have not discussed the vexed questions of the absolute rates of primary production, or the specific microzooplankton cycles in the upper layers. These are considered in other papers. There are important generalizations, proposed by Eppley and others, which could permit extra-polations over large open ocean and coastal areas. There is also the potential of the satellite color imagery to provide one surface parameter at a wide range of space and time scales. There are major challenges to the chemical oceanographer in the context of emerging physical theories of the "ventilated thermocline" which link near-surface fluxes and their seasonal variations with events in the main thermocline. Can we expect to combine satellite observations with sediment trap data using the "Eppley" functional relations for the upper layer, to provide the basis for descriptions of global fluxes of chemical constituents?

REFERENCES

Boyle, E.A., Huested, S.S., and Jones, S.P., 1981, On the distribution of copper, nickel, and cadmium in the surface waters of the North Atlantic and North Pacific Ocean, J. Geophys. Res., 86:8048.

Broecker, W.S., and Peng, T.-H., 1974, Gas exchange rates between air and sea, Tellus, 26:21.

Broecker, W.S., Cromwell, J., and Li, Y.H., 1968, Rates of vertical eddy diffusion near the ocean floor based on measurements of the distribution of excess ^{222}Rn, Earth Planet. Sci. Lett., 5:101.

Denman, K.L., and Mackas, D.L., 1978, Collection and analysis of underway data and related physical measurements, in: "Spatial Pattern in Plankton Communities", J.H. Steele, ed., pp.85-109, Plenum Press, New York.

Derenbach, J.B., Astheimer, H., Hansen, H.P., and Leach, H., 1979, Vertical microscale distribution of phytoplankton in relation to the thermocline, Mar. Ecol. Progr. Ser., 1:187.

Evans, G.T., 1978, Biological effects of vertical-horizontal interactions, in: "Spatial Pattern in Plankton Communities", J.H. Steele, ed., pp.157-179, Plenum Press, New York.

Fasham, M.J.R., 1978, The application of some stochastic processes to the study of plankton patchiness, in: "Spatial Pattern in Plankton Communities", J.H. Steele, ed., pp.131-156, Plenum Press, New York.

Gagosian, R.B., and Nigrelli, G.E., 1979, The transport and budget of sterols in the western North Atlantic Ocean, Limnol. Oceanogr., 24:838.

Gamble, J.C., 1978, Copepod grazing during a declining spring phyto-plankton bloom in the northern North Sea, Mar. Biol., 49:303.

Geophysical Institute, Bergen, 1976, Some preliminary results from a synoptic experiment in the Norwegian coastal current, Serial Rep. Mar. Inv., No.1, 55pp.

Goldman, J.C., 1984, Oceanic nutrient cycles, in: "Flows of Energy and Materials in Marine Ecosystems. Theory and Practice", M.J.R. Fasham, ed., pp. 137-170, Plenum Press, New York.

Jackson, G.A., 1980, Phytoplankton growth and zooplankton grazing in oligotrophic waters, Nature, Lond., 284:439.

Jenkins, W.J., 1980, Tritium and ^{3}He in the Sargasso Sea, J. Mar. Res., 38:533.

Jenkins, W.J., 1982, Oxygen utilization rates in North Atlantic subtropical gyre and primary production in oligotrophic systems, Nature, Lond., 300:246.

Lee, R.F., and Barnes, A.T., 1975, Lipids in the mesopelagic copepod, Gaussia princeps. Wax ester utilization during starvation, Comp. Biochem. Physiol., 52B:265.

Lee, R.F., Hirota, J., and Barnett, A.M., 1971, Distribution and importance of wax esters in marine copepods and other zooplankton, Deep-Sea Res., 18:1147.

Mackas, D.L., 1984, Spatial autocorrelation of plankton community composition in a continental shelf ecosystem, Limnol. Oceanogr., 29:451.

McCarthy, J.J., and Goldman, J.C., 1979, Nitrogenous nutrition of marine phytoplankton in nutrient-depleted waters, Science, N.Y., 203:670.

Platt, T., and Denman, K.L., 1975, Spectral analysis in ecology, Ann.Rev. Ecol. Systematics, 6:189.

Sargent, J.R., 1976, The structure, metabolism and function of lipids in marine organisms, in: "Biochemical and Biophysical Perspectives in Marine Biology", D.C. Malins and J.R. Sargent, eds, pp 149-212, Academic Press, New York.

Schmitt, R.W., 1979, Flux measurements on salt fingers at an interface, J. Mar. Res., 37:419.

Schulenberger, E., and Reid, J.L., 1981, The Pacific shallow oxygen maximum, deep chlorophyll maximum and primary productivity, reconsidered, Deep-Sea Res., 28:901.

Spencer, D.W., Bacon, M.P., and Brewer, P.G., 1981, Models of the distribution of ^{210}Pb in a section across the North Equatorial Atlantic Ocean, J. Mar. Res., 39:119.

Steele, J.H., 1978, Some comments on plankton patches, in: "Spatial Pattern in Plankton Communities", J.H. Steele, ed., pp 1-20, Plenum Press, New York.

Steele, J.H., and Henderson, E.W., 1977, Plankton patches in the northern North Sea, in: "Fisheries Mathematics", J.H. Steele, ed., pp. 1-19, Academic Press, London.

Yentsch, C.S., 1980, Phytoplankton growth in the sea: A coalescence of disciplines, in: "Primary Productivity in the Sea", P.G. Falkowski, ed., pp. 17-32, Plenum Press, New York.

THE CHEMISTRY OF NEAR-SURFACE SEAWATER

Peter S. Liss

School of Environmental Sciences
University of East Anglia
Norwich, NR4 7TJ U.K.

ABSTRACT

The top few hundred micrometres of the sea (often called the surface microlayer) is a difficult part of the oceans to study. The depth of the region is almost impossible to define in a meaningful way, and microlayer thicknesses are by default specified in terms of what the various sampling devices appear to collect. The organic composition of the microlayer is poorly characterized and its study suffers from similar problems to analogous work in bulk seawater. Some dissolved constituents in the microlayer appear to show small enrichments in concentration over subsurface waters, although such enrichments now seem smaller than was previously thought to be the case. Particulate material, on the other hand, does show significant microlayer enrichments.

What is clear is that processes in the water close to the sea surface are important not only for the chemistry of the underlying water but also for the marine atmosphere. Such processes include air-sea exchange of stable gases and particles, as well as free radicals and other unstable species. Furthermore, biological and photochemical activity near the ocean surface leads to the production of gases and to speciation changes for ionic forms. Photoreduction of multivalent elements in the presence of organic matter appears to be a particularly important sea surface mechanism for redox reactions.

INTRODUCTION

Although near—surface waters are acknowledged to be a very important zone with respect to chemical processes in the oceans, the chemistry of these waters has received rather little attention. For this and other reasons discussed later, our knowledge in this area is still somewhat rudimentary. In the present account some aspects of the topic will be discussed. No attempt has been made to be comprehensive; several review papers have appeared in recent years of which the most relevant in the present context include those by MacIntyre (1974a,b), Liss (1975), Wangersky (1976), Hunter and Liss (1981), Hardy (1982), and Lion (1984). Topics are dealt with in the form of questions. This correctly implies that subjects selected for inclusion are ones for which significant uncertainties still exist. In the final section the possible importance of

reactions and processes occurring in the near surface water for deeper
parts of the upper oceans is discussed. This serves as an introduction to
the more detailed treatment of some of these topics in other chapters of
this book.

HOW THICK IS THE SEA SURFACE REGION?

The answer to this question is about as indefinite as that to the
query "How long is a piece of string?" If one is dealing with such
phenomena as ion rejection, interfacial double layers, or true monolayers
the answer is several nanometres. However, for the oceanographically
important processes of, for example, gas exchange or heat conduction then
about 100 μm is probably the relevant thickness. These very different
depth scales are elegantly illustrated in Fig. 1, which is taken from
MacIntyre (1974a). It should be noted that in Fig. 1 the vertical axis is
logarithmic and, as MacIntyre points out, continuing the depth scale to
include the deepest ocean trenches would extend it by less than a factor
of two. He remarks that in terms of *processes* the sea surface region may
prove as rich a field for research as the deeper half of the "logarithmic"
ocean.

Thus, the only answer to the question posed above is that it depends
on the properties one is interested in. Since this chapter is about the
chemistry of near-surface seawater, here it is the thickness of water
required in order to obtain chemical information. Obviously, chemists
would like to be able to analyse material from as near the interface as
possible. However, the most widely used samplers collect material down to
depths of tens to hundreds of micrometres from the surface, so that most of

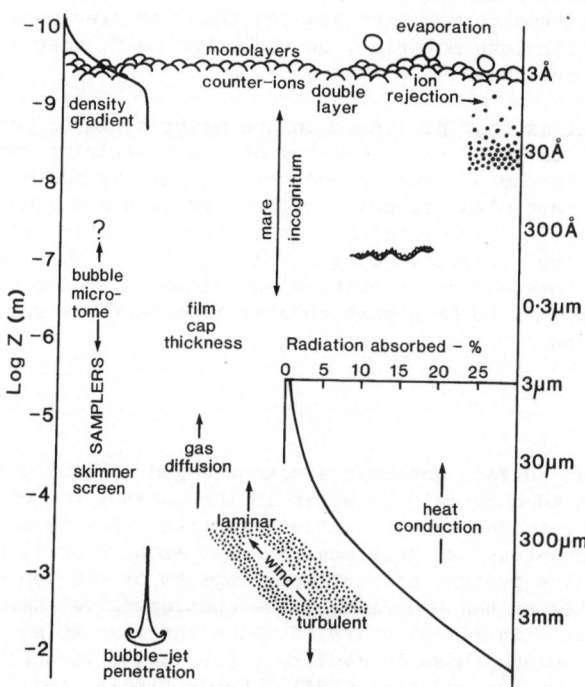

Fig. 1. Depth range of processes occurring
        in the top millimetres of the ocean
        surface (from MacIntyre, 1974a).

the presently available information on chemical composition is for this thickness of the sea surface. The term 'sea surface microlayer' is often used to describe samples harvested in this depth range. There are two sampling techniques which appear to collect thinner layers, but their utility is rather limited. These, together with the more conventional "microlayer" samplers, are discussed in the next section.

HOW IS THE SEA-SURFACE REGION SAMPLED?

The first practical device for sampling the sea surface microlayer was devised by Garrett (1965). It consists of a mesh screen of metal or plastic sample bottle for subsequent chemical analysis. With this type of device the top 100-300 μm are collected, and reasonably large volumes of water (hundreds of millilitres) can be harvested. Garrett screens have proved the most popular of the microlayer samplers so far invented. Other devices, collecting similar thicknesses and volumes, which have been found useful include a rotating ceramic drum (Harvey, 1966) and a glass plate (Harvey and Burzell, 1972). Since they can harvest reasonably large volumes, sophisticated chemical analysis of the water is possible, and these "microlayer" samplers have so far been the mainstay of research into the chemical composition of near-surface seawater.

A cryogenic device, in which a slice of water about 1 mm thick is frozen onto a flat plate collector cooled with liquid nitrogen, has been described by Hamilton and Clifton (1979). Although not widely used, it has found application for the collection of microlayer samples for dissolved trace gas analysis (Turner and Liss, 1985). Clearly, more conventional samplers in which the water is exposed to the atmosphere after collection are unsatisfactory for this type of work.

Layers nominally substantially thinner than those harvested by the samplers described above have been obtained by collecting the aerosol formed when artificially-produced bubbles burst at the sea surface. The bubble acts as a microtome and is thought to peel off a layer from the sea surface whose thickness is 0.05% of the diameter of the bursting bubble (MacIntyre, 1968). For a 1 mm bubble this corresponds to a "cut" of approximately 0.5 μm; i.e. two orders of magnitude thinner than can be achieved with the more traditional "microlayer" samplers. A practical realisation of this idea has been described by Fasching et al. (1974) and is called by them a Bubble Interfacial Microlayer Sampler (BIMS). With this device it is possible to collect tens of millilitres of water, so that quite detailed chemical analysis is possible. The device does suffer from a number of problems, including the need to deploy it under only the calmest of sea states, and high blanks (measured with the bubbler switched "off") particularly in the analysis of trace substances. More fundamental problems arise since the bubbles scavenge material from the subsurface water during their rise to the interface, and from the possibility of fractionation occurring during the actual bursting process. All in all, it would seem that the BIMS is more likely successfully to mimic and collect naturally produced marine aerosol then to sample an undisturbed slice of the order of 1 μm thickness from the sea surface.

The sampling device which appears to harvest the thinnest layers is the germanium prism described by Baier et al. (1974). This relies on the traditional Langmuir and Blodgett technique of using a hydrophilic material to collect monolayers from water surfaces. The thickness sampled appears to be approximately 30 nm when the collected material is in the dry state. It has been argued by Hunter and Liss (1981) that in the wet state (as at the sea surface) the material will be considerably expanded, possibly reaching dimensions of up to 1 μm. With such thin layers and considering

43

the small dimensions of the sampling prism, it is inevitable that only very small sample volumes are collected, so that the analysis which can be performed is strictly limited. In fact, by making the prism of germanium it is possible to obtain infra-red spectra of the adsorbed material directly, and this allows some of the major organic functional groups to be identified.

Further details of the samplers discussed above, as well as others not mentioned, are to be found in several of the review articles listed in the Introduction to this Chapter. A paper specifically devoted to sea surface samplers is that by Garrett and Duce (1980).

WHAT IS THE ORGANIC CHEMICAL COMPOSITION OF MATERIAL FROM NEAR THE SEA SURFACE?

In early work only lipid-type material was analysed (Garrett, 1967), so that it was widely accepted that the organic matter in the microlayer was largely "dry" surfactants (straight chain largely saturated hydrocarbons with a hydrophilic head group). By straightforward mass balance arguments it can be shown that lipids represent, at the most, only 10% of the total organic matter (Liss, 1975).

Infra-red analysis of the thinner layers harvested by the germanium prism technique indicates the presence of mainly polysaccharide and polypeptide chains, fixed to the sea surface by occasional hydrophobic functional groups (Baier et al., 1974). Material of this type is often referred to as "wet" surfactant.

The past decade has seen little further fundamental advance in our knowledge of the organic chemistry of the sea surface. Characterisation of this material is at least as difficult as the analogous problem in bulk seawater, indeed there are many reasons for thinking that the organic material in the microlayer and the bulk water organics is rather similar in nature. For many years considerable efforts have been made to elucidate the organic chemistry of subsurface seawater, but these studies have only recently begun to yield information on possible pathways to the formation of marine humic materials as well as to its structure (Harvey et al., 1983). Much of this work is likely to be directly applicable to similar studies of microlayer organics, so that significant steps forward can be expected in the near future.

WHAT SUBSTANCES OCCUR AT ELEVATED CONCENTRATIONS IN THE SEA SURFACE MICROLAYER?

Early studies of the sea surface microlayer seemed to indicate that, relative to bulk seawater, it was enriched substantially with respect to many components, e.g., total dissolved and particulate organics, bacteria and other microorganisms, pesticides and PCBs, trace metals, and plant nutrients (Liss, 1975). However, this assessment was based on very few analyses and, as more data have become available, the range of enriched substances and their degrees of enrichment have both decreased. For example, Chapman and Liss (1981) found little or no significant microlayer enrichment for any of the plant nutrients in a range of samples collected in U.K. coastal waters. These authors suggested that the apparent enrichments of nutrients found by others may have been due to depleted concentrations in the subsurface water, relative to the microlayer, rather than true microlayer enrichments. Carlson (1983) has compiled measurements of dissolved organic carbon (DOC) in microlayer and bulk seawater and these are shown in Fig. 2. Most points plot close to the 1:1 line indicating that

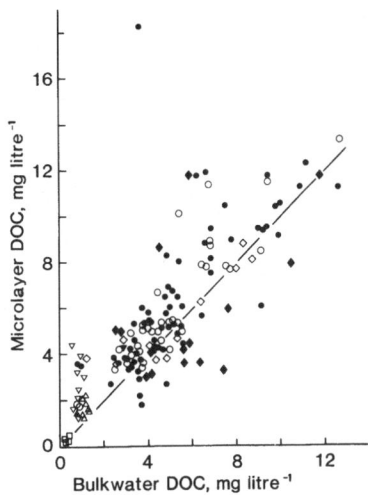

Fig. 2. Compilation of reported concentrations
of dissolved organic
carbon (DOC) in the
microlayer and bulk
seawater (from
Carlson, 1983).
Broken line indic-
ates equal concent-
rations in both
samples.

substantial microlayer enrichment for DOC is the exception rather than the
rule.

Small but significant enrichments for particular fractions of the
total organics are found. For example, measurements of ultra-violet absorbance, a measure of phenolic compounds, all plot above the 1:1 line,
as shown in Fig. 3. Furthermore, for this particular class of compounds
there is a clear relationship between degree of enrichment and the presence
or absence of visible slicks on the surface (Carlson, 1982).

One of the clearest examples of microlayer enrichment is shown by
particulate organic carbon (POC). Carlson (1983) summarises data he has
collected from estuarine, coastal and oceanic regions (his Table 1) and in
all cases the microlayer is enriched in POC, the extent of the enrichment
(ratio of microlayer to bulk sea-water concentrations) varying from 1.36 to
38.4. A plausible case can be made for the idea that the reason why POC
always shows microlayer enrichment whereas DOC generally does not is that
the surface active fraction of the DOC in bulk seawater achieves a lowering
of its free energy by adsorbing onto particulate material. The particles
are scavenged by bubbles and transported to the sea surface by them; the
adsorbed surface-active organic matter then helps the particles to remain
in the microlayer.

DOES THE MICROLAYER AFFECT THE CHEMISTRY OF THE UNDERLYING SEAWATER?

There are three things which point to the microlayer as a region of
the oceans where chemical transformations are likely to occur: (1) the flux

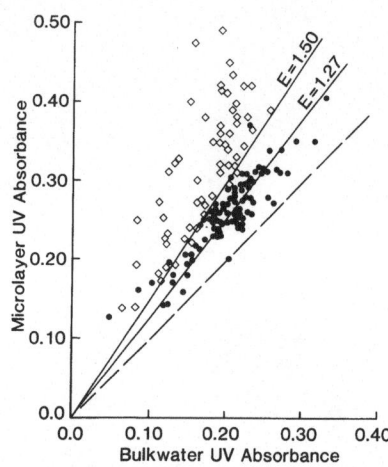

Fig. 3. Microlayer ultra—violet absorbance
(/280 nm) against that of bulk sea--
water (from Carlson, 1982). Dashed
line indicates equal absorbance in
both samples. Line of slope 1.27
(r^2 = 0.99, including 0,0) denotes
mean enrichment of clean—surface
samples in terms of ultra—violet
absorbance. Line representing
enrichment of 1.50 approximates
boundary between clean and slicked
samples. Lowest absorbances (<0.10)
of bulk seawater are from waters of
salinity <33‰. ● — Clean surfaces;
◇— Slicked surfaces. Several slicked
surfaces had absorbances >0.50.

of photons (and atmospherically-generated radicals and other reactive
species) is largely unattenuated; (2) there is an abundance of certain
classes of microorganisms (Sieburth, 1983, and this volume); and (3) the
presence of dissolved and particulate organic matter, the former not
necessarily enriched relative to subsurface water — see earlier discussion.
The formation or consumption of chemical species in the microlayer will
make it a potential source or sink of these species for the underlying
water, as well as for the atmosphere. Although conditions appear to favour
the microlayer as an important zone of chemical change, there is so far
rather little direct evidence (i.e., gathered at sea) for this contention.
Some possibilities are discussed below. Here, no consideration is given to
the processes of gas exchange across the air-sea interface; the topic is
dealt with in this book by Roether. A discussion of the possible role of
organic material (both natural and man-made) at the sea surface on inter—
facial gas transfer is given by Liss (1983).

Biologically Produced Gases

With respect to dissolved trace gases, it is well known that many of
them are formed by organisms in the marine photic zone, e.g. methane,
nitrous oxide, dimethyl sulphide (DMS) and other sulphur gases, hydrogen,
methyl iodide. However, in most cases it appears that the depth of maxi—
mum production of the gases is some metres below the surface. Of course,
such subsurface production will serve as a source of the gases for both the
underlying water and for air-sea exchange. There is one report of micro—

layer enrichment for DMS (Nguyen et al., 1978). These authors used a Garrett screen to sample the microlayer and found it to be up to five times enriched in DMS relative to subsurface waters. A more extensive series of measurements by Turner and Liss (1985), using the more acceptable cryogenic microlayer sampling technique described earlier, showed that there is often no measurable enrichment. On the occasions when significant enrichment is found it may well be due to the stress of the sampling procedure causing the microlayer organisms to release DMS into the sample. In any case, unless the rate of production in the microlayer is very large, it is hard to see how microlayer enrichment can be maintained in the face of both downwards mixing into the underlying water and escape across the sea surface.

Input of Reactive Atmospheric Species

Thompson and Zafiriou (1983) have tried to assess the effect of inputs of atmospherically-generated free radicals and other reactive species on both the chemistry of the microlayer and that of the underlying water. In the latter case they find that, in general, the air-to-sea flux is insufficient to compete with rates of *in-situ* production. However, for the microlayer region some interesting effects can occur. For example, input from the atmosphere of radicals such as HO_2 and CH_3O_2 is calculated to lead to a significant destruction in the microlayer region of NO produced photochemically in seawater (see later). Such 'flux capping' could lead to substantial overestimation of the sea-to-air flux of NO, when the conventional subsurface water concentrations are used to estimate the flux. Another example is the reaction of atmospheric O_3 with some dissolved components of the microlayer (principally I^-, NO_2^-, and organic matter). The reaction of O_3 with I^- is particularly interesting geochemically since it can lead to the formation of various volatile iodine compounds, which are then free to cross the air-sea interface. Such fluxes will add to the biologically mediated flux of methyl iodide (Rasmussen et al., 1982) and together they will help to explain the elevated I/Cl ratios found in the marine atmosphere and in terrestrial environments (Miyake and Tsunogai, 1963).

Photochemical Transformations

The chemistry of light-induced chemical reactions in near-surface seawater is a subject still very much in its infancy, but one which must prove a fertile area for research in the future. There are already indications of its importance, and some examples are given below. The current state of knowledge has been reviewed by Zika (1981) and Zafiriou (1983 and this volume).

One of the first examples of photochemistry in action in seawater was the report by Wilson et al. (1970) that carbon monoxide measured in surface seawater was produced photochemically as well as microbiologically. It has been suggested that one mechanism of CO formation is photolysis of formaldehyde to yield hydrogen gas together with carbon monoxide (Zika, 1981). As discussed by Zafiriou (1983), another possible route to CO is by photochemical breakdown of marine humic material.

Another light-induced reaction which leads to the formation of a gaseous product is photolysis of nitrite ions to yield nitric oxide and hydroxyl free radicals (Zafiriou et al., 1980). The former can then escape across the sea surface to the marine atmosphere, while the hydroxyl radical will react with many different species in surface seawaters.

Irradiation of organic material in seawater can lead to the formation of hydrated electrons, which will react rapidly to form, amongst other products, hydrogen peroxide. This substance is of considerable importance for the redox chemistry of seawater since it has been suggested by Breck (1974) that it is the O_2/H_2O_2 couple rather than the O_2/H_2O couple which controls the redox potential of the water. Recent measurements of H_2O_2 in surface seawaters give concentrations of about 10^{-7} mol l^{-1}. This implies a pE of 6.5 if the O_2/H_2O_2 couple is controlling (Moffett and Zika, 1983) as opposed to a pE of 12.5 if O_2/H_2O controls. This is clearly a field where considerable work still needs to be done, but a difference of six orders of magnitude in predicted electron activity gives a lot of scope for revision of our ideas on the redox chemistry of surface seawaters. For example, the substantially less oxidizing conditions at a pE of 6.5 would imply that for several redox couples the reduced rather than the oxidized form would become dominant.

Finally, there are several examples where photochemistry may alter the speciation of trace metals in seawater. In the laboratory, mercuric acetate solutions have been found to produce small but analytically significant amounts of monomethylmercury ion and dimethylmercury on irradiation using only ordinary fluorescent lighting (Jewett et al., 1975). It is not known whether this reaction also occurs in seawater. If it does, however, then since the dimethylmercury is volatile and will tend to escape to the atmosphere, it may help to explain the elevated mercury concentrations reported in air samples collected over some oceanic areas (Fitzgerald et al., 1984). A second example is from the work of Sunda et al. (1983) who found in laboratory experiments that Mn(IV) in seawater can be reduced to soluble Mn(II) when illuminated with sunlight or fluorescent light in the presence of marine humic material. Although not yet shown to be operative in the environment, this reaction would provide one plausible explanation for the elevated levels of dissolved manganese measured near the sea surface in oceanic vertical profiles (Landing and Bruland, 1980; Statham et al., 1985). A similar light-induced reduction of Fe(III) to soluble Fe(II) in the presence of organic matter has been demonstrated in laboratory studies using lake water (Finden et al., 1984).

It is clear from the above examples, which probably represent only the tip of the iceberg of possible reactions, that photochemical transformations are likely to play an important role in the chemistry of the upper oceans.

WHAT RESEARCH NEEDS TO BE DONE IN THE FUTURE?

Future research in the area of sea surface chemistry could profitably concentrate on the following aspects:

(1) Identification of those substances which are truly enriched at or near the sea surface. Substances considered should include both organic and inorganic components of dissolved and particulate fractions, the latter to include living microorganisms.

(2) Assessment of the role of the sea surface region in modifying the two-way transfer of energy and matter between the atmosphere and the oceans due to the presence of chemicals, particularly those that form films, at and near the interface.

(3) Examination of chemical (including photochemical) and biological processes in bringing about chemical transformations in near-surface seawater, and the significance of such changes for both atmospheric and oceanic chemistry.

REFERENCES

Baier, R.E., Goupil, D.W., Perlmutter, S., and King, R., 1974, Dominant chemical composition of sea surface films, natural slicks, and foams, J. Rech. Atmos., 8:571.

Breck, W. G., 1974, Redox levels in the sea, in: "The Sea, Volume 5, Marine Chemistry", E.D. Goldberg, ed., pp.153—179, Wiley-Interscience, New York.

Carlson, D.J., 1982, Surface microlayer phenolic enrichments indicate sea surface slicks, Nature, Lond., 296:426.

Carlson, D.J., 1983, Dissolved organic materials in surface microlayers: Temporal and spatial variability and relation to sea state, Limnol. Oceanogr., 28:415.

Chapman, P., and Liss, P.S., 1981, The sea surface microlayer: Measurements of dissolved iodine species and nutrients in coastal waters, Limnol.Oceanogr., 26:387.

Fasching, J.L., Courant, R.A., Duce, R.A., and Piotrowicz, S.R., 1974, A new surface microlayer sampler utilizing the bubble microtome, J. Rech. Atmos., 8:650.

Finden, D.A.S., Tipping, E., Jaworski, G.H.M., and Reynolds, C.S., 1984, Light-induced reduction of natural iron (III) oxide and its relation to phytoplankton, Nature, Lond., 309:783.

Fitzgerald, W.F., Gill, G.A., and Kim, J.P., 1984, An equatorial Pacific Ocean source of atmospheric mercury, Science, N.Y., 224:597.

Garrett, W.D., 1965, Collection of slick-forming materials from the sea surface, Limnol.Oceanogr., 10:602.

Garrett, W.D., 1967, The organic chemical composition of the ocean surface, Deep-Sea Res., 14:221.

Garrett, W.D., and Duce, R.A., 1980, Surface microlayer samplers, in: "Air-Sea Interaction: Instruments and Methods", F. Dobson, L. Hasse and R. Davis, eds, pp.471—490, Plenum Press, New York.

Hamilton, E.I., and Clifton, R.J., 1979, Techniques for sampling the air-sea interface for estuarine and coastal waters, Limnol. Oceanogr., 24:188.

Hardy, J.T., 1982, The sea surface microlayer: Biology, chemistry and anthropogenic enrichment, Progr.Oceanogr. 11:307.

Harvey, G.R., Boran, D.A., Chesal, L.A., and Tokar, J.M., 1983, The structure of marine fulvic and humic acids, Mar. Chem., 12:119.

Harvey, G.W., 1966, Microlayer collection from the sea surface: A new method and initial results, Limnol. Oceanogr., 11:608.

Harvey, G.W., and Burzell, L.A., 1972, A simple microlayer method for small samples, Limnol. Oceanogr., 17:156.

Hunter, K.A., and Liss, P.S., 1981, Organic sea surface films, in: "Marine Organic Chemistry", E.K. Duursma and R. Dawson, eds, pp.259—298, Elsevier, Amsterdam.

Jewett, K.L., Brinckman, F.E., and Bellama, J.M., 1975, Chemical factors influencing metal alkylation in water, in: "Marine Chemistry in the Coastal Environment", T.M. Church, ed., pp. 304—318, American Chemical Society, Washington, D.C.

Landing, W.M., and Bruland, K.W., 1980, Manganese in the North Pacific, Earth Planet. Sci. Lett., 49:45.

Lion, L.W., 1984, The surface of the ocean, in: "The Handbook of Environmental Chemistry", Volume 1, Part C, O. Hutzinger, ed., pp.79—104, Springer-Verlag, Berlin.

Liss, P.S., 1975, Chemistry of the sea surface microlayer, in: "Chemical Oceanography", Second edition, Volume 2, J.P. Riley and G. Skirrow, eds, pp.193—243, Academic Press, London.

Liss, P.S., 1983, Gas transfer: Experiments and geochemical implications, in: "Air-Sea Exchange of Gases and Particles", P.S. Liss and W.G.N. Slinn, eds, pp.241—298, Reidel, Dordrecht.

MacIntyre, F., 1968, Bubbles: A boundary-layer "microtome" for micron-thick samples of a liquid surface, J. Phys. Chem., 72:589.

MacIntyre, F., 1974a, Chemical fractionation and sea-surface microlayer processes, in: "The Sea, Volume 5, Marine Chemistry", E.D. Goldberg, ed., pp.245—299, Wiley-Interscience, New York.

MacIntyre, F., 1974b, The top millimeter of the ocean, Scient. Amer., 230:62.

Miyake, Y., and Tsunogai, S., 1963, Evaporation of iodine from the ocean, J. Geophys. Res., 68:3989.

Moffett, J.W., and Zika, R.G., 1983, Oxidation kinetics of Cu(I) in seawater: Implications for its existence in the marine environment, Mar. Chem., 13:239.

Nguyen, B.C., Gaudrey, A., Bonsang, B., and Lambert, G. 1978, Re-evaluation of the role of dimethyl sulphide in the sulphur budget, Nature, Lond., 275:637.

Rasmussen, R.A., Khalil, M.A.K., Gunawardena, R., and Hoyt, S.D., 1982, Atmospheric methyl iodide (CH_3I), J. Geophys. Res., 87:3086.

Sieburth, J. McN., 1983, Microbiological and organic-chemical processes in the surface and mixed layers, in: "Air-Sea Exchange of Gases and Particles", P.S. Liss and W.G.N. Slinn, eds, pp.121—172, Reidel, Dordrecht.

Statham, P.J., Burton, J.D., and Hydes, D.J., 1985, Cd and Mn in the Alboran Sea and adjacent North Atlantic: Geochemical implications for the Mediterranean, Nature, Lond., 313:565.

Sunda, W.G., Huntsman, S.A., and Harvey, G.R., 1983, Photoreduction of manganese oxides in seawater and its geochemical and biological implications, Nature, Lond., 301:234.

Thompson, A.M., and Zafiriou, O.C., 1983, Air-sea fluxes of transient atmospheric species, J. Geophys. Res., 88:6696.

Turner, S.M., and Liss, P.S., 1985, Measurements of various sulphur gases in a coastal marine environment, J. Atmos. Chem., 2:223.

Wangersky, P.J., 1976, The surface film as a physical environment, Ann. Rev. Ecol. Systematics, 7:161.

Wilson, D.F., Swinnerton, J.W., and Lamontagne, R.A., 1970, Production of carbon monoxide and gaseous hydrocarbons in seawater: Relation to dissolved organic carbon, Science, N.Y., 168:1577.

Zafiriou, O.C., 1983, Natural water photochemistry, in: "Chemical Oceanography", Second edition, Volume 8, J.P. Riley and R. Chester, eds, pp.339—379, Academic Press, London.

Zafiriou, O.C., McFarland, M., and Bromund, R.H., 1980, Nitric oxide in seawater, Science, N.Y., 207:637.

Zika, R.G., 1981, Marine organic photochemistry, in: "Marine Organic Chemistry", E.K. Duursma and R. Dawson, eds, pp.299—325, Elsevier, Amsterdam.

PROCESSES AFFECTING UPPER OCEAN CHEMICAL STRUCTURE

IN AN EASTERN BOUNDARY CURRENT

James J. Simpson

Scripps Institution of Oceanography
La Jolla, California 92093   U.S.A.

ABSTRACT

The primary transport and transformation processes which alter the
chemical speciation and structure of the upper (100-200 m) ocean within
1000 km of the Californian coast have been examined from the process-
oriented perspective of space and time scale analysis. In this region of
the ocean, upwelling, offshore mesoscale eddies and large-scale "El
Niño"-type processes dominate. Each of these physical processes has a
velocity field which transports chemical species into and out of the
euphotic zone in a unique way. Likewise, each of the resulting mass
transports produces a chemical signature which unambiguously characterizes
that physical process. Major alteration of chemical species which occurs
within the euphotic zone is associated with primary biological productivity. The results presented here for the California Current may have
general application for the interpretation of near-surface chemical variability observed in large areas of the world's oceans.

INTRODUCTION

This paper develops a conceptual framework for the interpretation of
the space-time structure of chemical variability observed in the upper
(0-200 m) ocean. A process-oriented perspective is used as the basis for
this conceptual model. Thus, dominant physical and biological processes
which significantly influence upper ocean chemical structure are identified
as forcing functions while the observed chemical structure is interpreted
as a response function. Each of the physical and biological processes in
turn is associated with a range of characteristic space and time scales.
Such a framework forms a convenient basis for both the design of chemical
oceanographic experiments and the subsequent analysis of data in a theor-
etically sound context.

SCALES OF VARIABILITY

Scale analysis was introduced to physical oceanography twenty years
ago by Stommel (1963). The Stommel diagram is a three-dimensional
representation of the 'spectral' distribution of physical variables (e.g.,
sea level, ocean currents) plotted as a function of time and space on a

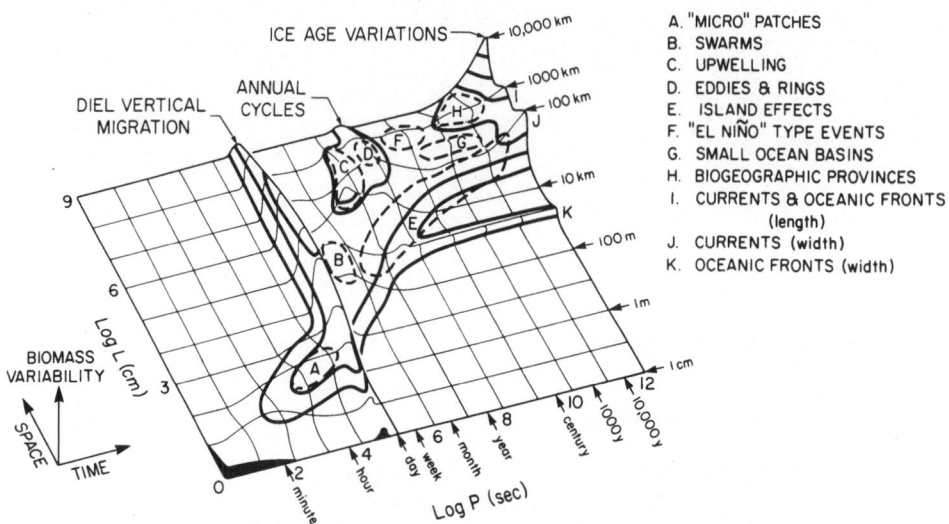

Fig. 1. Conceptual model of the time-space variability of zooplankton biomass and the processes contributing to it (from Haury et al., 1978).

logarithmic scale. The vertical axis represents the amount of variance of the physical variable within a given space-time (or frequency-wavelength) window. Haury et al. (1978) adapted the Stommel diagram to study the variability of oceanic biological processes; their Stommel diagram for zooplankton biomass is shown in Fig. 1. Although there are some scale differences betwen phytoplankton and zooplankton biomass, Haury et al. (1978) concluded that this diagram was also representative of phytoplankton biomass because most of the zooplankton variability shown in Fig. 1 is related to physical processes which would affect phytoplankton similarly. Likewise, the physical and biological processes which most significantly affect chemical variability in the upper ocean (e.g., Simpson and Zirino, 1980; Traganza et al., 1981; Barber and Smith, 1981; Simpson, 1983, 1984a; McGowan, 1983; Lynn, 1983a) are represented in Fig. 1. Hence, we interpret the model shown in Fig. 1 as an approximation of chemical variability in the upper ocean.

Both Stommel (1963) and Haury et al. (1978) have addressed some of the inherent problems associated with construction of Stommel diagrams. First, the shape of the diagram is dependent upon the oceanographic area under study and the smaller-scale features may be related to the phase of the larger-scale processes. For example, patchiness can greatly reduce the uniformity and spatial homogeneity of biological structures and hence chemical structures (Cox et al., 1982). The construction of a Stommel diagram, however, necessarily assumes that the statistical and spectral properties of the physical and biological processes are independent of both space and time. Secondly, the lack of sufficient data for a given process over all space scales, and especially over long time scales, imposes a severe limitation. For example, the proper analysis of barotropic ocean currents requires *very* long time series. Such long-term time series of oceanographic data generally are not available. Hence, some of the information displayed in the diagram represents a "best guess" rather than a quantitative measure of variability (Haury et al., 1978). Finally, as pointed out by Haury et al. (1978), the construction of a Stommel diagram for a biological property includes not only an estimate of the biological variance of that property at any scale, but also an estimate of its biological importance. Thus, the height of the peaks in the Stommel diagram for zooplankton biomass (Fig. 1) depends not only on the actual biomass variance in a given frequency-wavelength window but also on the effect this variance may have on the ecosystem. As Haury et al. (1978) indicate, small fluctuations in biomass over short space and time scales may have equal or greater importance in determining ecosystem structure than have larger-scale fluctuations.

Clearly, the Stommel diagram is not a true spectral representation of the variance of a given property because the assumptions of time series analysis (e.g. stationarity, uniformity, time invariance) all are violated in its construction. Also, the vertical axis is not a true measure of natural variance and hence, unlike spectra, the Stommel diagram is not a variance-conserving representation. Nonetheless, the Stommel diagram provides a simple conceptual tool for determining the qualitative relations between the scales of variability of a given set of oceanographic observables and for the design of oceanographic experiments. Here, this tool is used to investigate the physical, chemical and biological structures associated with three of the processes illustrated in Fig. 1. These are upwelling, mesoscale eddy systems and their interaction with coastal streamers, and "El Niño"—type phenomenon. They were chosen for detailed discussion because they are the dominant processes which affect lateral and vertical transport of chemical species and transmutation of chemical species within the upper 200 m of eastern boundary currents (Simpson and Zirino, 1980; Barber and Smith, 1981; Traganza et al., 1981; McGowan, 1983; Simpson, 1984a). Moreover, these same three processes also affect the distribution and *in situ* alteration of chemical species within western boundary currents (Cheney and Richardson, 1976; The Ring Group, 1981; Scott, 1981) and within the equatorial regions of the worlds's oceans (Smith, 1968; Donguy et al., 1983).

UPWELLING

Coastal upwelling generally occurs along eastern boundaries of the ocean (California, Peru, and Northwest Africa) where the predominant equatorward winds are part of a semi—stationary, mid—ocean high pressure system (Barber and Smith, 1981). These northwesterly or southeasterly winds, combined with the Earth's rotation, produce a net offshore transport of surface water whose direction is 90° to the right of the wind direction in the northern hemisphere or 90° to the left in the southern hemisphere (Ekman, 1905). The depth of the surface layer directly affected by the winds (i.e., the Ekman layer) typically is between 10—20 m. Although the large—scale atmospheric wind system provides the driving force and the large—scale wind—driven oceanic circulation determines the distributions of water properties available for upwelling, the actual vertical transport of subsurface nutrient—rich water into the Ekman layer (i.e., the upwelling) only occurs within a narrow zone adjacent to the coast. Theoretical studies (Yoshida, 1955, 1967; Charney, 1955) show that the width of this upwelling zone is given by the baroclinic radius of deformation, $R = hN/f$, where h is the water depth, N is the mean Brunt–Väisälä frequency, and f is the Coriolis parameter. Typically, R is 10—15 km in coastal upwelling areas (Smith, 1981). Thus, coastal upwelling is a mesoscale dynamical response to large—scale atmospheric forcing. Coastal upwelling ecosystems differ from most other oceanic ecosystems because upwelling circulation can produce optimal conditions for nutrient supply through the vertical transport of subsurface water into the euphotic zone and maintain optimal light conditions for photosynthesis through the resultant stabilized horizontal divergent flow (Barber and Smith, 1981). Hence, the usual limits on biological primary productivity are reduced. A more detailed discussion of upwelling ecosystems is given by Smith (1981) and by Barber and Smith (1981).

A region of the southern California Current off Pt. Conception and a region of the northern California Current between Pt. Reyes and Pt. Arena (see Fig. 2 in Simpson, 1984b) were sampled continuously along cross-shelf transects from the coast to about 150 km offshore. Pt. Arena frequently is the center of active coastal upwelling (Cox et al., 1982). Upwelling is less frequent off Pt. Reyes. Upwelling does occur off Pt. Conception (Cox et al., 1982), although often it is less intense than off Pt. Arena. Differences in the strength and frequency of upwelling events off Pt. Conception and off Pt. Arena result from differences in coastline topography, bathymetry, local wind field, variation of the Coriolis parameter with latitude, and the interaction of the local circulation with the mean flow of the California Current (Barber and Smith, 1981).

Simultaneous measurements of temperature, salinity, pH, dissolved oxygen, chlorophyll a, nitrate, phosphate and silicon were made continuously from R. V. New Horizon while underway during July, 1979. Water for the measurements was drawn from about 3 m. Data were sampled from all instruments at intervals of 20 seconds and recorded digitally. For a ship's speed of 18 km hr^{-1}, this sampling interval in time corresponds to about 100 m in space. From these data, sequential one-minute temporal averages (300 m spatial averages) were made. Discrete samples were drawn from the flow for calibration of the underway data about every 15 minutes. Details of the analytical, calibration and computational procedures are given by Simpson (1985).

Potentiometric determinations of total inorganic carbon dioxide and of titration alkalinity were made with the GEOSECS titrators (Bradshaw et al., 1981). Discrete values of pH and pCO_2 were calculated from these data using the methods given by Broecker and Takahashi (1978) and by Takahashi (1981). These values were used to calibrate the continuous pH measurements. Continuous pCO_2 and ΣCO_2 were calculated from the continuously measured pH and from alkalinity, calculated from a continuous measurement of salinity, using the equations given by Skirrow (1975). Simpson (1985) has shown that the salinity-derived alkalinities agree to within 0.8% of the measured titration alkalinities and that the discrete and continuous pCO_2 distributions typically agree to within ± 10-15 ppm.

Surface distributions of temperature, nitrate, chlorophyll a and percent saturation of dissolved oxygen for the northern region are shown in Fig. 2; similar maps for the southern region (Pt. Conception) are given by Simpson (1985). Percent saturation of dissolved oxygen was calculated using the solubility equation for oxygen in seawater (Weiss, 1970). The cold plume of nutrient-rich water, which extends eastward from Pt. Arena, confirms that Pt. Arena is a center of active upwelling. These distributions show that the subsurface waters reach the surface only within a narrow (~10 km) band adjacent to the coast. This is consistent with theoretical models (Yoshida, 1955) and observations from other major upwelling areas of the world's oceans (Smith, 1981). Temperature also shows that the waters adjacent to Pt. Reyes are more typical of warm oceanic water than of freshly upwelled water. While nitrate shows some of the gross features of the temperature distribution (e.g., upwelling at Pt. Arena and mean southward advection parallel to the coast), it also shows significant alteration by in situ biological processes. For example, the concentration of nitrate is greatly reduced from its source concentration (Pt. Arena) along a narrow band of inshore water parallel to the coast and in a larger region near Pt. Reyes. The areas of reduced nitrate concentration are spatially coherent with regions of high chlorophyll a (and hence, implied high phytoplankton biomass) and regions of high supersaturation of dissolved oxygen.

Statistical analyses of the data (Simpson, 1985) show that the mean

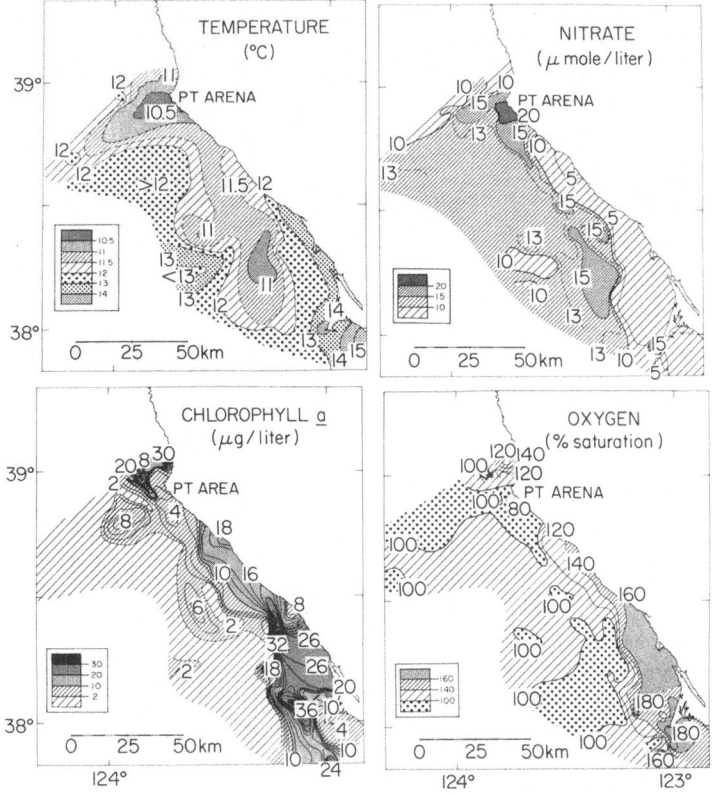

Fig. 2. Surface distributions of temperature, nitrate,
chlorophyll a and percent saturation of dissolved oxygen, for a northern region of the
California Current, July, 1979 (from Simpson,
1984b).

cross-shelf spatial gradients in sampled properties and the small-scale
fluctuations about these mean gradients are consistent with the conclusion
that phytoplankton photosynthesis, not air-sea exchange, controlled the *in
situ* concentrations of CO_2 and O_2 in this region of the California Current
during this experiment. Moreover, analysis of the vertical structure
beneath these surface patterns indicates that maximum plant biomass and
maximum primary productivity occur close to the surface in these regions of
the California Current and that these near-surface biologically active
areas produce the regions of O_2 supersaturation shown in Fig. 2.

Percent saturation of dissolved oxygen is plotted as a function of the
departure of pCO_2 from atmospheric equilibrium (taken as 330 ppm) in
Fig. 3. Inshore data in Fig. 3 refer to samples inshore of the upwelling
fronts shown in Fig. 2. The pCO_2 distribution has values within the range
± 200 ppm of atmospheric equilibrium while the O_2 concentration range is as
high as 185% supersaturated and as low as 70% undersaturated with respect
to atmospheric equilibrium. If the *in situ* concentrations of O_2 and CO_2
were in equilibrium with respect to their atmospheric counterparts, then
the data in Fig. 3 would be clustered around the point (0,100). Inshore
deviations from equilibrium are larger than offshore deviations. The
offshore deviations, however, are large compared to the open-ocean near-
surface equilibrium conditions generally reported (Keeling, 1968; Weiss
et al., 1982). These data clearly show that the ocean-atmosphere system
in this region of the California Current was not in equilibrium with
respect to the air-sea exchange of O_2 and CO_2 during this experiment.

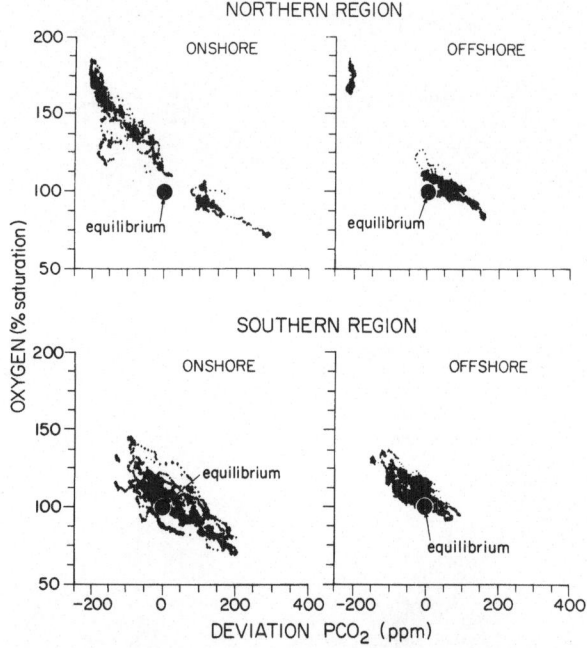

Fig. 3. Percent saturation of dissolved
oxygen *versus* the deviation from
atmospheric equilibrium of pCO_2 for
northern (see Fig. 2) and southern
(Pt. Conception) regions of the
California Current. Samples were
taken at 3 m depth. Data are sub-
divided into those for inshore and
offshore regions. The large dot
(0,100) indicates the equilibrium
coordinate (from Simpson, 1985).

Upwelling brings cold, nutrient-rich, oxygen-depleted and pCO_2 supersaturated waters to the surface at the coast of eastern boundary currents
and at the equator (Smith, 1968). Rapid equilibration of these freshly
upwelled waters with the atmosphere would tend to reduce the pCO_2 concentration to 330 ppm and increase the O_2 concentration to about 100%
saturation. Further, equilibration of mid-latitude waters with the
atmosphere should produce a very narrow pH distribution centred near pH 8.1
(Skirrow, 1975). The range of pH observed during this experiment, however,
was approximately 7.9-8.4 (Simpson, 1985). Summer heating cannot account
for the degree of O_2 supersaturation (~185%) because the observed temperature range (Fig. 2) is too small to affect significantly the solubility
of oxygen (Weiss, 1970). Mixing of oxygen-saturated waters of different
temperatures and salinities also is an inadequate explanation for the
levels of O_2 (Shulenberger and Reid, 1981). In addition, plots of ΣCO_2
(μmol kg^{-1}) *versus* dissolved oxygen concentration (μmol kg^{-1}) for both
the inshore and offshore sections of the southern region have mean slopes
of -0.82 and -0.83, respectively (Simpson, 1985). The negative sign shows
that ΣCO_2 is consumed as O_2 is produced, a result consistent with the
stoichiometry of photosynthesis. Also, these mean slopes are in good
agreement with the stoichiometric slope of -0.78 expected from the Redfield
et al. (1963) model of photosynthesis. Only *in situ* photosynthesis can
explain those regions of the curves (Fig. 3) supersaturated with respect to

O_2 and undersaturated with respect to CO_2, and the ΣCO_2 and O_2 relations reported by Simpson (1985). Finally, respiration would tend to reduce the O_2 concentration. The analysis of Simpson (1985), however, indicates that the effects of respiration are small compared to those of photosynthesis, at least on time scales of several days in active upwelling areas. The analysis of Shulenberger and Reid (1981), moreover, suggests that this is true for even longer time scales and in other areas of the ocean.

The distribution of phytoplankton primary productivity for the Pacific Ocean is shown in Fig. 4. This projection, preserving equal-area, was produced from data originally obtained by the Food and Agriculture Organiz- ation of the United Nations (1981). Shown in Fig. 5a is the percent of ocean surface area for each of the five primary productivity zones shown in Fig. 4; the corresponding total daily rates of carbon fixation are shown in Fig. 5b. While the surface area of the central Pacific gyres greatly exceeds the surface area of active upwelling regions (Figs 4, 5a), the total daily amount of fixed carbon (i.e., product of zone primary product- ion rate and zone surface area shown in Fig. 5b) for active upwelling areas almost balances the daily amount of photosynthetically fixed carbon prod-

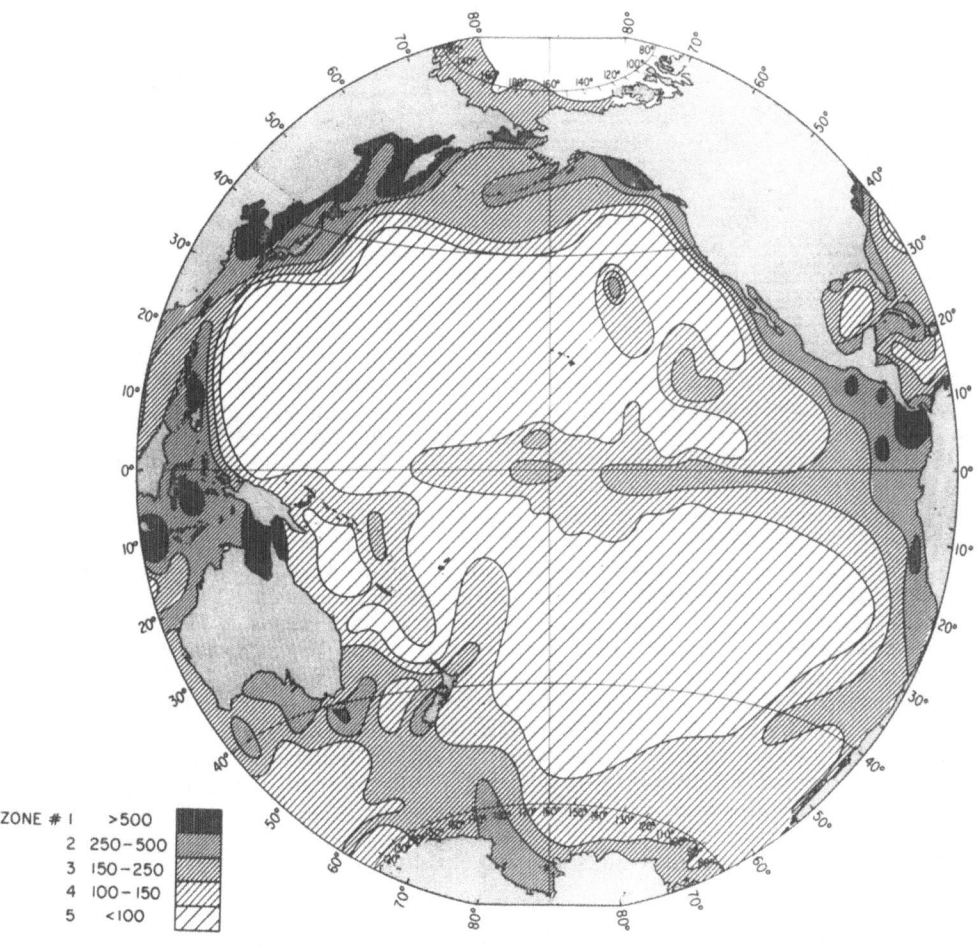

ZONE # 1   >500
2   250-500
3   150-250
4   100-150
5   <100

Fig. 4. Area preserving projection of the distribution of phytoplankton primary productivity (mg C m^{-2} day^{-1}) for the Pacific. Original data from Food and Agriculture Organization of the United Nations (1981).

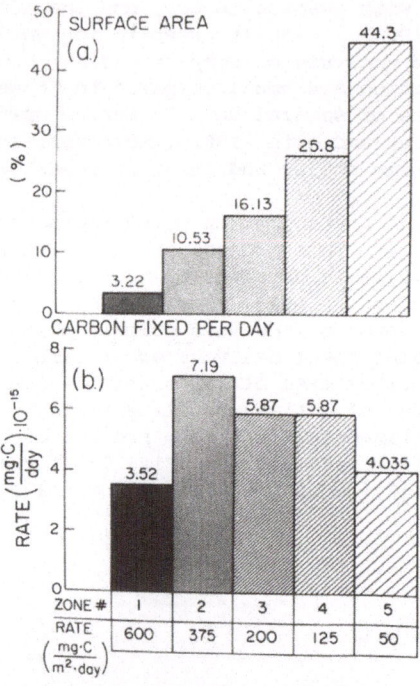

Fig. 5. Histograms of (a) percent
of surface area of the
Pacific, and (b) carbon
fixation (mg C day^{-1}) for
the five productivity
zones shown in Fig. 4.

uced in the central gyres. In fact, the combination of biomass fixed in
the near coastal zones (zones 1-3 in Fig. 4) exceeds that fixed in the
much larger open ocean areas (zones 4 and 5 in Fig. 4). This pattern of
primary productivity for the Pacific (Figs 4, 5) is found throughout all
the world's oceans (Ryther, 1969).

The estimates used here, like those of Ryther, however, have two major
sources of uncertainty. The first is the estimation of the percent of
surface area in a given productivity zone. Presently, the best estimates
of this quantity are based upon available data grouped by regions in
accordance with the breakdown adopted by FAO. Significant improvement in
these estimates is likely to be made in the future by mapping the global
distribution of phytoplankton biomasss with remote-sensing techniques,
i.e., improved versions of the Coastal Zone Color Scanner (CZCS). Current
CZCS data, however, are consistent with the large-scale pattern of primary
productivity shown in Figs 4 and 5. The second source of uncertainty is
the assumed rates of primary productivity. Estimates of biomass fixation
(Fig. 5b) based on these assumed rates, however, are probably conservative
because many recent studies (e.g. Shulenberger and Reid, 1981) suggest that
primary productivity rates seriously underestimate the actual *in situ*
productivity.

The large—scale patterns (Figs 4 & 5) of primary production (with
implied CO_2 consumption and O_2 production) are consistent with the small—
and mesoscale observations reported herein (Figs 2 & 3), observations of pH
off Peru (Simpson and Zirino, 1980), observations of oxygen off Washington
and Oregon (Stefánsson and Richards, 1964) and observations of plant

biomass and oxygen off northwest Africa (Minas et al., 1982). The data (Figs 2-5) suggest that over the central gyres the air–sea exchange of O_2 and CO_2 will tend to approach equilibrium conditions because near–surface biological production is low, the time scales of the dominant physical processes which affect the circulation are long, and "event"–type processes are infrequent. Over most coastal regions, however, both biological and physical processes which affect *in situ* CO_2 and O_2 concentrations are likely to produce large departures from air–sea equilibrium (e.g., Fig. 3). The time scales over which these departures from equilibrium can be maintained or regenerated by physical processes (upwelling and stratification) and biological processes (photosynthesis) against the equilibrium restoring forces of air–sea exchange is uncertain. It is indeed likely that eastern boundary currents (California Current, Peru Current, Benguela Current), and other biologically productive areas of the world's oceans, are frequently out of equilibrium with the atmosphere. In general, the equatorial region may also be in disequilibrium because upwelling is a semi–permanent feature of the equatorial circulation.

Water mass formation occurs at the surface in many of the highly productive areas of the world's oceans (Fig. 4). Simpson (1985) has shown that near–surface ΣCO_2 in these areas is significantly affected by biological processes. These latter analyses support the conclusion of Shiller and Gieskes (1980) that uncertainties in direct estimates of oceanic CO_2 transport, based upon total CO_2, titration alkalinity and AOU relationships (Brewer, 1978; Chen and Pytkowicz, 1979) partially result from an incomplete knowledge of the effects of biological processes on total CO_2 in the biologically productive areas of the world's oceans. This occurs because phytoplankton photosynthesis can cause large departures from *in situ* near surface equilibrium concentrations of O_2 and CO_2 with respect to their atmospheric counterparts, while the direct oceanic transport models (e.g., Brewer, 1978) explicitly assume that the near surface concentration of O_2 is spatially and temporally in equilibrium with the atmosphere.

The data in Figs 2-5 also show that a careful distinction must be made between the air–sea equilibrium concentrations of CO_2 and O_2 and equilibrium between *in situ* processes which produce these concentrations. The physical process of upwelling is a non–equilibrium process because it produces abnormally high (or low) concentrations of chemical species in the euphotic zone. Analysis of phytoplankton biomass and of the activity of the enzyme laminarinase in zooplankton samples (Cox et al., 1982) shows that the approach to equilibrium conditions between phytoplankton production rates and zooplankton grazing rates in an upwelling system is a function of both zooplankton localization (i.e., behavioral aggregation; see Peterson et al., 1979) and the upwelling circulation. A complex interplay between these physical and biological processes determines if biological equilibrium (balance between grazing and primary production rates) is ever achieved. The non–equilibrium concentrations of O_2 and CO_2 (Fig. 3) are the chemical indicators of *in situ* non–equilibrium physical and biological processes. The buffering mechanisms of the carbonate system in seawater (Skirrow, 1975) respond to these physical and biological reactive stresses in order to maintain an *in situ* quasi–equilibrium among carbonate species and hydrogen ions. This *in situ* quasi–equilibrium of the carbonate system, while related to the air–sea equilibrium concentrations of CO_2 and O_2, is distinct from it.

Finally, different control mechanisms have been suggested to explain observed changes in the pCO_2 of natural waters. For example, Weiss et al. (1982) showed that the seasonal oscillation of pCO_2 in the central Pacific gyre (15°S, 70°W) was controlled almost completely by thermodynamic processes; the temperature–induced seasonal change in the amplitude of pCO_2 was 20 ppm whereas the biologically–induced change was only about 0.6 ppm.

In contrast, Simpson (1985) measured near surface (3 m) concentrations of dissolved oxygen as high as 13 ml l^{-1} and values of pCO_2 as low as 150 ppm at temperatures between 14–15°C in the biologically productive waters of the inshore California Current. In this case, however, the data for solubility (Weiss, 1970) predict saturation concentrations for dissolved oxygen of only about 5.8 ml l^{-1}. Clearly, these two experiments represent extremes of the control conditions on the concentrations of O_2 and CO_2 in natural water systems, control conditions, however, remarkably consistent with the large-scale distributions of primary productivity in Figs 4 and 5. Hence, the design of both regional and local experiments to determine the processes affecting the structure and distribution of chemical species in the upper ocean must be made carefully. Further, the results of such experiments must be scaled properly, if accurate global distributions of CO_2 and O_2 are to be inferred from them.

MESOSCALE EDDIES

Most observations of rings and eddies formed near oceanic boundary currents (Vastano et al., 1980; Nilsson and Cresswell, 1981) show that these features are mesoscale dynamical structures with cores of water of non-local origin. Because the cores carry non-local water, the transfer of water properties between neighboring water masses can occur within the interior of the water masses, and not simply at their boundaries. Hence, several authors (e.g., Cheney and Richardson, 1976) have emphasized the potential role of eddies as transport mechanisms in major oceanic current systems. This transport mechanism not only affects the physical dynamics of the interacting water masses but also the distribution and structure of chemical properties (Scott, 1978, 1981; Brandt et al., 1981) and biological populations (Cox and Wiebe, 1979; The Ring Group, 1981; Brandt, 1981; Brandt et al., 1981).

From 9 to 17 January, 1981, two intersecting, orthogonal vertical sections, each about 300 km in length, were made through a warm-core eddy centred near 32.4°N, 124.0°W, southwest of Pt. Conception, California. Vertical sections of temperature made during this experiment (see Fig. 2, Simpson et al., 1984a) show a complex three-layer system. A subsurface warm-core eddy, which extends from about 200 to 1400 m, is the dominant dynamical feature. The eddy diameter is about 150 km at the 7°C isotherm and its geometric center is well-defined. A warm surface layer extending to about 75 m lies over the eddy. Between the surface layer and the sub-surface warm-core eddy, there is a cold-core region which extends to about 175 m. A region of minimum vertical shear between the bottom of the cold-core region and the top of the warm-core eddy (~200 m) is thus implied. A conceptual diagram of the three-layer system is shown in Fig. 6.

The observed distributions of water properties (temperature, salinity, density, and dissolved oxygen; see Fig. 8 in Simpson et al., 1984a) are inconsistent with a single, local generation process for the three-layer system. These observations show that the core of the eddy (between 200–600 m) consists of shelf-slope water which originated in the California Undercurrent (CU). Local waters from the Deep Poleward Flow (DPF) are found in the three-layer system just below the core of the warm-core eddy (~700 m). Distributions of potential density and potential vorticity (see Fig. 23, Simpson et al., 1984a) show that lateral exchange of fluid between local waters and the eddy system is severely inhibited between about 200–600 m depth. Regions of possible fluid exchange consistent with these distributions of potential density and potential vorticity are shown in Fig. 6 of this paper.

Intense physical, chemical and biological fronts, which extend from

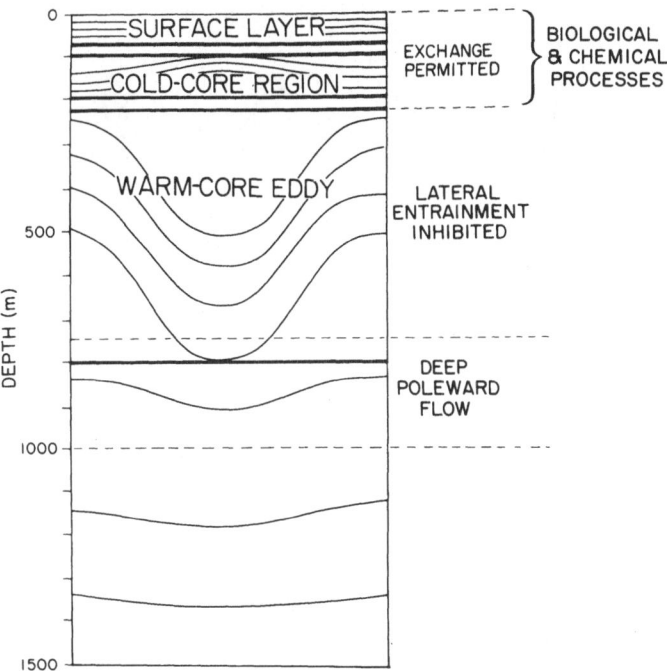

Fig. 6. Conceptual model of the observed three-
layer offshore mesoscale eddy system.
Regions of possible fluid exchange, based
on the vorticity dynamics of the system,
between exterior waters and the interior
waters of the three-layer system are also
shown (from Simpson, 1984a).

the surface to greater than 100 m in depth, partially surround the upper
layers of the three-layer system. A strong thermal front (Fig. 7a) is
located along the northeastern quadrant of the eddy system, while a strong
haline front (Fig. 7b) is located along the southwestern quadrant. Simpson
(1984a) has shown that the structure of the associated biological and
chemical fronts (e.g., chlorophyll a, Fig. 7c; nitrate, Fig. 7d) was formed
by the same physical processes that formed the thermal fronts, and that
subsequently their structure was modified by local biological processes,
primarily photosynthesis. These frontal structures support the conclusion
that waters of coastal origin were entrained into and around the upper two
levels of the northeastern quadrant of the eddy system, while waters of
oceanic origin were entrained along the southwestern quadrant (Simpson,
1984a; Haury, 1984).

The interaction of these surface frontal structures with the three-
layer system was studied with satellite observations of sea surface temperature (Koblinsky et al., 1984). Their analysis showed that, for several
months, bands of warm, salty, oceanic water from the southwestern salinity
front (feature E in Fig. 1 in Koblinsky et al., 1984) wrapped around cooler
water (feature D in Fig. 1 in Koblinsky et al., 1984); part of this coastal
water was entrained into the center of the surface expression of the eddy.
A conceptual diagram of this interleaving of surface water, based upon the
sea surface temperature data, is shown in the xy plane of Fig. 8. Simpson
(1984a) hypothesized that the lateral interleaving and mixing processes
which occurred in the vicinity of these surface fronts produced highly
diffusive water types, some of which were gravitationally unstable. Some

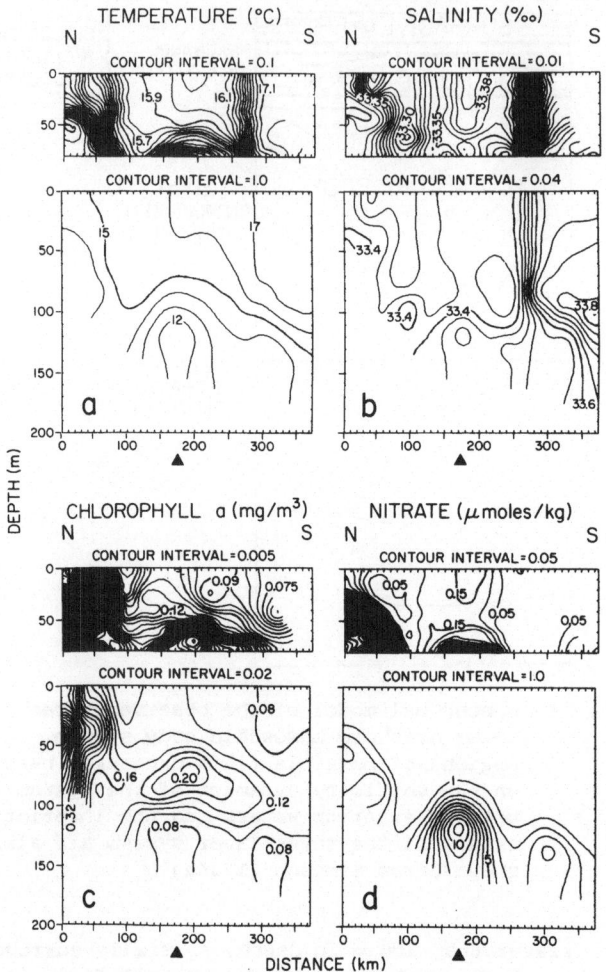

Fig. 7. Vertical sections of temperature,
salinity, chlorophyll a and nitrate
through the N-S section of the eddy
system. Only data from the upper two
layers of the eddy system are shown.

of these water types sank along density surfaces to produce the deeper-
lying fronts shown in Fig. 7. That this phenomenon can occur in eddies is
supported by Traganza et al. (1981), who studied inshore cyclonic eddies
and their interactions with upwelling sources in the California Current
using both *in situ* and remote sensing techniques. Additional support for
the lateral entrainment hypothesis is found in studies of warm-core eddies
off Australia (Scott, 1978, 1981).

The dominant feature in Fig. 7 is a dome-shaped structure of upward-
sloping properties. The diameter of this cold-core region is about 150 km
at the 12.5°C isotherm (Fig. 7a). Simpson (1984a) suggested that this
cold-core region was produced by a forced-secondary circulation at depth.
The forced-secondary circulation was, in turn, a dynamical consequence of
the lateral entrainment of non-local waters discussed above. Independent
biological evidence (Haury, 1984) also supports lateral entrainment and
forced-secondary circulation as the generation mechanism of the cold-core
region.

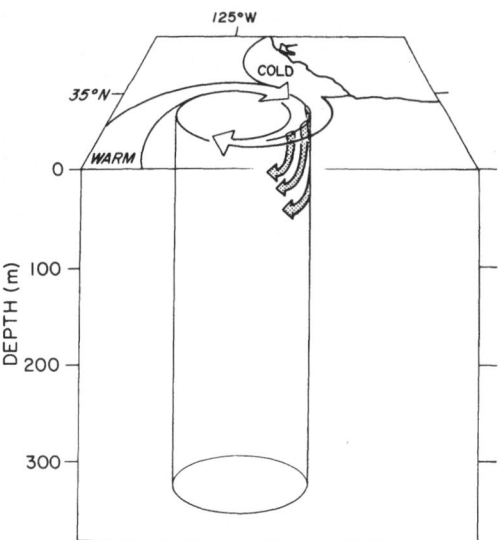

Fig. 8. Conceptual diagram of the interleaving of inshore (cold) and
oceanic (warm) surface waters
based upon infrared satellite
imagery. The sinking of unstable, highly diffusive water
types into and around the
upper two layers of the three-
layer system also is shown
(from Simpson, 1984a).

The distributions of chemical properties, chlorophyll a, and phaeo-
phytin pigment, shown in Fig. 7 and in Figs 6—11 of Simpson (1984a), are
atypical of those expected of a warm—core eddy (Scott, 1978; eddies E and
F, September, 1978 in Tranter et al., 1980; eddy J in Brandt et al., 1981)
or of a cold—core ring (The Ring Group, 1981). Warm—core eddies are
expected to have low plankton concentrations at their center because low
nutrient waters generally form their cores (Jitts, 1965; Scott, 1978);
high plankton and nutrients generally are found at their frontal boundaries
(Tranter et al., 1980; Lukashev and Chernyakova, 1979). Cold—core rings,
on the other hand, are expected to have high plankton and nutrient concentrations at their center (The Ring Group, 1981; Yoder et al., 1981)
and low plankton and nutrients at their frontal boundaries (The Ring Group,
1981). The three—layer system reported here has relatively high chlorophyll a concentrations (Fig. 7c) and relatively high nutrients (Fig. 7d),
both in the center and around some of its frontal boundaries. Hence, its
chemical and biological distributions possess simultaneously some of the
attributes associated with both warm—core eddies and cold—core rings. The
data show that the observed chemical structure resulted from a complex
interplay among non—local eddy generation processes at the time the
three—layer system was formed and a continuous set of interactions among
the three—layer system, both inshore (cold) and offshore (warm) waters of
the California Current and coastal and local biological processes. The
continued interactions of the three—layer system with the coastal and
offshore waters of the California Current make offshore eddies in the
California Current System fundamentally different, chemically and biologically, from cold—core Gulf Stream rings because these rings are thought to
be isolated vortices embedded in the relatively homogeneous water of the
Sargasso Sea (The Ring Group, 1981).

Satellite infrared imagery of the California Current (Bernstein
et al., 1977; Traganza et al., 1981; Koblinsky et al., 1984) consistently
shows coastal streamers of cold, nutrient-rich water wrapping around warmer
offshore water. Moreover, recent observations in the California Current
(Peláez, 1984) of chlorophyll concentration derived from remotely-sensed
ocean color using the CZCS, coupled with *in situ* observations of chlorophyll concentration, show entrainment of plankton-rich coastal waters by
offshore eddies. These data imply that entrainment (e.g., stirring
associated with geostrophic turbulence) may occur with some frequency in
the offshore California Current. Haury (1984) has suggested that this
lateral entrainment of coastal water into the offshore domain of the
California Current might help to explain the two offshore maxima in zooplankton biomass observed by Bernal and McGowan (1981) in the thirty-year
time series of zooplankton data collected by the California Cooperative
Oceanic Fisheries Investigations (CalCOFI). This latter study and the work
of Simpson et al. (1984b) support the conclusion that the overall biological productivity, water property mixing, heat transport and circulation of
at least the offshore California Current may be largely determined by
processes associated with the offshore mesoscale eddy field. Thus, unlike
Gulf Stream cold-core rings, offshore California Current eddies transport
chemical and biological species by three separate mechanisms: interior core
transport of non-local water (the mechanism of Cheney and Richardson,
1976), continued lateral entrainment of non-local water (the mechanism of
Simpson, 1984a), and vertical transport which results from density-driven
flows and double diffusive phenomena associated with the lateral entrainment of different water types into the eddy system (the mechanism of
Simpson, 1984a). Hence, such eddy systems provide complex natural environments for the transformation of chemical species.

EL NIÑO PHENOMENA

A major El Niño (i.e., El Niño Southern Oscillation (ENSO) event)
occurred in the tropical Pacific during 1982 (Donguy et al., 1983). It is
generally agreed that the abnormally warm waters associated with El Niño
are part of a large-scale air-sea interaction; Quinn (1974) has shown that
El Niño is preceded by large peaks in the twelve-month running means of the
Southern Oscillation Index.

Measurements made in the California Current during 1982-83 (Simpson,
1983) showed several anomalous conditions: warm sea surface temperature
anomaly (1-3°C), depression of the thermocline by 50m or more, and monthly
mean sea levels 20-24 cm above the corresponding long-term (56-year)
monthly means. During this same period, the inshore poleward flow of the
California Countercurrent was enhanced (Lynn, 1983a). These analyses show
that a major Californian "El Niño" occurred coincident with the 1982-83
equatorial ENSO event.

Vertical sections of temperature, temperature anomaly, salinity and
salinity anomaly, taken along CalCOFI line 90 (a line heading 240°T from
the California coast at 33°25'N, 117°54.3'W) during August 1983 are shown
in Fig. 9. They were calculated relative to the long-term (28 year) August
CalCOFI mean for line 90, using the method of Lynn (1967). These anomalies
are typical of the subsurface structures observed in the California Current
during the 1982-83 El Niño event (Simpson, 1984c) and their climatological
significance, in terms of standard deviations above (for temperature) and
below (for salinity) the long-term monthly means for August, is given in
these papers. Simpson (1984c) also showed that the positive temperature
and negative salinity subsurface anomalies (Fig. 9) were accompanied by
statistically significant positive dissolved oxygen anomalies. This
combination of subsurface anomaly structure, characteristic diagrams

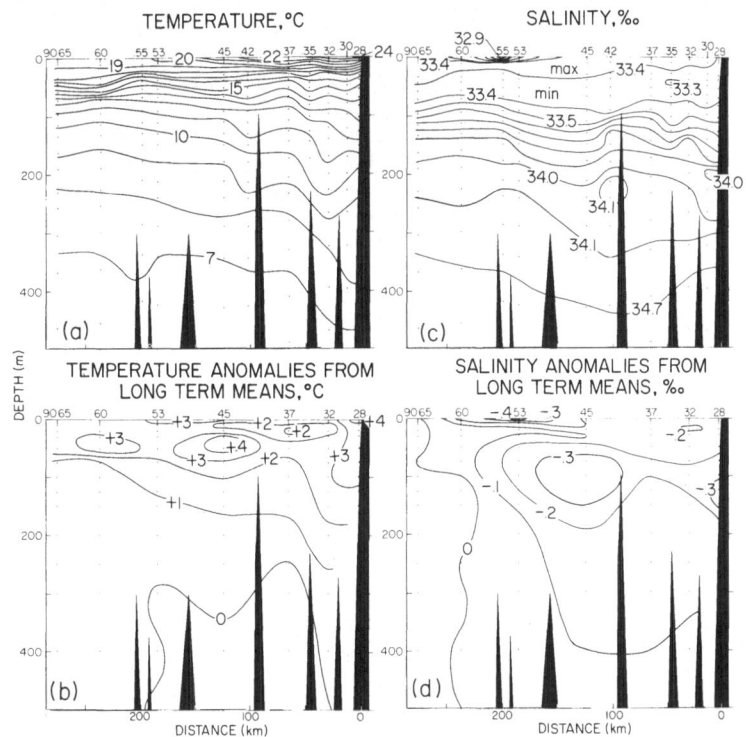

Fig. 9. Vertical sections of temperature, temperature anomaly, salinity and salinity anomaly along CalCOFI
line 90 for August 1983.

(Simpson, 1984c) and offshore sign reversals in the anomalies of salt and
dissolved oxygen (Simpson, 1984c) taken along line 90 is consistent only
with enhanced onshore transport of Pacific Subarctic water from the
offshore California Current, i.e., primarily from the west and bounded by
the west—northwest and west—southward directions. Moreover, large—scale
observations of salinity made during March 1983 (Lynn, 1983b) and during
November—December 1983 confirm that the onshore transport of Pacific
Subarctic water occurred over the entire domain of the California Current.

    Theoretical studies (McCreary, 1976; Hurlburt et al., 1976) have
suggested that the anomalous poleward geostrophic flow, anomalous high
coastal sea levels and positive sea surface temperatures associated with
mid—latitude Pacific "El Niño" events are induced by a poleward propagating
coastally—trapped wave which originates in the eastern tropical Pacific.
Onshore transport of mass, however, excludes poleward propagating coastally
—trapped waves as a generation mechanism for the 1982—83 Californian "El
Niño" because both Kelvin waves and coastally—trapped waves have a cross—
shelf component of velocity which is zero (Mysak, 1980; Gill, 1982).
Further, the space scales of the observed anomaly structures (Fig. 9) are
at least an order of magnitude greater than the cross—shelf length scale of
a coastally—trapped wave as determined by its baroclinic Rossby radius of
deformation. Again, the observations fail to support coastally—trapped
waves as a generation mechanism for the 1982—83 Californian "El Niño".
These data, however, support the conclusion (Simpson, 1983) that the
expansion and intensification of the Aleutian low and the decrease in
strength of the North Pacific high produced an anomalous basin—wide

atmospheric circulation which coupled directly to the large—scale wind—
driven oceanic circulation to produce the Californian "El Niño" of 1982—83.
Further, the observed space and time scales of the anomalous atmospheric
forcing are consistent with the observed space and time scales of the
mid—latitude oceanic response (Simpson, 1983, 1984d).

El Niño—induced onshore transport of Pacific Subarctic water produced
an anomalous, yet dynamically very stable, vertical configuration in the
California Current. A quantitative measurement of this anomalous stability
is shown in the vertical section of static stability (Fig. 10a) taken along
CalCOFI line 90 during March 1983. Here, the subsurface maximum in static
stability is larger in magnitude and occurs at a greater depth than in the
corresponding long—term seasonal mean distribution of static stability
(Lynn et al., 1982). Moreover, the distributions of static stability for
May and August (Figs 10b,c) show that the subsurface stability maximum
continued to intensify during most of 1983. In fact, by August (Fig. 10c),
and to some extent as early as May (Fig. 10b), anomalously high values
of static stability were observed even at the surface. These anomalous

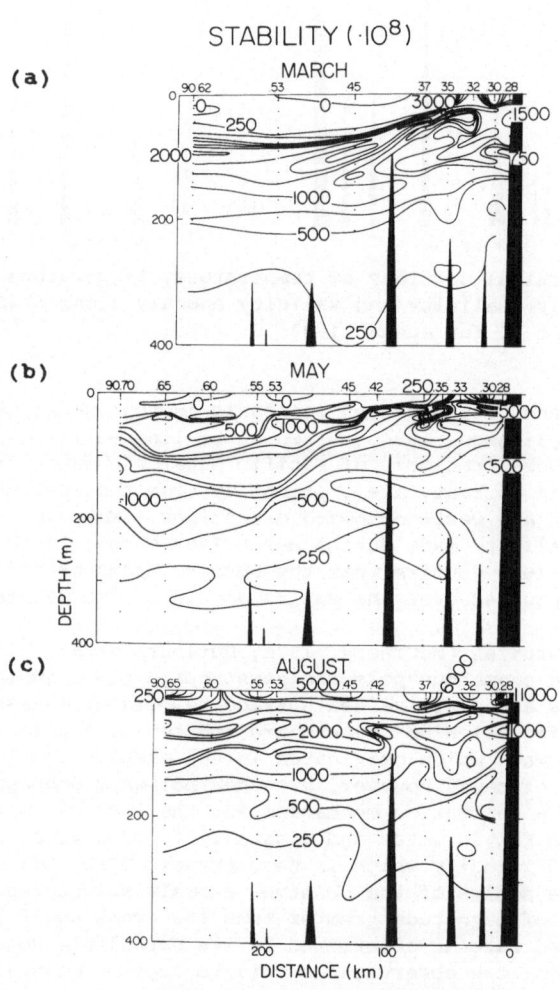

Fig. 10. Vertical section of static
stability along CalCOFI line
90 for (a) March, (b) May, and
(c) August, 1983.

high near-surface values of static stability were partially produced by local thermodynamic processes (i.e., enhanced near surface solar absorption during El Niño; see Simpson, 1984d).

McGowan (1983) compared the distributions of chlorophyll a measured during March and April 1983 with the corresponding long-term measurements and found that seaward of 100 km offshore the maximum in chlorophyll a occurred much deeper in the water column and that its magnitude was much smaller than normal. He suggested that the deep chlorophyll a maximum (DCM) was responding to the depression of the inshore thermocline discussed by Simpson (1983). The offshore DCM usually forms below a subsurface stability maximum because such a maximum inhibits vertical mixing, a physical constraint needed for the accumulation of biomass in time at a given depth zone. Hence, the deeper and stronger subsurface stability maxima (Fig. 10) observed during the 1982-83 El Niño event necessarily imply a deeper DCM. The potential effects of a deeper DCM on phytoplankton production are difficult to predict, particularly in such a hydrographic-ally complex region as the California Current. In general, the response of the offshore DCM to "El Niño" conditions will be a compromise between potentially greater production due to increased nutrient concentrations with increasing depth of the DCM, offset by a reduction in light available for photosynthesis.

During this same period (March-April, 1983), however, the nearshore (≤ 100 km) chlorophyll pattern was fairly normal (McGowan, 1983). This suggests that the ecological and chemical structure of the inshore domain of the California Current responded differently to that of the offshore domain, in response to the anomalous "El Niño" forcing. The stability sections (Fig. 10) show that inshore of 100 km the surface mixed layer is much shallower than in the offshore region. Typically (Cox et al., 1982; Simpson, 1985), a shallower near surface region of maximum chlorophyll is found near the base of these inshore shallow surface layers. There is considerable observational evidence (Venrick, 1979; Cullen and Eppley, 1981) that this inshore, shallow near-surface chlorophyll maximum develops into the DCM with increasing distance from the coast. Such a development of the offshore DCM is also consistent with the cross-shelf depth depen-dence of the subsurface stability maximum shown in Fig. 10. Moreover, Longhurst (1976) and Cullen and Eppley (1981) suggested that the DCM is neither a plant biomass maximum nor a productivity maximum. Their results showed that offshore both these parameters occur at depths shallower than that of the DCM and that the difference in depth between the various measures decreases nearshore. The near-surface spatial distributions of chlorophyll and oxygen (Fig. 2), which reflect normal inshore upwelling conditions in the California Current, are consistent with the data of Longhurst (1976) and Cullen and Eppley (1981) and with the inference that regions of maximum plant biomass and primary productivity occur very close to the surface in the inshore California Current. This inference also is consistent with measurements of temperature, chlorophyll and ^{14}C primary productivity (Dengler, 1981) which show that in the inshore upwelling region off Peru the maximum in surface chlorophyll coincides with the maximum in primary productivity.

Simpson (1984d) showed that the anomalous subsurface structure observed during the 1982-83 California El Niño was produced dynamically by a depression of the thermocline. He also showed that the corresponding anomalous surface structure was produced by a combination of dynamical and local thermodynamical processes. If the inshore near-surface biological and chemical structure responded to El Niño forcing similarly to the in-shore near-surface physical structure, then one would expect the offshore biological response to be governed primarily by dynamical processes and the inshore biological response to be governed by a combination of dynamical

and local thermodynamic processes. Further, the inshore biological response to El Niño—type forcing should be most pronounced during summer and
early fall when the combination of maximum near—surface irradiance absorption and reduced coastal upwelling produce the very large, inshore, near—
surface values of static stability seen in Fig. 10c.

Onshore transport of Pacific Subarctic water, subsequent development
of the deeper, stronger subsurface stability maximum, and development of
anomalous high near—surface inshore static stabilities provided the dynamical basis for the ecological and chemical modification of the structure of
the California Current which occurred during the 1982—83 "El Niño" event.
This can be seen by comparing vertical sections of temperature, salinity,
chlorophyll a and nitrate (Fig. 11) taken during typical upwelling conditions in the California Current (May, 1981) with similar distributions
(Fig. 12) taken during a major "El Niño" event (May, 1983). Data for May,
1983 were taken on two separate cruises about six days apart, This
interval of time, however, is very small compared to the El Niño time

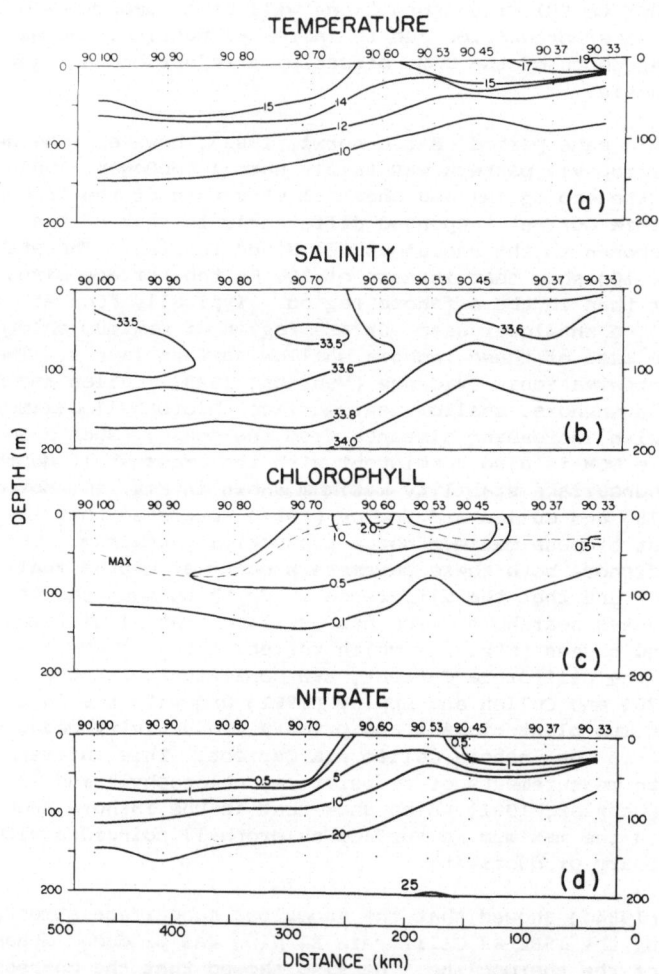

Fig. 11. Vertical sections along CalCOFI line 90
during May, 1981, of (a) temperature, (b)
salinity, (c) chlorophyll a, and (d)
nitrate, typical of upwelling conditions
in the California Current. (Data provided
by T. Hayward.)

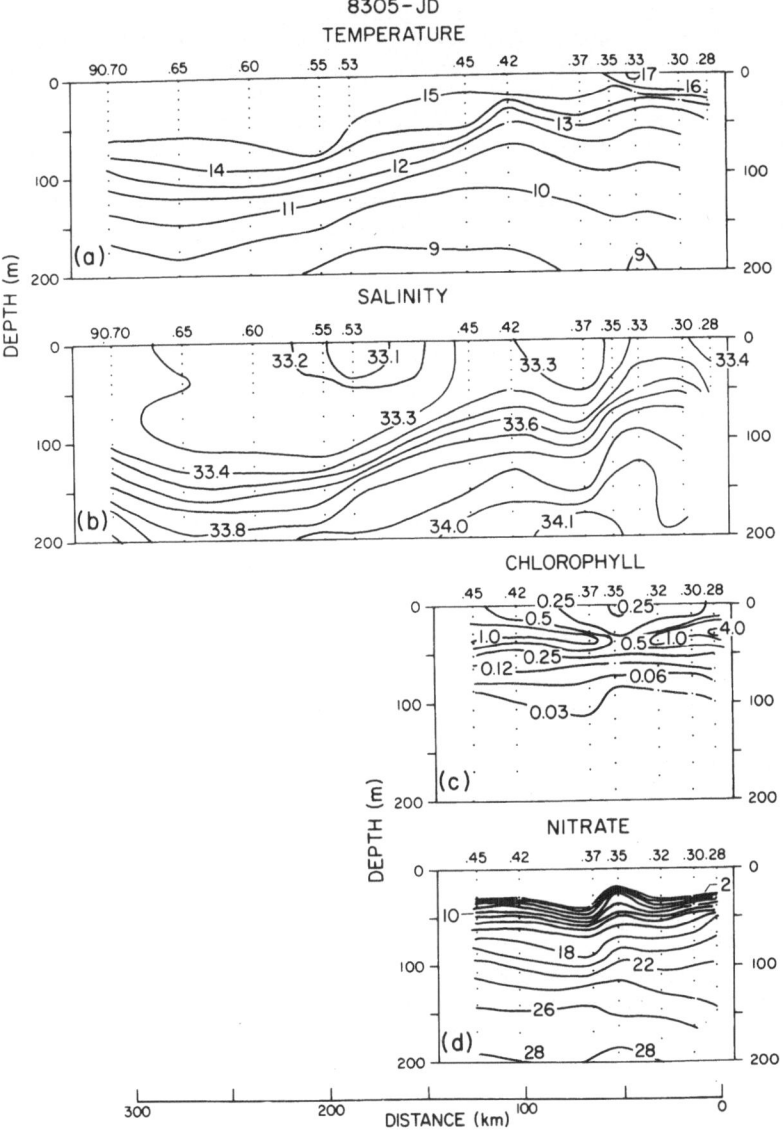

Fig. 12. Sections analogous to those shown in Fig. 11, but
under El Niño conditions in May 1983.

scale (18–24 months). Hence, the two cruises can be treated as a single
observational period in the context of the present discussion. Under
upwelling conditions, the surface temperature is relatively cool
(Fig. 11a), the inshore salinity is relatively high (Fig. 11b), the
magnitude of the offshore subsurface chlorophyll a maximum is large (Fig.
11c), the near-surface chlorophyll a concentrations (Fig. 11c) and the
near-surface nutrient concentrations (Fig. 11d) also are relatively high.
During El Niño conditions, however, the offshore surface temperature is
relatively warm (Fig. 12a), the inshore salinity is relatively low (Fig.
12b), the magnitude of the offshore subsurface chlorophyll a maximum is
smaller and occurs deeper in the vertical than usual (Fig. 12c), and the
near-surface distribution of chlorophyll a (Fig. 12c) and of nutrients
(Fig. 12d) is relatively low. The offshore phytoplankton standing crop

measured in August (Fig. 13a) was depleted even more than that measured during either March–April by McGowan (1983) or May (Fig. 12c) and the corresponding nitrate distribution for August (Fig. 13b) is substantially more depleted than that of May (Fig. 12d). In fact, by August even the near–shore chlorophyll pattern, which was fairly normal during March–April, 1983, was severely depleted compared to inshore chlorophyll concentrations observed during typical summertime upwelling conditions (e.g., compare Fig. 13 herein with Fig. 2 in Cox et al., 1982). Moreover, these latter changes in the ecological (Fig. 13a) and chemical (Fig. 13b) structure of the California Current are consistent with the temporal development of the vertical stability of the water column (Fig. 10).

The differences in the chemical and ecological structure of the California Current cited above occurred because El Niño–induced onshore transport of low salinity Pacific Subarctic water from the offshore California Current produced a stabilized horizontally convergent flow and downwelling in the inshore domain of the California Current. The El Niño data (Figs 12, 13) clearly show downwelling at the coast and a major depression of the inshore thermocline, instead of the usual uplift of isotherms (Fig. 11a) and of the nutricline (Fig. 11d) which results from coastal upwelling. In addition, the deeper, stronger subsurface stability maximum of El Niño inhibits the upward vertical transport of deeper–lying nutrients into the euphotic zone. The combination of the effects of reduced coastal upwelling and enhanced near–surface solar absorption are most pronounced during summer and in the inshore regions of the California Current (Fig. 13). These data clearly imply that during major "El Niño" events the biologically–induced disequilibrium in the air–sea exchange of CO_2 and O_2 over an eastern boundary current (Figs. 3–5) is replaced by conditions which more closely resemble the near equilibrium open–ocean situation because near–surface phytoplankton production is greatly reduced.

CONCLUSIONS

The primary transport and transformation processes which alter the chemical speciation and structure of the upper 200 m of the ocean within 1000 km of the coast of California were examined from the process–oriented perspective of space and time scale analysis. Upwelling, offshore meso–scale eddies, and El Niño were identified as the dominant physical

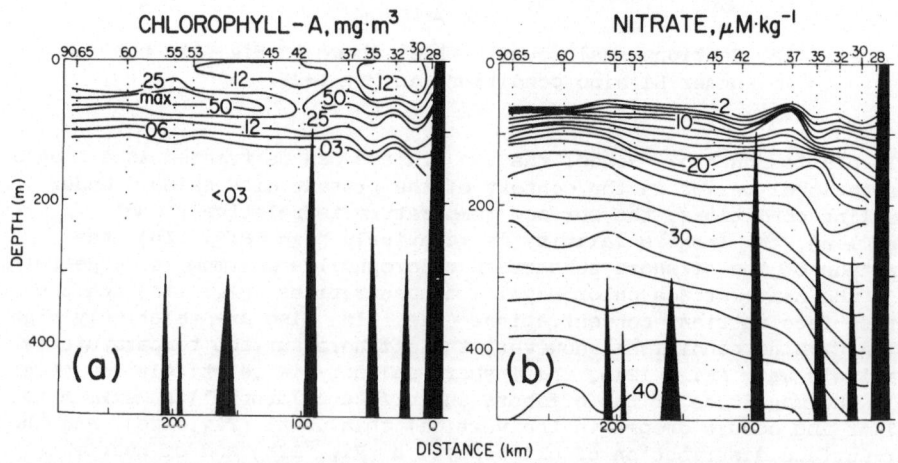

Fig. 13. Vertical sections along CalCOFI line 90 of (a) chlorophyll a, and (b) nitrate, under El Niño conditions, August 1983.

transport processes. Upwelling produces upward vertical and offshore transport of chemical species. Mesoscale eddies produce both onshore and offshore transport of chemical species through lateral entrainment of non-local waters and both upward and downward vertical transport of chemical species through double diffusive phenomena and density-driven flows which can give rise to subsurface secondary circulations. El Niño produces large-scale onshore transport of chemical species and downward vertical transport of chemical species within a relatively narrow (~200 km) wide band adjacent to the coast. The dominant biological process responsible for *in situ* transformation of chemical species within the upper 100 m was identified as phytoplankton photosynthesis. These physical and biological processes combine to produce complex oceanic environments for the alteration of chemical species. Experiments to measure the fluxes of chemical species between ocean and atmosphere must be designed carefully and scaled properly, with due attention paid to processes such as those cited herein, if accurate global chemical budgets are to be inferred from them.

ACKNOWLEDGEMENTS

This work was funded over a period of several years by the State of California through the Marine Life Research Group (MLRG) of the Scripps Institution of Oceanography, the California Space Institute, the Foundation for Ocean Research, the Office of Naval Research, and Sea Grant. Data were collected and processed by the technical support staffs of MLRG and PACODF. Special thanks to S. Anderson, G. Anderson, W. Bryan, G. Hemingway, A. Mantyla and R.T. Williams. Sharon McBride typed the manuscript, and the MLRG Illustrations Group prepared the figures. Special thanks are due to Fred Crowe, Nancy Hulbirt, René Wagemakers and Guy Tapper. The author is deeply indebted to Prof. J.L. Reid, Chairman of the Marine Life Research Group, for his continued support and encouragement. Dr. T. Hayward generously provided the data for Fig. 11. Dr. L.R. Haury carefully reviewed the manuscript.

REFERENCES

Barber, R.T., and Smith, R.L., 1981, Coastal upwelling ecosystems, in: "Analysis of Marine Ecosystems", A.R. Longhurst, ed., pp.31-68, Academic Press, London.

Bernal, P.A., and McGowan, J.A., 1981, Advection and upwelling in the California Current, in: "Coastal Upwelling", F.A. Richards, ed., pp.381-399, American Geophysical Union, Washington, D.C.

Bernstein, R.L., Breaker, L., and Whritner, R., 1977, California Current eddy formation: ship, air and satellite results, Science, N.Y., 195:353.

Bradshaw, A.L., Brewer, P.G., Shafer, D.K., and Williams, R.T., 1981, Measurements of total carbon dioxide and alkalinity by potentiometric titration in the GEOSECS program, Earth Planet. Sci. Lett., 55:99.

Brandt, S.A., 1981, Effects of a warm-core eddy on fish distributions in the Tasman Sea off East Australia, Mar. Ecol. Progr. Ser., 6:19.

Brandt, S.A., Parker, R.R., and Vaudrey, D.J., 1981, Physical and biological description of warm-core eddy J during September-October, 1979, Report No. 126, CSIRO Division of Fisheries and Oceanography, Cronulla.

Brewer, P.G., 1978, Direct observations of the oceanic CO_2 increase, Geophys. Res. Lett., 5:997.

Broecker, W.S., and Takahashi, T., 1978, The relationship between lysocline depth and in situ carbonate ion concentration, Deep-Sea Res., 25:69.

Charney, J.G., 1955, The generation of ocean currents by wind, J. Mar. Res., 14:477.

Chen, C.T., and Pytkowicz, R.M., 1979, On the total CO_2 - titration alkalinity - oxygen system in the Pacific Ocean, Nature, Lond., 281:361.

Cheney, R.E., and Richardson, P.L., 1976, Observed decay of a cyclonic Gulf Stream ring, Deep-Sea Res., 23:143.

Cox, J., and Wiebe, P.H., 1979, Origins of oceanic plankton in the Middle Atlantic Bight, Estuarine Coastal Mar. Sci., 9:509.

Cox, J.L., Haury, L.R., and Simpson, J.J., 1982, Spatial patterns of grazing-related parameters in California coastal surface waters, July 1979, J. Mar. Res., 40:1127.

Cullen, J.J., and Eppley, R.W., 1981, Chlorophyll maximum layers of the Southern California Bight and possible mechanisms of their formation and maintenance, Oceanol. Acta., 4:23.

Dengler, A.T., Jr., 1981, Spatial distributions of phytoplankton: Limitations of power spectrum techniques, Ph.D. Thesis, University of California, San Diego.

Donguy, J.R., Dessier, A., and Meyers, G., 1983, The 1982 Niño-like event, Trop. Ocean-Atmos. Newsl., 16:7.

Ekman, V.W., 1905, On the influence of the earth's rotation on ocean currents, Ark. Math. Astr. Fys. (Stockholm), 2:1.

Food and Agriculture Organization of the United Nations, Department of Fisheries, 1981, "Atlas of the Living Resources of the Sea", pp.13-19, FAO, Rome.

Gill, A.E., 1982, "Atmosphere-Ocean Dynamics", Academic Press, New York.

Haury, L.R., 1984, An offshore eddy in the California Current System, Part IV: Plankton distributions, Progr. Oceanogr., 13:95.

Haury, L.R., McGowan, J.A., and Wiebe, P.H., 1978, Patterns and processes in the time-space scales of plankton distributions, in: "Spatial Pattern in Plankton Communities", J.H. Steele, ed., pp.277-327, Plenum Press, New York.

Hurlburt, H.E., Kindle, J.C., and O'Brien, J.J., 1976, A numerical simulation of the onset of El Niño, J. Phys. Oceanogr., 6:621.

Jitts, H.R., 1965, The summer characteristics of primary productivity in the Tasman and Coral Seas, Aust. J. Mar. Freshw. Res., 20:65.

Keeling, C.D., 1968, Carbon dioxide in surface ocean waters, 4. Global distribution, J. Geophys. Res., 73:4543.

Koblinsky, C.J., Simpson, J.J., and Dickey, T.D., 1984, An offshore eddy in the California Current System, Part II: Surface manifestation, Progr. Oceanogr., 13:51.

Longhurst, A., 1976, Interactions between zooplankton and phytoplankton in the Eastern Tropical Pacific Ocean, Deep-Sea Res., 23:729.

Lukashev, Yu.F., and Chernyakova, A.M., 1979, Variability of nitrate fields associated with the passage of eddy formations, Mar. Chem., 19:409.

Lynn, R.J., 1967, Seasonal variation of temperature and salinity at 10 m in the California Current, CalCOFI Reports, 11:157.

Lynn, R.J., 1983a, The 1982-83 warm episode in the California Current, Geophys. Res. Lett., 10:1093.

Lynn, R.J., 1983b, Anomalous steric height in the California Current during the 1983-83 warm episode, Trop. Ocean-Atmos. Newsl., 21:23.

Lynn, R.J., Bliss, K.A., and Eber, L.E., 1982, "Vertical and horizontal distributions of seasonal mean temperature, salinity, sigma-t, stability, dynamic height, oxygen and oxygen saturation in the California Current, 1957-58", CalCOFI Atlas 30, University of California, San Diego.

McCreary, J.P., 1976, Eastern tropical ocean response to changing wind systems: with application to El Niño, J. Phys. Oceanogr., 6:632.

McGowan, J.A., 1983, El Niño and biological production in the California Current, Trop. Ocean-Atmos. Newsl., 21:23.

Minas, H.J., Packard, T.T., Minas, M., and Coste, B., 1982, An analysis of the production-regeneration system in the coastal upwelling area off N. W. Africa based on oxygen, nitrate and ammonium distributions, J. Mar. Res., 40:615.

Mysak, L.A., 1980, Topographically trapped waves, Ann. Rev. Fluid Mech., 12:45.

Nilsson, C.S., and Cresswell, G.R., 1981, The formation and evolution of East Australian Current warm-core eddies, Progr. Oceanogr., 9:133.

Peláez, J., 1984, Phytoplankton pigment concentrations and patterns in the California Current as determined by satellite, Ph.D. Thesis, University of California, San Diego.

Peterson, W.T., Miller, C.B., and Hutchinson, A., 1979, Zonation and the maintenance of copepod populations in the Oregon upwelling zone, Deep-Sea Res., 26:467.

Quinn, W.R., 1974, Monitoring and predicting El Niño invasion, J. Appl. Meteorol., 13:825.

Redfield, A.C., Ketchum, B.H., and Richards, F.A., 1963, The influence of organisms on the composition of sea water, in: "The Sea", Volume 2, M.N. Hill, ed., pp.26-77, Wiley-Interscience, New York.

Ring Group, The, 1981, Gulf Stream cold-core rings: Their physics, chemistry and biology, Science, N.Y., 212:1091.

Ryther, J.H., 1969, Photosynthesis and fish production in the sea, Science, N.Y., 166:72.

Scott, B.D., 1978, Hydrological features of a warm-core eddy and their biological implications, Report No.100, CSIRO Division of Fisheries and Oceanography, Cronulla.

Scott, B.D., 1981, Hydrological structure and phytoplankton distribution in the region of a warm-core eddy in the Tasman Sea, Aust. J. Mar. Freshw. Res., 32:479.

Shiller, A.M., and Gieskes, J.M., 1980, Processes affecting the oceanic distributions of dissolved calcium and alkalinity, J. Geophys. Res., 85:2719.

Shulenberger, E., and Reid, J.L., 1981, The Pacific shallow oxygen maximum, deep chlorophyll maximum, and primary productivity, reconsidered, Deep-Sea Res., 28:901.

Simpson, J.J., 1983, Large-scale thermal anomalies in the California Current during the 1982-1983 El Niño, Geophys. Res. Lett., 10: 937.

Simpson, J.J., 1984a, An offshore eddy in the California Current System, Part III: Chemical structure, Progr. Oceanogr., 13:71.

Simpson, J.J., 1984b, On the exchange of oxygen and carbon dioxide between ocean and atmosphere in an eastern boundary current, in: "Gas Transfer at Water Surfaces", W. Brutsaert and G.H. Jirka, eds, pp. 505–514, Reidel, Dordrecht.

Simpson, J.J., 1984c, El Niño-induced onshore transport in the California Current during 1982-83, Geophys. Res. Lett., 11:233.

Simpson, J.J., 1984d, A simple model of the 1982-83 Californian El Niño, Geophys. Res. Lett., 11:237.

Simpson, J.J., 1985, Air–sea exchange of carbon dioxide and oxygen induced by phytoplankton: Methods and interpretation, in: "Mapping Strategies in Chemical Oceanography", A. Zirino, ed., pp.409–450, American Chemical Society, Washington, D.C.

Simpson, J.J., and Zirino, A., 1980, Biological control of pH in the Peruvian coastal upwelling area, Deep-Sea Res., 27:733.

Simpson, J.J., Dickey, T.D., and Koblinsky, C.J., 1984a, An offshore eddy in the California Current System, Part I: Interior dynamics, Progr. Oceanogr., 13:5.

Simpson, J.J., Koblinsky, C.J., Haury, L.R., and Dickey, T.D., 1984b, An offshore eddy in the California Current System, Preface, Progr. Oceanogr., 13:1.

Skirrow, G., 1975, The dissolved gases: carbon dioxide, in: "Chemical Oceanography", Second Edition, Volume 2, J.P. Riley and G. Skirrow, eds, pp.1–181, Academic Press, London.

Smith, R.L., 1968, Upwelling, Oceanogr. Mar. Biol. Ann. Rev., 6:11.

Smith, R.L., 1981, A comparison of the structure and variability of the flow field in three coastal upwelling regions: Oregon, northwest Africa and Peru, in: "Coastal Upwelling - Coastal and Estuarine Science", Volume 1, F.P. Richards, ed., pp.107—118, American Geophysical Union, Washington, D.C.

Stefánsson, U., and Richards, F.A., 1964, Distributions of dissolved oxygen, density and nutrients off the Washington and Oregon coasts, Deep-Sea Res., 11:355.

Stommel, H., 1963, Varieties of oceanographic experience, Science, N.Y., 139:572.

Takahashi, T., 1981, GEOSECS Carbonate Chemistry, in: "GEOSECS Atlantic Expedition, Volume 1, Hydrographic Data", A.E. Bainbridge, ed., pp.61—63, U.S. Government Printing Office, Washington, D.C.

Traganza, E.D., Conrad, J.C., and Breaker, L.C., 1981, Satellite observations of a cyclonic upwelling system and "giant plume" in the California Current, in: "Coastal Upwelling - Coastal and Estuarine Science", Volume I, F.P. Richards, ed., pp.228—241, American Geophysical Union, Washington, D.C.

Tranter, D.J., Parker, R.R., and Cresswell, G.R., 1980, Are warm-core eddies unproductive?, Nature, Lond., 284:540.

Vastano, A.C., Schmitz, J.E., and Hagan, D.E., 1980, The physical oceanography of two rings observed by the Cyclonic Ring Experiment, Part 1: Physical structures, J. Phys. Oceanogr., 10:493.

Venrick, E.L., 1979, The lateral extent and characteristics of the North Pacific Central environment at $35^{\circ}N$, Deep-Sea Res., 26:1153.

Weiss, R.F., 1970, The solubility of nitrogen, oxygen, and argon in water and seawater, Deep-Sea Res., 17:721.

Weiss, R.F., Jahnke, R.A., and Keeling, C.D., 1982, Seasonal effects of temperature and salinity on the partial pressure of CO_2 in seawater, Nature, Lond., 300:511.

Yoder, J.A., Atkinson, L.P., Lee, T.N., Kim, H.H., and McClain, C.R., 1981, Role of Gulf Stream frontal eddies in forming phytoplankton patches on the outer southeastern shelf, Limnol. Oceanogr., 28:1103.

Yoshida, K., 1955, Coastal upwelling off the California coast, Rec. Oceanogr. Works Jpn, 2:8.

Yoshida, K., 1967, Circulation in the eastern tropical oceans with special reference to upwelling and undercurrents, Japan. J. Geophys., 4:1.

STRUCTURE AND EVOLUTION OF GULF STREAM WARM-CORE RINGS:

A PHYSICAL CHARACTERIZATION

Terrence M. Joyce

Woods Hole Oceanographic Institution
Woods Hole, Massachusetts 02543 U.S.A.

ABSTRACT

A major field study was undertaken in 1981 and 1982 to examine the structure of Gulf Stream warm-core rings, the changes in structure with time, and the processes responsible for the observed changes. The research program was multi-disciplinary and included components from physical, biological and chemical oceanography, remote sensing, and mathematical modeling. This report focuses on the physical structure of warm-core rings and its change with time. Some ramifications of physical processes upon biology and chemistry are briefly discussed.

INTRODUCTION

Warm-core rings (WCRs) are found in the Slope Water region between the Gulf Stream and the North American continental shelf. These rings or eddies are bodies of water 100 to 200 km in diameter that result when a northward meander separates from the Gulf Stream (Fig. 1). The WCR is comprised of a central core, the initial physical, chemical and biological properties of which originate in the Sargasso Sea. The core of the ring is surrounded and contained by a clockwise-rotating remnant of the Gulf Stream with surface current speeds of 50-200 cm s^{-1}. Cold-core rings are formed to the south of the Gulf Stream. Both types of rings maintain their identity for a substantial period of time (a few months to a year) and are the most energetic form of variability in the Northwest Atlantic Ocean (cf., Richardson, 1983). Typically, in any given year, eight pairs of warm and cold rings will be formed. They generally drift to the west after formation with speeds of 3-5 cm s^{-1}.

In 1981-82 a major field effort was undertaken to study the physical, biological, and chemical structure and evolution of WCRs (Joyce and Wiebe, 1983). This program involved more than 25 principal investigators from 13 different research institutions (Warm Core Rings Executive Committee, 1982). The location of the region of experimental study and the various rings examined in 1981-82 is shown in Fig. 2. Some specific research questions addressed were:

(1) What is the physical structure and motion? How are physical processes affected by interactions with surroundings? How does a ring change as it ages?

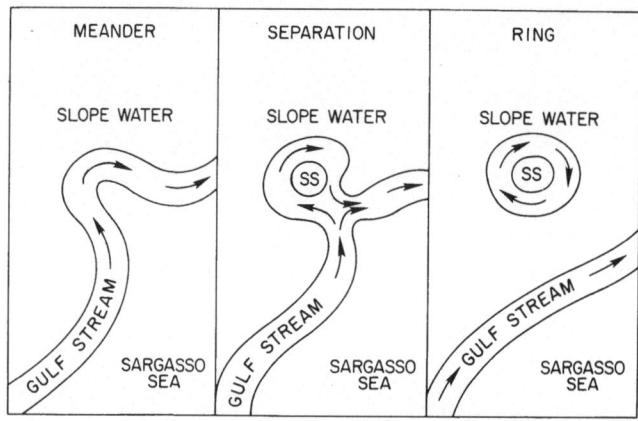

Fig. 1. Three stages in the formation of a warm-
core ring from a Gulf Stream meander.

(2)   What effect do the rings have on the large-scale chemical transport in the region? Do they alter vertical fluxes and affect
biological and chemical interactions?

(3)   What are the population distributions? How do they relate to
environmental factors in the ring, particularly as it matures?
What controls primary production and activity at higher trophic
levels within the ring?

This report focuses on the experimental results concerning the physical
characteristics of warm-core ring structure and evolution. First, the
vertical and horizontal structure is described. This is followed by a
discussion of the nature of the physical structure and the mechanisms
responsible for changes in it. Finally, some of the chemical and biological ramifications are considered.

PHYSICAL STRUCTURE

Nine stations were occupied on a section across ring 81D (the fourth
ring to form in 1981; Fig. 2). The station spacing was nominally 20
nautical miles. The sections (Fig. 3) illustrate the vertical and horizontal structure of potential temperature, salinity, potential density,
dissolved oxygen, dissolved silica, and nitrate. The temperature signature
of the warm-core can be seen throughout the water column below the depth of
the seasonal thermocline (50-100 db). The near-surface signature of the
ring is demarcated by warm and cold streamers at the ring periphery. The
central core contains a thick lens of 18°C water of Sargasso Sea origin. In
wintertime this water mass extends from the surface to depths of 500 m in
the Sargasso Sea and has a salinity of 36.5‰. These water type characteristics can be seen in both the temperature and salinity sections. While
the surface layer of the ring is still more saline than its surroundings,
its Sargasso Sea signature has been substantially freshened by mixing with
the fresher Slope Water outside the ring. The ring is denser than its
surroundings near the surface but less dense at pressures below 100 db.
This produces a sub-surface velocity maximum. The baroclinic signature of
the ring extends to the bottom with maximum horizontal density gradients
between station pairs 4 and 5, and 8 and 9. One persistent feature of the
near-surface oxygen distribution is a lens of low oxygen concentration
between the freshened surface layer and the 18°C water core. This low

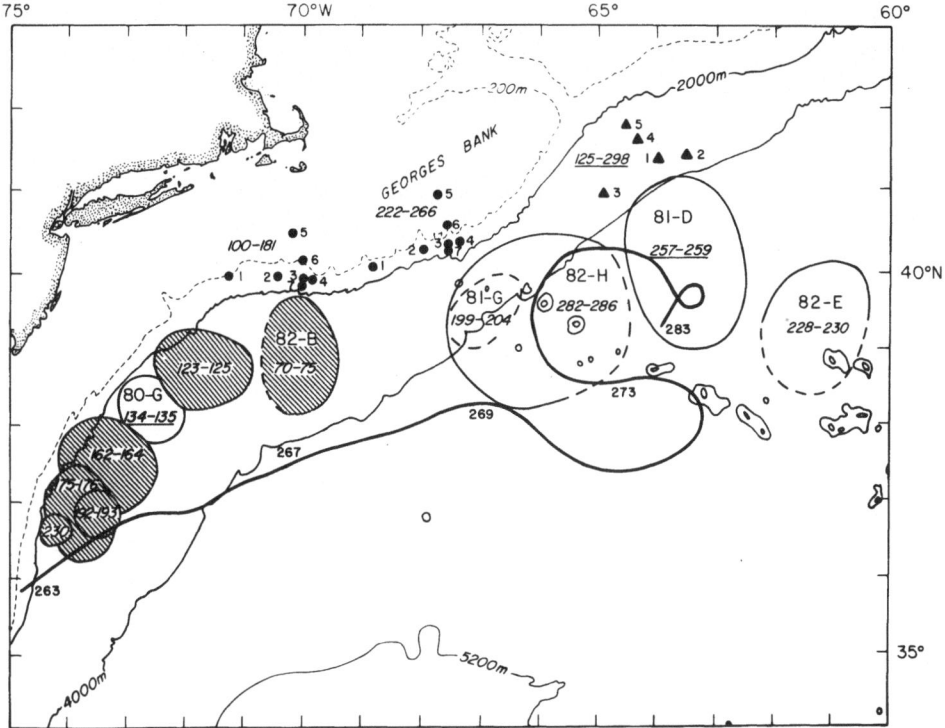

Fig. 2. Warm-core rings studied in 1981-82. Julian days, numbered from
1 January, are shown for each survey with 1981 days underlined.
Current meter mooring locations (dots and triangles) and measurement periods are also indicated. Shaded circles portray the
changing shape and location of ring 82B. The heavy line is the
track of a satellite-tracked drifter which was in ring 82B when
it was reabsorbed into the Gulf Stream. (From Joyce and Wiebe,
1983.)

oxygen, high salinity feature is thought to be a remnant of the Gulf Stream
which has spread across the interior of the ring. Samples for nutrient
measurements were collected at discrete depths with a rosette sampler.
Higher nitrate concentrations were observed at thermocline depths on the
periphery of the ring and under the ring near the bottom. Similar results
were obtained for dissolved silica.

Water mass analysis suggests that the ring was anomalously high in
salinity and low in oxygen relative to its surroundings, from the surface
to pressures of 1000 db. This is illustrated for salinity by examining
salinity anomalies of the ring with respect to a 'standard' reference
curve. The reference potential temperature/salinity curve was constructed
by Armi and Bray (1982) using spline fits to potential temperature/salinity
curves for the Northwest Atlantic. The existence of a detectable water
mass contrast between the ring and its surroundings (Fig. 4a) is a result
of water trapped in the ring during its formation from Sargasso Sea
components. The Slope Water south of Nova Scotia contains measurable
property differences compared with the Sargasso Sea down to temperatures of
3.4°C. Vertically isolated intrusions of fresher water can be detected on
the periphery of the ring. These are suspected to be signatures of lateral
mixing between the ring and its surroundings. The distributions of oxygen,
dissolved silica and nitrate are consistent with this water mass signature

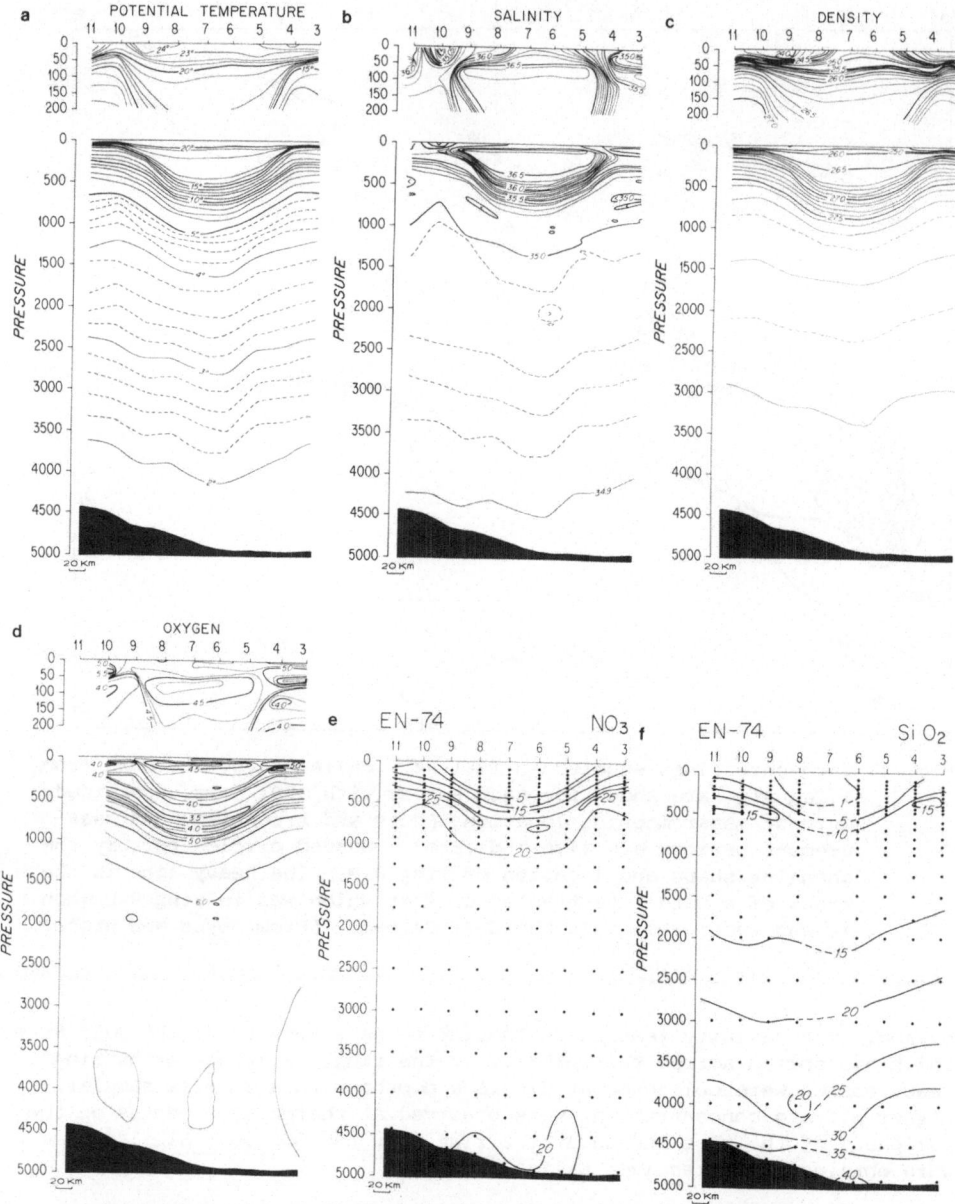

Fig. 3. Hydrographic sections of (a) potential temperature (°C); (b) salinity (‰); (c) potential density (kg m⁻³); (d) dissolved oxygen
(ml l⁻¹); (e) nitrate (μmol kg⁻¹); (f) dissolved silica (μmol kg⁻¹)
from CTD stations 3-11 made from R/V ENDEAVOR, September 1981 (from
Joyce, 1984).

of the ring: more saline water, low in oxygen and nutrients, of Sargasso
Sea origin, is present in the ring core in varying amounts to pressures in
excess of 1000 db.

In order to determine the velocity structure using hydrographic data
alone, physical oceanographers have used the "thermal wind" relationship
for geostrophic motion. In a ring this must be augmented to include a
centripetal acceleration because the ring currents are rapidly rotating

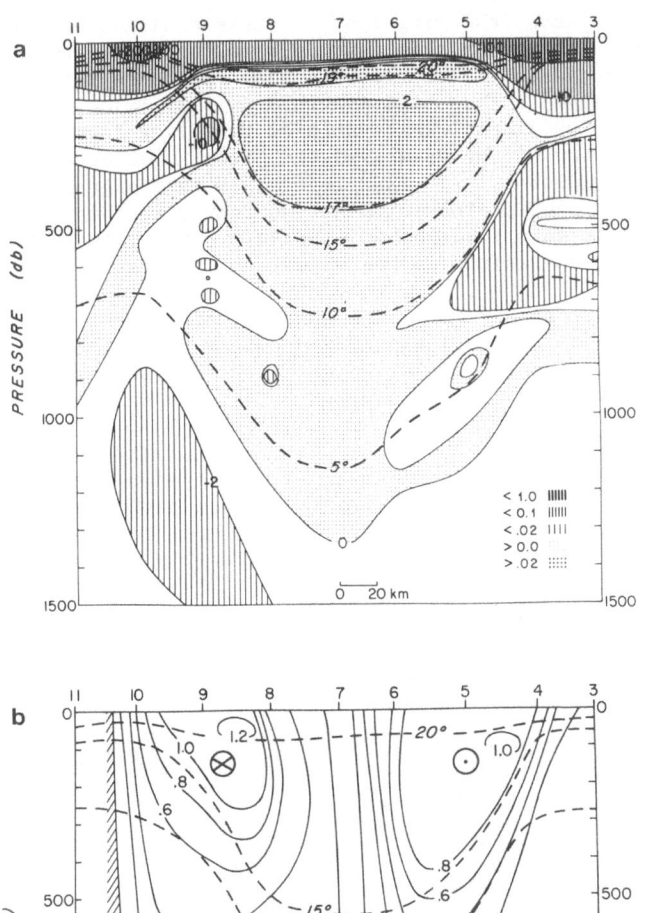

Fig. 4. For the section in Fig. 3, values for
(a) salinity anomaly (‰ x 100) from a
reference curve, and selected potential
temperature isotherms (dashed); (b) azimuthal velocity (x - 1; m s^{-1}) together
with potential temperature isotherms.
Cross-hatched regions have positive
azimuthal velocities. (From Joyce,
1984.)

near the ocean surface about the ring center. This modified balance, called the "gradient current" balance, yields azimuthal currents as a function of depth and radius with respect to some reference level. In our study of WCRs, we employed a shipboard acoustic-doppler velocity profiling system (Joyce et al., 1982) that was capable of profiling currents from the surface to a depth of 100 m. We have used the directly measured currents at 100 m depth as a reference level, and together with the gradient current balance, have inferred the azimuthal current structure of ring 81D. This has been discussed more fully by Joyce (1984). The results are depicted in Fig. 4b. The maximum currents near the surface are found at a radius of 70 km. The position of the velocity maximum becomes sub-surface as the ring center is approached. On the southwestern side of the ring (stations 3-7), the velocity maximum is found closer to ring center with increasing depth. The circulation to the northeast of ring center is confused by a small cyclonic eddy, discussed more fully by Joyce (1984). Below 1000 db, the circulation is weak and confined to a small annulus. Thus the distributions of the velocity and salinity anomaly support the view that the trapped fluid within the ring is contained by the circulation in a cone-shaped region which narrows with increasing depth and vanishes below approximately 1500 db. Within the core of the ring, the circulation near ring center is similar to a solid body, rotating in a clockwise direction with decreasing rates as depth increases.

The topography of the main thermocline can be mapped in plan view

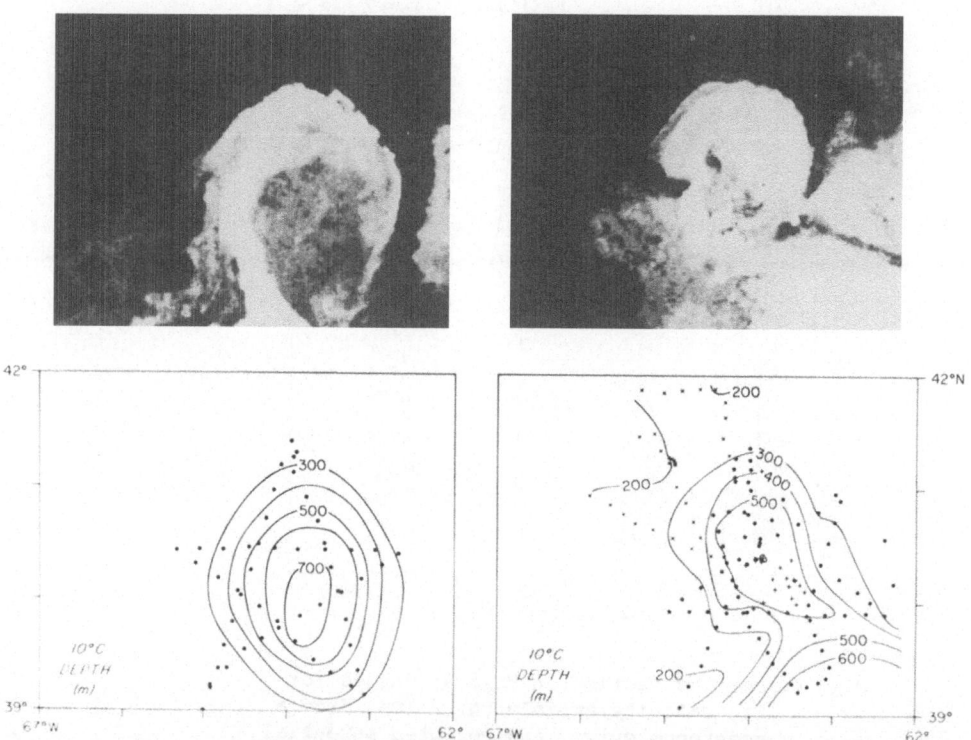

Fig. 5. Plan view of warm-core ring 81D for the days 257-259 (14-16 September) (left) and days 267-270 (24-27 September) (right). Top panels are satellite SST images produced at the University of Miami from the NOAA-7 AVHRR radiometer. Bottom panels are maps of the topography of the 10°C isotherm in metres. (From Joyce et al., 1984.)

using the depth of the 10°C isotherm. This has been done for ring 81D and the results plotted (Fig. 5) for two different surveys of the ring. A pair of satellite images of sea surface temperature (SST) during each of the two surveys is included to illustrate the close relationship between the deep hydrographic structure and SST. In the first survey, the depression of the ring's thermocline reached depths in excess of 700 m. The ring was elliptical in shape with the major axis oriented nearly north-south. The ring is demarcated in SST by a warm streamer of Gulf Stream water on the west and a cold streamer of Shelf Water east of ring center. During the second survey, the ring was partially attached to a Gulf Stream meander to the southeast and the maximum depth of the 10°C isotherm was reduced to less than 600 m. Interactions with the Gulf Stream are responsible for the ultimate death of all WCRs. In the ten day interval between the two surveys, the ring suffered a glancing but not fatal blow from the Gulf Stream. This will be discussed more fully in the next section.

The close correspondence between SST and the deep structure in WCRs (Brown et al., 1983) permits them to be tracked remotely and enables *in situ* measurements to be located with respect to ring center and other major features such as warm and cold streamers. The existence of the streamers illustrates the effectiveness of WCRs in the stirring of the water masses of the Northwest Atlantic Ocean. Reference to Fig. 4 shows that the streamers observed in SST are limited in depth extent to the upper 100 m. Because the circulation of the ring extends throughout the upper 1000 m, it can be inferred that rings are active stirring agents for the entire upper ocean.

The physical structure of WCRs has been illustrated with data from a single ring, 81D. While other rings can have different sizes and shapes, most of their essential features are the same. The core waters of rings may be quite different, however, depending on their life histories. Wintertime cooling can reduce the core temperature in WCRs from values near 18°C, as in ring 81D, to 13°C. Observations of two other rings are contrasted with ring 81D in Fig. 6. Each has undergone different degrees of winter cooling.

EVOLUTION OF PHYSICAL STRUCTURE OF WARM-CORE RINGS

Flierl (1984) has recently reviewed models for the translation of isolated eddies. Depending on the details of the model, translation can be in any direction. Isolated, strong warm-core rings, with no initial motion in the deep ocean, will propagate to the west leaving a trail of planetary waves in their wake. The WCRs observed thus far move to the west and are strongly constrained by the continental slope to the north and the Gulf Stream to the south. Further, Joyce and Kennelly (1985) have shown that a warm-core ring is advected by upper level currents in the surrounding Slope Water. Evans et al. (1985) have also demonstrated that the translation rate of ring center can be influenced by changes in the local bathymetry as well as by interaction with the Gulf Stream and other WCRs.

Gulf Stream interactions are the most intense, episodic agents for WCR evolution. As was noted above with reference to Fig. 5, a glancing encounter between a WCR and a Gulf Stream meander rapidly altered the size and shape of ring 81D. The loss of available potential energy during this strong interaction was 7.7×10^9 W: more than an order of magnitude greater than has been observed for "isolated" rings. During these strong interactions, the surface expression of the ring in SST can vanish entirely and the near surface layer can be substantially replaced (Joyce et al., 1984).

Interactions with the atmosphere can also be both intense and inter-

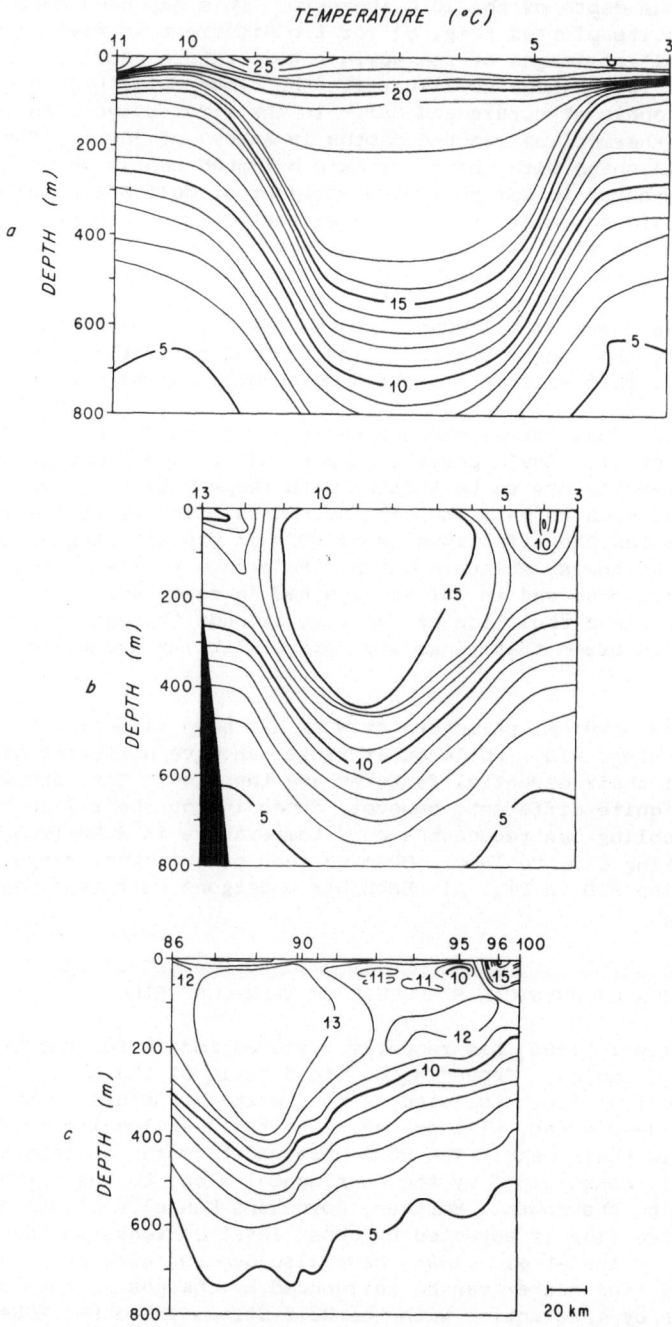

Fig. 6. Temperature sections of three different WCRs:
(a) 81D, formed in July, observed in late
September, 1981; (b) 82B, formed in February,
observed in late April, 1982; (c) 80C/F,
formed in May, 1980, observed in May, 1981.
Each ring has been subjected to differing
amounts of winter cooling. Station numbers
(CTD for (a) and (b); XBT for (c)) are shown
at the top of each section. (From Joyce and
Stalcup, 1985.)

mittent, especially in winter. The passage of atmospheric cyclones over the Northwest Atlantic Ocean can be accompanied by outbreaks of cold, dry continental air over the ocean. Because the surface temperature of a WCR is greater than that of the surrounding water in winter, large sensible and latent heat losses can occur which can, in a matter of a few days, induce vigorous convection in the ring core to depths of a few hundred metres. These cooling events occur intermittently, but their integrated effect will produce a thermal signature in the core of WCRs similar to that shown in Fig. 6. Dewar (1985) has studied this process using a one-dimensional mixed layer model. Measurements of the convective overturn in WCRs, however, suggest that one-dimensional models will produce neither the correct temperature (Joyce and Stalcup, 1985) nor salinity (Schmitt and Olson, 1985) in the core of WCRs. Lateral mixing is required between the ring and the cooler, fresher surrounding waters.

The WCR most extensively studied in our multi-disciplinary research program was ring 82B (see Fig. 2). It was first visited in March of 1982. Three multi-ship cruises followed in April, June and August as this ring drifted towards Cape Hatteras, North Carolina, where it eventually was reabsorbed into the Gulf Stream in mid-September (Joyce and Wiebe, 1983). We were fortunate in being able to study this ring during most of its seven-month life cycle.

When ring 82B formed in late February, 1982, it probably contained a core of 18°C water. When first observed by a ship in March, this layer had cooled slightly to 17.7°C. Between March and our first main cruise in April, late winter cooling had reduced the core temperature of the ring to 15.7°C. Subsequently, a seasonal thermocline developed with surface temperatures exceeding 27.7°C by August. The evolution of the thermal struct-

POTENTIAL TEMPERATURE

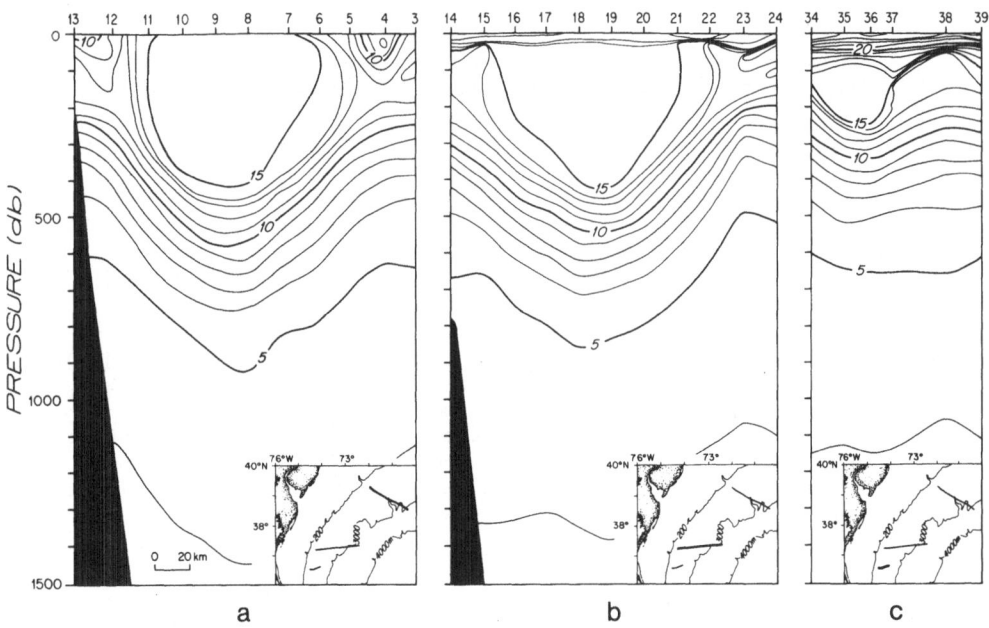

Fig. 7. Temperature sections across ring 82B made by R/V ENDEAVOR in 1982 on days 111-113 (April), 166-168 (June) and 228-229 (August). Section locations are shown in the inserts. (From Joyce and Kennelly, 1985.)

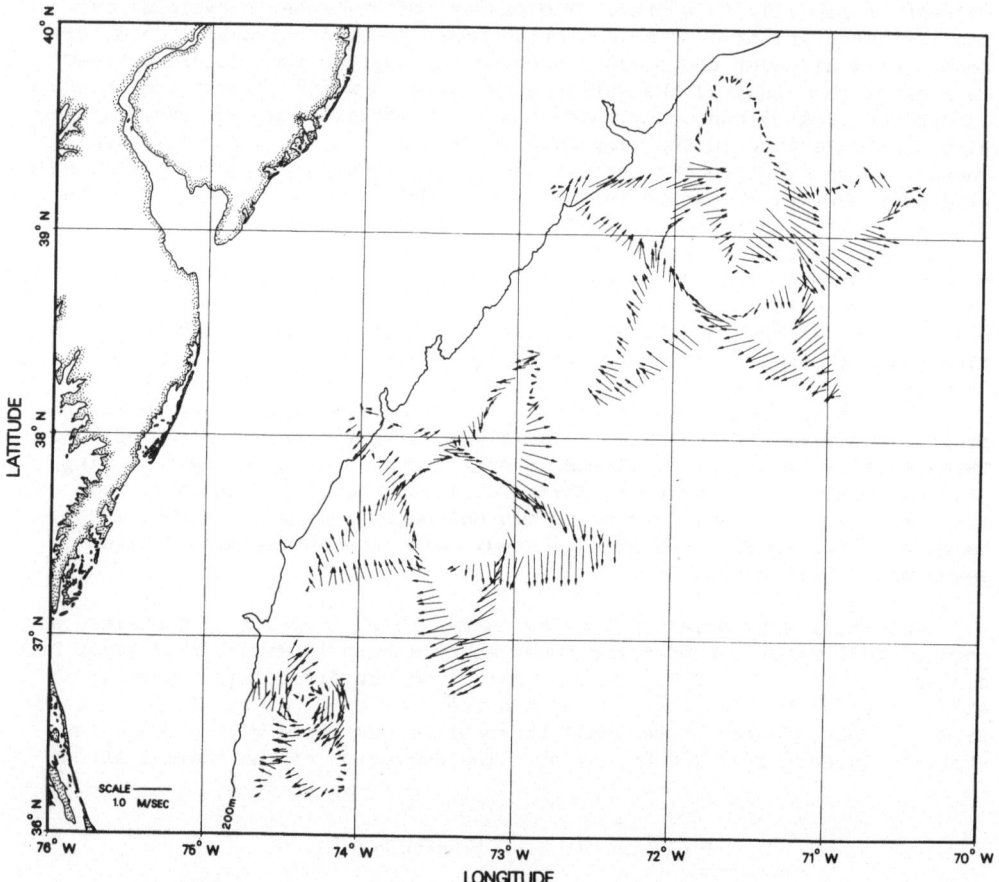

Fig. 8. Composite of 100 m current maps following ring 82B in April, June
and August, as it drifted to the west-southwest. The coastline and
200 m isobath are indicated. (From Joyce and Kennelly, 1985.)

ure of the ring (Fig. 7) from April through August shows the development of
the seasonal thermocline and the diminishing size of the ring. The near-
surface circulation of ring 82B (Fig. 8) illustrates the diminution in ring
size as the ring drifted to the west-southwest in close proximity to the
North American continental shelf.

The depth of the 10°C isotherm can be used as an index of ring
strength, while the areal extent of the region over which this isotherm is
deeper than 300 m is a useful measure of ring size. Both of these quantities, along with the thickness of the thermostad core and the surface
salinity, have been plotted for the ring center *versus* time in Fig. 9. Soon
after ring formation, the core temperature decreased to 15.7°C. Thereafter
the core temperature remained at this value while the layer thickness
decreased with time, as did the 10°C isotherm depth and the ring diameter.
The change in ring volume and energetics has been discussed in Olson et al.
(1985). Throughout its life cycle, the surface waters of ring 82B mixed
with fresher waters of the Slope Water, causing the anomalously salty
surface layer of the ring to become less saline. The most rapid change in
the physical characteristics of the ring began about Julian day 180 when it
suffered a glancing blow from the Gulf Stream. Between days 110 and 170,
the ring evolved without further cooling and interactions with the Gulf
Stream (Evans et al., 1985). This period is of great biological and chemical interest because of the occurrence of a "spring bloom" in the ring.

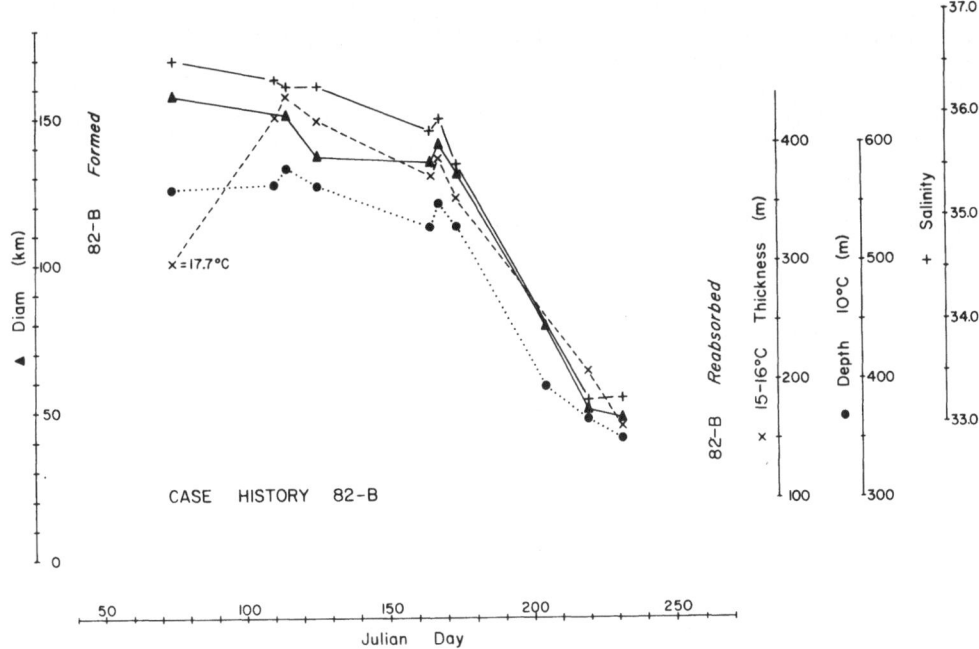

Fig. 9. History of ring 82B observations, showing the changes in the diameter of the ring, depth of the 10°C isotherm, the thickness of
the thermostad (the layer with a temperature between 15 and 16°C),
and the surface salinity at the ring center (adapted from Joyce
and Wiebe, 1983).

DISCUSSION

It is perhaps a fortuitous coincidence of circumstances that the
period for major biological changes in ring 82B coincided with both the
cessation of winter cooling and the isolation of the ring from interactions
with the Gulf Stream. The period from days 110 to 170 (April-June) will be
the focus for many of the "budget-type" calculations which will attempt to
unravel the degree to which observed changes in the ring ecosystem can be
explained by advection, mixing, or *in situ* processes.

From the observed change in depths of isopycnal surfaces, we can infer
that during this time interval the vertical velocity in the center of the
ring was upward throughout much of the upper 1000 m with velocities of
about 0.5–1.5 m day^{-1}. This slow decay of the thermocline depression of
the WCR is similar to observations of WCRs in the East Australian Current
(Niilson and Cresswell, 1981) and equal in magnitude but opposite in
direction to vertical velocities in decaying Gulf Stream cold-core rings
(Cheney and Richardson, 1976). The convergence and divergence pattern of
the vertical motion at the center of ring 82B will be used to infer the
slow radial advection velocities of the ring. These, together with the
observed changes in the salinity, should enable us to estimate the magnitude of lateral mixing which erodes the positive salinity anomaly of the
ring. This work is in progress and will be reported elsewhere. Preliminary indications are that between April and June, *in situ* processes may
dominate the observed changes in the near surface nutrients and biota.

The fact that vertical motion in the ring center is upwards implies
that the flux of nutrients will be into the photic zone and that vertical
gradients in the nutricline will be increased. Whether or not this flux is

sufficient to fuel the observed biological growth remains to be answered. It is clear, however, that the initial growth upon cessation of convective cooling was aided by high levels of nitrate ($6\mu mol\ kg^{-1}$) at the surface (Fox *et al.*, 1984). This nitrate was brought up to the surface from the main thermocline as it was eroded by convective cooling during late winter and early spring

The April-June period may be the most useful in developing an understanding of the evolution of semi-isolated rings and thus of processes which occur in the world ocean, but which can be best studied in a "semi-controlled" environment. The entire life history of ring 82B provides a comprehensive framework in which we can examine the dynamics of energetic eddies. In discussing Fig. 9, we stressed ring properties that changed with time. While these changes give insight into causal mechanisms, much can also be learned from properties which remain nearly constant in time. For example, once late winter convection stopped, the temperature, salinity, and potential vorticity of the core of the ring changed very little with time. The latter quantity was estimated by Joyce and Kennelly (1985) to be $(f + 2\omega)/h_{10}$, where f is the Coriolis parameter, ω is the angular rotation rate of the ring at 100 m depth, and h_{10} is the depth of the $10°C$ isotherm. Thus, despite the rapid ring decay due to a Gulf Stream interaction, these sub-surface properties of the ring core reservoir remain relatively constant, suggesting that mixing of mass and momentum are not essential to a basic description of interactions between rings and the Gulf Stream.

ACKNOWLEDGEMENTS

This research was supported by grants to the Woods Hole Oceanographic Institution from the National Science Foundation (OCE80-16983) and the National Aeronautics and Space Administration (NAGW-272) in the U.S.A. The author is grateful to many of his colleagues in the Warm-Core Rings Program for stimulating discussions and early access to their scientific results.

REFERENCES

Armi, L., and Bray, N., 1982, A standard analytic curve of potential temperature versus salinity for the western North Atlantic, J. Phys. Oceanogr., 13:384.

Brown, O., Olson, D., Brown, J., and Evans, R., 1983, Satellite infrared observations of the kinematics of a warm-core ring, Aust. J. Mar. Freshw. Res., 34:535.

Cheney, R.E., and Richardson, P.L., 1976, Observed decay of a cyclonic Gulf Stream ring, Deep-Sea Res., 23:143.

Dewar, W., 1985, Mixed layers in Gulf Stream rings, Dyn. Atmos. Oceans, in press.

Evans, R., Baker, K., Brown, O., and Smith, R., A chronology of warm-core ring 82B, J. Geophys. Res., 90:8803.

Flierl, G.R., 1984, Model of the structure and motion of a warm-core ring, Aust. J. Mar. Freshw. Res., 35:9.

Fox, M.S., Bates, P.F., and Kester, D.R., 1984, Nutrient data for warm-core ring 82B, from R/V Knorr Cruise 93, Technical Report 84-1, University of Rhode Island Graduate School of Oceanography, Narragansett, Rhode Island.

Joyce, T.M., 1984, Velocity and hydrographic structure of a Gulf Stream warm core ring, J. Phys. Oceanogr., 14:936.

Joyce, T.M., and Kennelly, M.A., 1985, Upper ocean velocity structure of Gulf Stream warm-core ring 82-B, J. Geophys. Res., 90:8839.

Joyce, T.M., and Stalcup, M.C., 1985, Wintertime convection in a Gulf Stream warm-core ring, J. Phys. Oceanogr., 15:1032.

Joyce, T., and Wiebe, P., 1983, Warm-core rings of the Gulf Stream, Oceanus, 26:34.

Joyce, T.M., Bitterman, D.S., Jr., and Prada, K.E., 1982, Shipboard acoustic profiling of upper ocean currents, Deep-Sea Res., 29:903.

Joyce, T., Backus, R., Baker, K., Blackwelder, P., Brown, O., Cowles, T., Evans, R., Fryxell, G., Mountain, D., Olson, D., Schmitt, R., Smith, P., Smith, R., and Wiebe, P., 1984, Rapid evolution of a Gulf Stream warm-core ring, Nature, Lond., 308:837.

Niilson, C.S., and Cresswell, G.R., 1981, The formation and evolution of East Australian Current warm-core eddies, Progr. Oceanogr., 9:133.

Olson, D., Schmitt, R., Kennelly, M., and Joyce, R., 1985, A two layer diagnostic model of the long term physical evolution of warm-core ring 82-B, 1984, J. Geophys. Res., 90:8813.

Richardson, P.L., 1983, Gulf Stream Rings, in: "Eddies in Marine Science", A.R. Robinson, ed., pp.20-45, Springer-Verlag, Berlin.

Schmitt, R., and Olson, D., 1985, Winter convection in warm rings: thermocline ventilation and the formation of mesoscale lenses, J. Geophys. Res., 90:8823.

Warm Core Rings Executive Committee, 1982, Multidisciplinary program to study warm core rings, Trans. Amer. Geophys. Un., 63:834.

TURBULENCE IN THE UPPER LAYER

Thomas Osborn

Chesapeake Bay Institute
Shady Side, Maryland 20764 U.S.A.

ABSTRACT

The role is examined of small scale turbulence in determining the distribution of chemical species and other scalar quantities. Turbulent stirring is identified as the cumulative effect of fluctuations in the velocity components combined with fluctuations in the concentrations. Recent work enables the vertical eddy diffusivity coefficient to be para- meterized in terms of small scale parameters for two portions of the water column; the region just below the surface and the region that is stratified. Much more information is necessary before parameterizations can be related directly to bulk parameters such as wind speed or Väisälä frequency. Recent measurements from a submarine of turbulent fluctuations and salt fingers are shown as examples of the small scale processes in the ocean.

INTRODUCTION

Turbulence is classically defined as random, fluctuating, three- dimensional motion that is both dissipative and diffusive. The diffusive nature of turbulent flows is due to an accumulation of fluctuating advect- ive events. Consider a region of anomalously warm water. The fluctuating advective motions shear, stretch, and distort this region, increasing its surface area and the temperature gradient at its surface. Both of these effects increase the molecular transport of heat from the region, which by now can be quite convoluted in shape. As the heat diffuses away from the water molecules that were originally warmer, the temperature feature disappears. Eckart (1948) distinguished between stirring by the turbulent motion and the final mixing process that is due to molecular diffusion. The effect of the turbulent stirring motion is important in chemical processes where the rates of molecular diffusion are often quite small.

Mixed layer and upper layer modeling have been most successful in predicting the oceanic response to synoptic scale storms. The air-sea gas exchange process is different from the air-sea buoyancy transfer phenomena and, as well, the sources and sinks for chemical species are distributed throughout the upper layer. One can expect that modeling the distributions of chemical species is a different and possibly a more exacting test of our understanding of the physics of the upper layer.

The last fifteen years have seen a strong development in the field of oceanic turbulence. We are now able to measure the small scale temperature and velocity fluctuations associated with the smallest scales of the turbulent stirring process. These measurements allow us to estimate the variance of the temperature gradient fluctuations and the variance of the velocity shear. The former is used to determine the rate of dissipation of the variance of the temperature fluctuations while the latter is used to calculate the rate of turbulent energy dissipation, a fundamental parameter in the energetics of the processes. Measurements have shown turbulent patches in the seasonal thermocline that have vertical extents from as little as 1 m to as much as 50 m. Time scales of the order of many times the Väisälä period have been observed. Features have been traced horizontally for distances of 1 km or more. These measurements have provided estimates of vertical mixing rates and some insight into the forcing mechanisms for the turbulence. There is strong evidence that the internal wave field, both the gravitational and the inertial waves, play major roles in the energy transfer.

Our present understanding of the physical processes that occur in the upper ocean is still insufficient. Recent work on doubly diffusive convection and turbulence gives valuable insights into these processes, which are important in determining chemical concentrations in the upper portion of the ocean. This account looks at some of the recent information about these different phenomena in order to see their effect on the transport of a chemical species. It concentrates on small-scale turbulence which causes mixing across isopycnal surfaces. The discussion will seem to emphasize vertical mixing as a one-dimensional process, but that is only because of our limited understanding of the role of two-dimensional variability and the details of the process of stirring by intrusions along isopycnal surfaces.

MODELING OF TURBULENT MIXING

When dealing with turbulent processes the normal approach is to decompose the flow field into a mean flow and a fluctuating flow. A major problem is in determining what motions are part of the mean flow and what motions are fluctuations about the mean. We are accustomed to determining means from time or space averages of the appropriate quantities. The theoretical derivations are based on ensemble averages of the flow-averages over many realizations of the flow with the same initial conditions and the same boundary conditions.

Recent work on eddies and large scale flow variations shows that there is a lot of low-frequency, large-scale variability in the ocean. Where do we separate the variation of the mean field from the fluctuations about that mean state? The best separation occurs when there is a definitive separation in terms of the time and space scales. For example, the surface wave field can be separated from the tidal variation because of the great difference in frequencies and wavelengths. Unfortunately, the many processes that contribute to scalar distributions in the upper layer are not so easily separated, since the space and time scales can be comparable.

In spite of these problems, let us use the Reynolds decomposition of the fields into their mean and fluctuating parts to study the problem of turbulent transport. Scalar quantities such as chemical species, and to some extent temperature, obey conservation equations that can be exemplified by the equation

$$\frac{DC}{Dt} = - \nabla \cdot \vec{M}_C + SO - SI \qquad (1)$$

where $\frac{D}{Dt} = \frac{\partial}{\partial t} + \vec{U} \cdot \nabla$, C is the concentration of the scalar quantity, $\vec{M}_c$ is the molecular flux, and SO and SI are the sources and sinks of C. This equation is exact and complete. We set the concentration $C = \langle C \rangle + c'$ and the velocity $\vec{U} = \langle \vec{U} \rangle + \vec{u}'$, where the process of averaging is denoted by the pair of brackets $\langle \ \rangle$. The averages of c' and $\vec{u}'$, the fluctuating part of the concentration and the velocity, are zero, i.e., $\langle c' \rangle \equiv 0$ and $\langle \vec{u} \rangle \equiv 0$. With these assumptions we can derive an equation for this mean concentration:

$$\frac{\partial \langle C \rangle}{\partial t} + \langle \vec{U} \rangle \cdot \nabla \langle C \rangle = -\nabla \circ \langle (\vec{U}'c') \rangle + \kappa_c \nabla^2 \langle C \rangle + \overline{SO} - \overline{SI} \tag{2}$$

We assume that $\langle \vec{M}_c \rangle$ is proportional to the gradient of $\langle C \rangle$ and κ_c is the molecular diffusivity of C. The important term is the divergence of the turbulent diffusion, also called the turbulent flux, $\langle \vec{u}'c' \rangle$. This term represents the net advective effect of the velocity fluctuations $\vec{u}'$, which average to zero under our definition of the averaging process.

The modeling of the flow as a time variable mean and a fluctuating part produces the term $\langle u'c' \rangle$ in the equation for the mean scalar concentration. In numerical modeling parlance, these terms are the "sub-grid scale processes". They are the physical processes that exist on scales that are too small to be resolved by the grid on which the calculations are performed. Another perspective is to accept that the specific details of the flow are not important in describing the final configuration. The turbulent flux is a net advective effect by the fluctuations in the flow, and we wish to average over these fluctuations.

The turbulent fluxes in the ocean are impossible to measure directly. Unstable platforms support the sensors and the orbital velocities from surface waves and internal waves contaminate the velocity measurements. Sensors always suffer from limitations in their response characteristics and many chemical species cannot be measured easily, if at all, in situ. Even the heat flux, $\langle \vec{u}'\theta' \rangle$, and the vertical component of the momentum flux or stress, $\langle u'w' \rangle$, cannot be measured directly in the ocean. These measurements may become possible in the future but probably not in the next ten or twenty years.

Some problems of turbulent transport are simplified because horizontal variation is relatively unimportant. In these cases the problem becomes one-dimensional with just the vertical variation and transport contributing. Most of the early mixed layer models were one-dimensional; let us look for some insight from the results. This work gives us estimates for the turbulent diffusion in terms of the vertical eddy diffusivity coefficient which we assume to be the same for all passive scalars.

Since direct flux measurements are not possible, some indirect methods for estimating fluxes have been developed. For the heat and mass flux there are some models that enable us to estimate the turbulent transport. A model due to Osborn and Cox (1972) permits estimates of the vertical heat flux in situations where horizontal intrusions are not important. The flux is estimated from measurements of the variance of the small scale temperature gradient fluctuations:

$$\langle w'\theta' \rangle = -\kappa \frac{\langle (\nabla \theta')^2 \rangle}{\frac{\partial \langle \theta \rangle}{\partial z}} \tag{3}$$

where κ is the coefficient of molecular diffusion and $d\theta/dz$ is the mean vertical temperature gradient. It is convenient to model the heat flux as an eddy coefficient K_θ times the mean vertical gradient $d\theta/dz$. Under these assumptions the formula for the eddy coefficient possesses a certain symmetry:

$$K_\theta = \kappa <(\nabla\theta')^2>/(\partial<\theta>/\partial z)^2 \qquad (4)$$

The model requires a mean temperature gradient and is not appropriate if the upper layer is well mixed. This model has been used extensively by Gregg (1975, 1976), Gargett (1976), Elliott and Oakey (1976, 1980) and Caldwell (1978). The results suggest that the eddy coefficient is of the order of 0.03 cm^2 s^{-1} in the seasonal thermocline, a value which is substantially below the canonical value of 1 cm^2 s^{-1} many had expected. The values for the eddy coefficients for other passive scalar quantities are normally the same as the values for heat since it is the same turbulent velocity fluctuations that are performing the transport. The same formalism that leads to equations (3) and (4) can be applied to any passive scalar (salt, oxygen, etc.) for which the variance of the gradient can be measured. It is most appropriate for temperature since the small scale variance of the temperature gradient is presently measurable, unlike those for the other passive scalars in the ocean.

Another model for estimating the vertical diffusion coefficient in the stratified portion of the water column is based on the turbulent kinetic energy equation (Osborn, 1980). This model calculates an upper limit on the diffusion rate from measurements of the turbulent kinetic energy dissipation rate, ϵ:

$$K_m \leq 0.2\epsilon/N^2 \qquad (5)$$

where N is the Väisälä frequency. This estimate is useful for several reasons. First, it gives support to the low values estimated from the earlier temperature gradient measurements. Secondly, this formula uses the dissipation data that are now more prevalent than the temperature gradient data, and it allows a connection to ideas about the relationships of internal waves and turbulence in the ocean (Gargett and Holloway, 1984) as well as a prediction of the variation of K as a function of the Väisälä frequency (Gargett, 1984).

The eddy coefficients in equations (4) and (5) are based on restrictive assumptions and require a one-dimensional regime so that lateral transport effects can be neglected. Quite obviously such a situation does not exist in many interesting regions, such as frontal boundaries and around intrusions.

The nature of the turbulence at the very surface of the ocean has been examined by Dillon et al. (1981), who indicate that the surface of the ocean can sometimes be modelled as a constant stress layer. While they caution against universal application of this idea, since their result is for a 7 m s^{-1} wind over a 6 m thick upper layer in a lake, the result is appealing since a constant stress layer can be modeled by an eddy coefficient. In cases where there is no stratification

$$K_c = 0.4U_* D \qquad (6)$$

where U_* is the friction velocity and D is the depth. Figure 1 shows the fit of the measured dissipation data to the theoretical log normal profile based on the local wind speed. This work on the details of the surface layer turbulence gives a model for the turbulence at the surface of the ocean. The constant stress layer model must be coupled to the physics of the gas transfer through the surface and to another model for transfer through the rest of the upper layer.

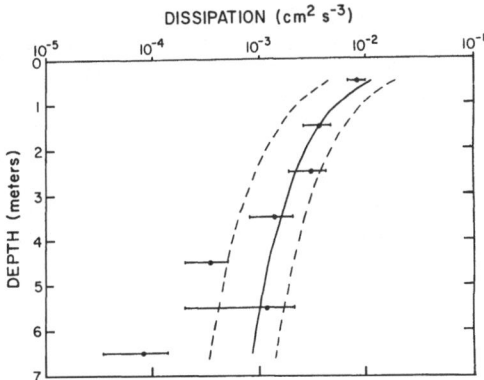

Fig. 1. Dissipation profile showing the
fit of the dissipation rate
near the surface to the predictions of the constant stress
model (from Dillon et al., 1981).

A complete interface model will include buoyancy flux through the
layer and the effect of stratification on the diffusion rate. Even including these details, the physical regime associated with the surface
forced constant stress layer probably does not extend very far into the
upper layer. Oakey and Elliott (1982), who studied a 20 m thick upper
layer, remarked that: "Even in periods of high wind speeds, the mixed
layer is not being continually overturned by turbulence generated at the
surface. Generation of turbulence, at least in the lower half of the mixed
layer, appears to be through shear instability".

The details of the response of the upper layer to atmospheric forcing
are not yet known. Thorpe et al. (1977) and Dillon and Caldwell (1978)
show examples of the sporadic events that occur in the upper layer to cause
mixing and entrainment of fluid from below. Modeling the frequency and the
effects of such occurrences is not yet possible. The turbulent stress and
the velocity shear cannot be predicted so the turbulent energy production
cannot be predicted. The problem is essentially circular for the turbulent
production generates the velocity fluctuations whose correlation determines
the stress which, in conjunction with the boundary conditions, determines
the velocity profile. At present the best approach appears to be to treat
the upper layer as a well mixed reservoir, unless there is enough of a mean
gradient for one to assign an eddy coefficient to the turbulent transport
process.

Thorpe (1984) shows that the diffusivity near the surface of the ocean
can be estimated from inverted echo-sounder measurements of the scattering
cross-section from bubbles. This work is especially promising because the
method can produce information about temporal variation of the diffusivity
as a function of depth and factors such as wind speed and wave breaking.
The work also provides information on the role of bubbles in gas transfer.

DOUBLY-DIFFUSIVE CONVECTION

The ocean is not a simple horizontally homogeneous system. There are
lateral differences in temperature and salinity. Sometimes the lateral
changes are concentrated in what are called frontal regions, and the
different water masses form intrusions that extend into each other. In
these regions we often find relatively warm salty water on top of colder

less saline water. In such cases there is a special kind of convection pattern, called salt fingers, which can form due to the difference in the rate of molecular diffusivity between salt and heat. Salt fingers drive a convective pattern by removing gravitational potential energy from the salinity (by lowering the center of mass of the salt molecules). About half of the potential energy is dissipated by viscous drag on the flow. About half of the potential eneregy removed from the salinity field is returned as potential energy in the temperature field due to the downward advection of heat in the fingers. (See Turner (1973) and Huppert and Turner (1981) for details of the formation and pictures of the regular pattern of the convection cells.) Salt fingers only occur under the specific conditions when the salinity and the temperature are decreasing with depth.

While turbulence raises the center of mass of the fluid, salt fingers lower the center of mass. Turbulent stirring and the associated mixing move temperature, salinity, and density down their respective gradients, from regions of high values to regions of lower values. Salt fingers also transfer heat and salt from regions of relatively higher concentration to regions of relatively lower concentration. However, since they do not transfer heat and salt with the same efficiency, the fingers appear to transfer density against its mean gradient, that is from regions of lower density to regions of higher density.

Oceanic observations of salt fingers are limited but fairly convincing (Williams, 1974; Magnell, 1976; Gargett and Schmitt, 1982). In the ocean, salt fingers can occur as vertically aligned square packed cells with sides of about 0.04 m. The situation is a set of cells of alternate upflow and downflow. Heat is transferred laterally from the downflowing warm (relatively salty) water, making it heavier, to the adjacent upflowing cooler cell, making it lighter (because it is less salty). The exact dimensions depend on the local stratification as well as the age and nature of the fingers. The thickness of the fingering layer is of the order of 1 m. If the water column is vertically sheared by the mean velocity profile, then the fingers will appear as sheets parallel to the velocity vector with a 40% greater thickness. The fingers and the interfaces between them are so small because the process depends on the difference in the rates of molecular diffusion. Molecular diffusion is only effective in the ocean over short scales of the order of a few centimetres.

Laboratory studies of salt fingers (Turner, 1973; Schmitt, 1979) show that the fingers grow very rapidly. Time scales are of the order of one Väisälä period if the value of $R = \alpha T/\beta S$ is between 1 and 2. Thus, the fingers have the possibility of being very important in frontal regions where the values of R can be very low due to the intrusive nature of the flow. Another region of potential importance is in the central gyres where evaporation leads to salinity maxima at the surface. The fingers that occur are a strong mechanism for transferring chemical species vertically. Because the molecular diffusion rates for chemical species vary, the effect of salt fingers on the individual species is different. This effect will need to be considered in the eventual incorporation of salt fingering into the problem of cross-isopycnal transport in the ocean. Present work is focused on the role of salt fingers in the ocean but there is no present capability to estimate their importance quantitatively.

There is a second type of doubly diffusive convection which arises when the warmer and saltier water is under the relatively cooler, fresher water mass. In this case heat is transferred across the horizontal interface by molecular conduction while little salt transfer occurs. Thus water parcels above the interface tend to get warmer and rise away from the interface, while those below tend to get cooler and sink from the interface, thereby maintaining its sharpness. This type of doubly diffusive convection does not involve transfer of fluid vertically between the two reservoirs but rather away from their common interface.

By their very nature, intrusions are amenable to one type of doubly diffusive convection on their upper surface and the other type on their lower surface. The relative efficiency of the two processes tends to drive the intrusion across isopycnal surfaces as it extends into the fluid (Turner, 1978). There may well be turbulence associated with the intrusion also. Here is a critical problem: what is the role of the horizontal processes in the vertical mixing of the ocean? Gregg (1976,1980) gives examples of the complicated situation that arises due to intrusive features. Gargett (1976) details the existence of small frontal regions even in the central gyre regions where horizontal uniformity is most likely. Much work is needed on this problem before the answer is clear.

EXAMPLES OF OCEANIC PROCESSES

In this section we will examine some recent measurements of small scale processes in the ocean. The data will show turbulent patches, salt fingering, surface forced convection, and intrusions. The objective is to give the reader some feeling for the present work in small scale turbulence, the phenomena observed, and the processes that are active in the ocean. The examples are selected from some of our recent measurements using the submarine USS Dolphin (Osborn and Lueck, 1984). Many other examples are available in the literature and a listing of some of the papers is provided in the bibliography.

The measurements from the submarine differ from much of the previous work since the submarine moves in a predominantly horizontal direction, compared to the free fall vehicles which profile vertically. There are complications; the submarine tends to change depth, even when the crew attempts to maintain a uniform depth; moreover, the isopycnal surfaces in the ocean are not level or stationary in time. The net result of these effects ensures that the data acquired are not a simple horizontal, level transect, parallel to the density surfaces. For this reason one operational technique has been to cycle the submarine over a vertical depth range, but keep the vertical speed much smaller than the horizontal speed which is of the order of 1.2 m s^{-1}.

It is customary to begin a dive with a transect from the surface to 100 m depth to get an idea of the turbulent activity. Figure 2 shows two transects through the upper 100 m of the water column. The first was made in December 1980, the second in April 1982. The data show the temperature profile and the local rate of turbulent energy dissipation estimated directly from the shear of the small scale velocity fluctuations. The upper layer is well-mixed in December with relatively large dissipation rates. There is evidence of entrainment at the base of the mixed layer, where the dissipation rate decreases before it increases to the dissipation maximum in the thermocline. Below the maximum, the dissipation rate decreases sharply to the noise level. There are intermittent active patches further down in the water column.

In April 1982 there is no mixed layer at the surface, but rather a substantial temperature gradient extending to the surface. The surface is at the top of the temperature trace, about 3 m depth, the offset arising because the pressure transducer is lower than the temperature sensor. There are several turbulent patches in the upper 100 m. One must bear in mind with these profiles that, while they are shown as vertical transects, the submarine went several kilometres horizontally between the top and the

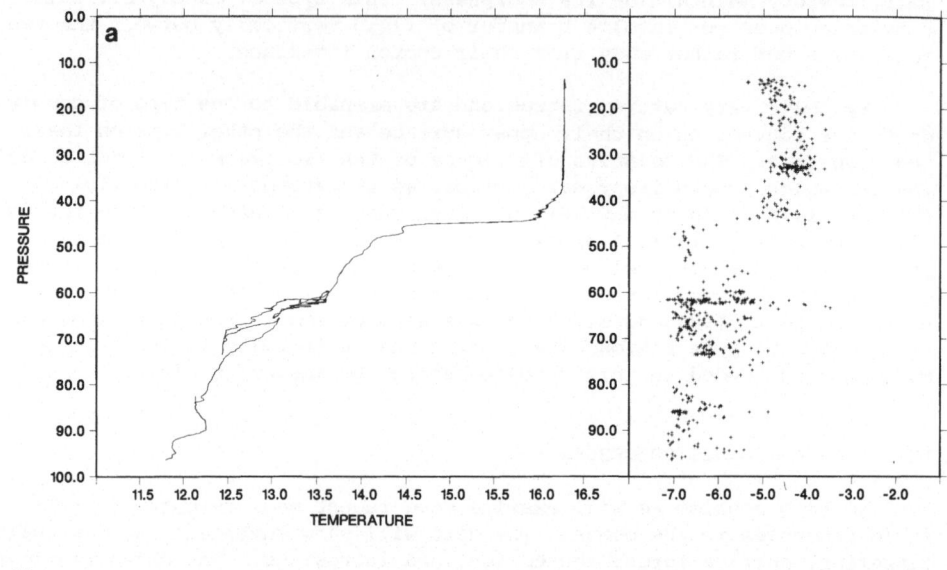

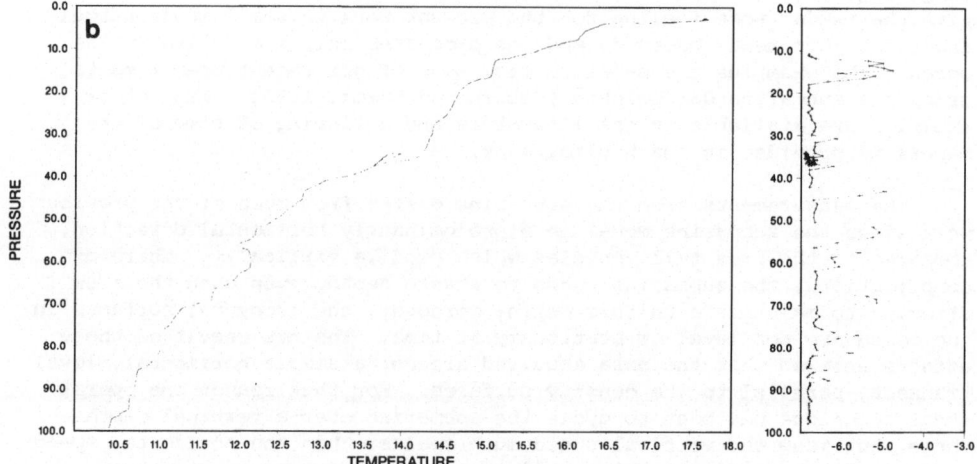

Fig. 2. Dissipation profiles off San Diego California: (a) 4 December 1980
near midnight, (b) 22 April 1982 about 15.00 local time. Temperature scales are in °C and dissipation scales are in W m⁻³. (From
Osborn and Lueck, 1984.)

Fig. 3. CTD profiles of the upper 20 m off San Diego, 19 April 1982 at
10.00 hours: (a) a downward profile by the submarine, (b) an upward
profile. Note the strong temperature gradient just below the surface, the saline surface layer, and the relatively small salinity
variation. The T-S diagram shows the dominance of temperature in
the sensitivity variation. The sloping lines are lines of constant
sigma-T. The surface is at approximately 3 decibar pressure
because the pressure transducer was mounted that much lower than
the temperature and conductivity transducers. The offset is not
removed because it varies slightly in time. The top of each trace
is truncated at our best estimate of the location of the sea surface. The letter labels refer to features shown in the turbulence
data in Fig. 4.

bottom of the profile. An intrusion of warm water occurs below 55 decibar. The two profiles in Fig. 2 are typical of the vertical intermittency seen in much of the data. The averaging techniques used in estimating the eddy coefficients from equations (4) and (5) must smooth over the details of any single profile and establish mean values representative of the time and space scales that determine the distributions.

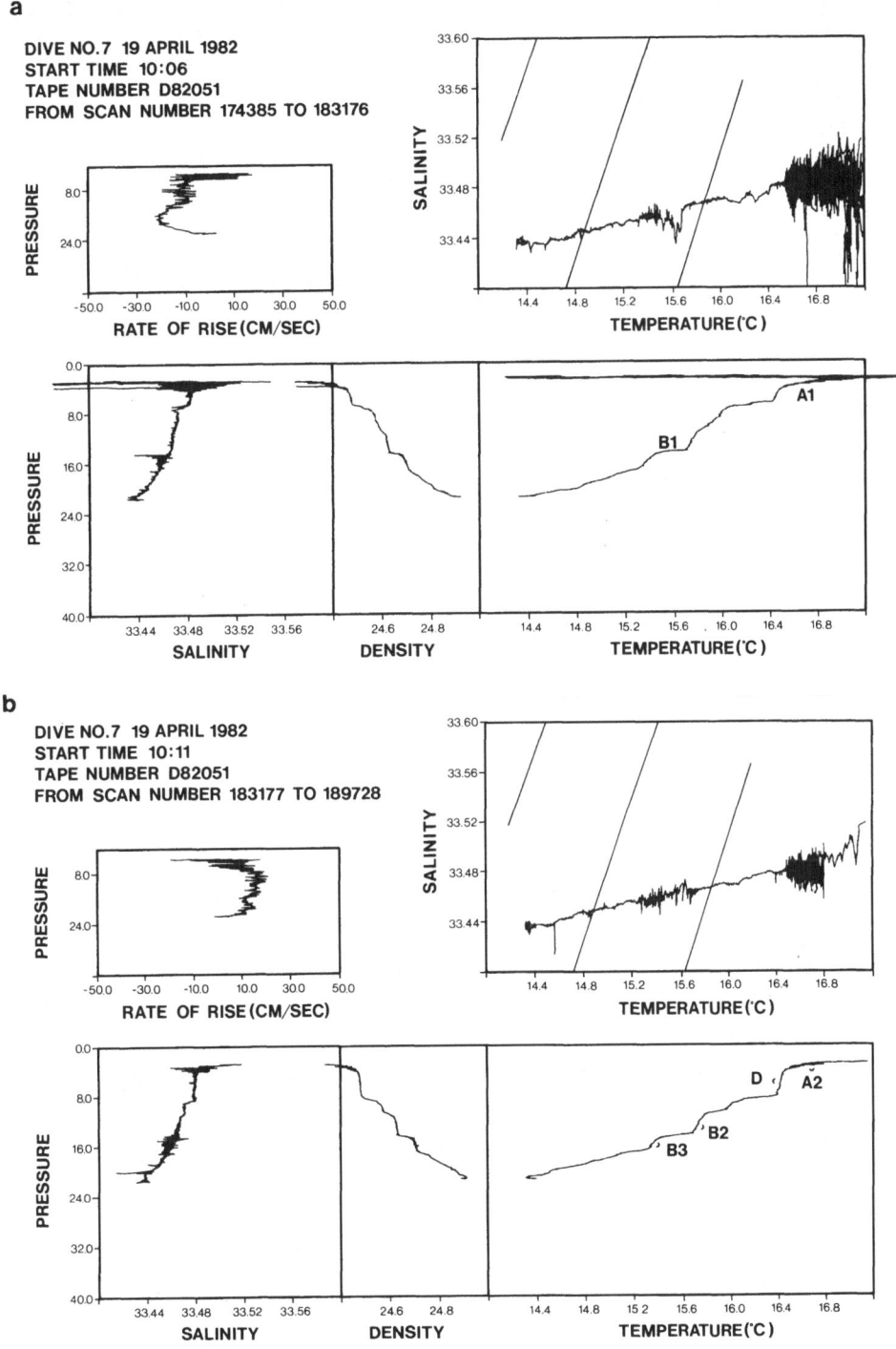

Of special interest is the very surface of the ocean. Here the exchange processes can be limited by the processes associated with the interface. The probes on the submarine are mounted at the top of a tripod over 4 m above the deck and just aft of the bow. This location allows us to profile through the surface of the ocean. Figure 3 shows the CTD data acquired on two consecutive profiles through the surface on 19 April 1982 about 10.00. There is a strong temperature gradient just below the water surface, with an indication of a saline layer in the upper 3 m. The T-S diagrams are similar for the two profiles except for the small amount of salinity spiking associated with the sharp temperature gradients. The temperature profiles are similar but distorted by lateral variations in thickness of various features.

The time series plots for the turbulent velocity and temperature gradient signal for these transects from the surface down to depth are shown in Fig. 4. The velocity shear, the temperature, and the temperature gradient are plotted as a function of time and not depth. Since the vertical speed varies, the horizontal scale does not relate directly to depth. Rather the temperature features must be related to the CTD trace and depth deduced from that plot. The velocity shear signals show turbulent patches at the surface in both transects (these features are labelled A1 and A2 in Fig. 4). Turbulent features labelled B1, B2, and B3 are identified in Figs 3 and 4. The features are different in the two transects. There is a single patch superimposed on the temperature step in the first transect. In the second profile, the feature appears to be two patches, one on each side of the temperature step.

The striking difference between the two transects is the temperature microstructure (labelled D) that appears in the temperature gradient channel in the mixed layer just below the surface of the upward transect. There is no associated velocity signal above the noise level. Spectral calculations give results similar to those reported by Gargett and Schmitt (1982). The wavelength of the phenomenon is about 0.06 m. The signal is presumably due to salt-fingers. The situation is a little unclear since the CTD data show no strong source region. The value of R (the ratio of the relative

Fig. 4. Profiles of the turbulence parameters during two sequential cycles through the surface: (a) the down cycle, (b) the immediately subsequent up cycle. Total elapsed time for the cycle is 6 min. and the distance covered was 0.65 km. Temperature, its derivative, and the velocity shear in the frequency range from 3 hz to 20 hz, are shown. The velocity shear channel shows the presence of turbulent velocity fluctuations. Data are plotted as a function of time not depth. Features of the temperature profile can be related to the temperature profile from the CTD in order to determine the position of the temperature gradient and velocity shear patches. Figures are cut from one continuous time trace.

The down trace shows a turbulent patch (marked A1) at the surface and another about 16 decibar (marked B1). On the up cycle the turbulence is still found in that region but there are two patches, B2 and B3, above and below the temperature step. Note on the up cycle the large amount of temperature gradient signal in the 3 m thick region just below the surface labelled D. There is no velocity signal accompanying the temperature microstructure. This region is believed to be salt-fingering. The convective velocities of the fingers are below the noise level of the velocity sensing system. Above the salt fingers there is still the surface turbulent region labelled A2.

contributions of the temperature and salinity gradients to the density gradient) is greater than 10, well away from the value for fast growth of the fingers. Atmospheric conditions were very calm, with slight wind, and the morning fog was just burning off. The waves were only 0.1 m in

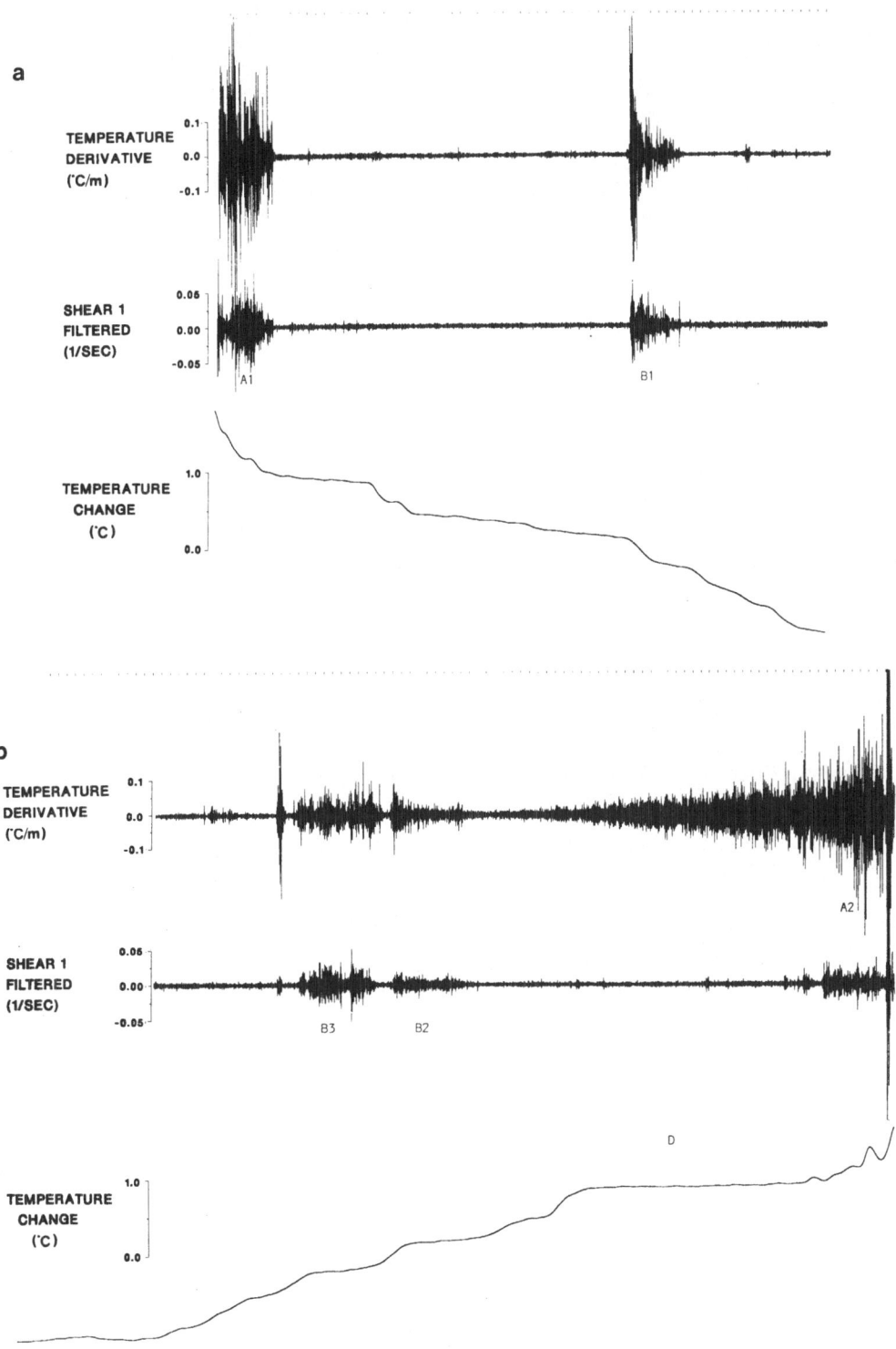

amplitude. The phenomenon occurred in 12 out of 13 profiles at varying
intensity over a horizontal distance of 7 km. The combination of a thin
turbulent layer at the surface with a fingering region below could mean
that there was significantly more flux than one would estimate from the
wind stress and equation (6).

CONCLUSION

While much remains to be done in the field of oceanic turbulence,
there is a growing body of information on the mechanisms and the rates for
the cross isopycnal transport of properties. There is an understanding of
turbulence in the upper layer both in terms of mechanisms and intensities.
The next step is to relate the small scale measurements to the mean field
and thus develop improved schemes to parameterize the turbulence.

ACKNOWLEDGEMENTS

This work has been supported by the Office of Naval Research.

REFERENCES

Caldwell, D.R., 1978, Variability of the bottom mixed layer on the Oregon
    Shelf, Deep-Sea Res., 25:1235.

Dillon, T.M., and Caldwell, D.R., 1978, Catastrophic events in the
    surface mixed layer, Nature, Lond., 276:601.

Dillon, T.M., Richman, J.G., Hansen, C.G., and Pearson, M.D., 1981,
    Near-surface turbulence measurements in a lake, Nature, Lond.,
    290:390.

Eckart, C., 1948, An analysis of the stirring and mixing processes in
    incompressible fluids, J. Mar. Res., 7:265.

Elliott, J.A., and Oakey, N.S., 1976, Spectrum of small-scale oceanic
    temperature gradients, J. Fish. Res. Bd Can., 33:2296.

Elliott, J.A., and Oakey, N.S., 1980, Average microstructure levels and
    vertical diffusion for Phase III, Gate, in: "Oceanography and
    Surface Layer Meteorology in the B/C-Scale", G. Siedler and J.D.
    Woods, eds, pp.273–294, Pergamon Press, Oxford.

Gargett, A.E., 1976, An investigation of the occurrence of oceanic
    turbulence with respect to finestructure, J. Phys. Oceanogr.,
    6:139.

Gargett, A.E., 1984, Vertical eddy diffusivity in the ocean interior,
    J. Mar. Res., 42:359.

Gargett, A.E., and Holloway, G., 1984, Dissipation and diffusion by
    internal wave breaking, J. Mar. Res., 42:15.

Gargett, A.E., and Schmitt, R.W., 1982, Observations of salt fingers in
    the central waters of the eastern North Pacific, J. Geophys. Res.,
    87:8017.

Gregg, M.C., 1975, Microstructure and intrusions in the California
    Current, J. Phys. Oceanogr., 5:253.

Gregg, M.C., 1976, Finestructure and microstructure observations during the passage of a mild storm, J. Phys. Oceanogr., 6:528.

Gregg, M.C., 1980, The three-dimensional mapping of a small thermohaline intrusion, J. Phys. Oceanogr., 10:1468.

Huppert, H.E., and Turner, J.S., 1981, Double-diffusive convection, J. Fluid Mech., 106:299.

Magnell, B., 1976, Salt fingers observed in the Mediterranean Outflow region using a towed sensor, J. Phys. Oceanogr., 6:511.

Oakey, N.S., and Elliott, J.A., 1982, Dissipation within the surface mixed layer, J. Phys. Oceanogr., 12:171.

Osborn, T.R., 1980, Estimates of the local rate of vertical diffusion from dissipation measurements, J. Phys. Oceanogr., 10:83.

Osborn, T.R., and Cox, C.S., 1972, Ocean fine structure, Geophys. Fluid Dyn., 3:321.

Osborn, T.R., and Lueck, R.G., 1985, Turbulence measurements from a submarine, J. Phys. Oceanogr., 15:1502.

Schmitt, R.W., 1979, The growth rate of super-critical salt fingers, Deep-Sea Res., 26:23.

Thorpe, S.A., 1984, On the determination of K_v in the near-surface ocean from acoustic measurements of bubbles, J. Phys. Oceanogr., 14:855.

Thorpe, S.A., Hall, A.J., Taylor, C., and Allan, J., 1977, Billows in Loch Ness, Deep-Sea Res., 24:371.

Turner, J.S., 1973, "Buoyancy Effects in Fluids", Cambridge University Press, London.

Turner, J.S., 1978, Double-diffusive intrusions into a density gradient, J. Geophys. Res., 83:2887.

Williams, A.J., III, 1974, Salt fingers observed in the Mediterranean Outflow, Science, N.Y., 185:941.

CONVECTION IN THE UPPER OCEAN

Theodore D. Foster

Division of Natural Sciences
University of California
Santa Cruz, California 95064   U.S.A.

ABSTRACT

    Convective phenomena in the upper ocean are important to an under-
standing of the flow patterns that occur in the mixed layer. Small scale
convection can occur just below the sea surface whenever there is a transfer of heat from the ocean to the atmosphere. The horizontal scale of this
convection will typically be a few centimetres. Usually this convection
will be intermittent with a characteristic period of the order of a minute.
The small scale convection can transfer energy to larger spatial and temporal scales until the spatial scale reaches a size of the same order as
the depth of the mixed layer.

INTRODUCTION

    In the atmosphere convective phenomena are clearly visible in the
various types of cumulus clouds. In the ocean convective phenomena are not
readily apparent and their observation usually requires special visualization techniques. Using an infrared scanner from aircraft, McAlister and
McLeish (1965) have observed structures on the sea surface that closely
resemble cumulostratus cloud forms (Fig. 1, upper) or convection cells in
laboratory experiments (Foster, 1965b) using dye techniques for visualization (Fig. 1, lower). Using an underwater shadowgraph system, Williams
(1974) has observed double-diffusive convection in the ocean. Most of our
knowledge of convection in the ocean, however, has been from temperature
and salinity profiles and theoretical extrapolation from laboratory
experiments.

    Buoyancy-driven convection in the upper ocean is often masked by
wind-induced stirring. Nevertheless, buoyancy provides considerable
structure to the flow field even when wind stresses dominate buoyancy
forces, and under calm and low wind conditions (speeds less than about
1 m s^{-1}) buoyancy-driven effects can dominate. McAlister and McLeish
(1965) found that as the wind speed increased, the effect of the wind
stress was to distort the sea surface infrared pattern so that the cells
became elongated in the direction of the wind.

    In the upper ocean three basic types of convection are found: thermal,
haline and thermohaline. Thermal convection in an almost pure case occurs

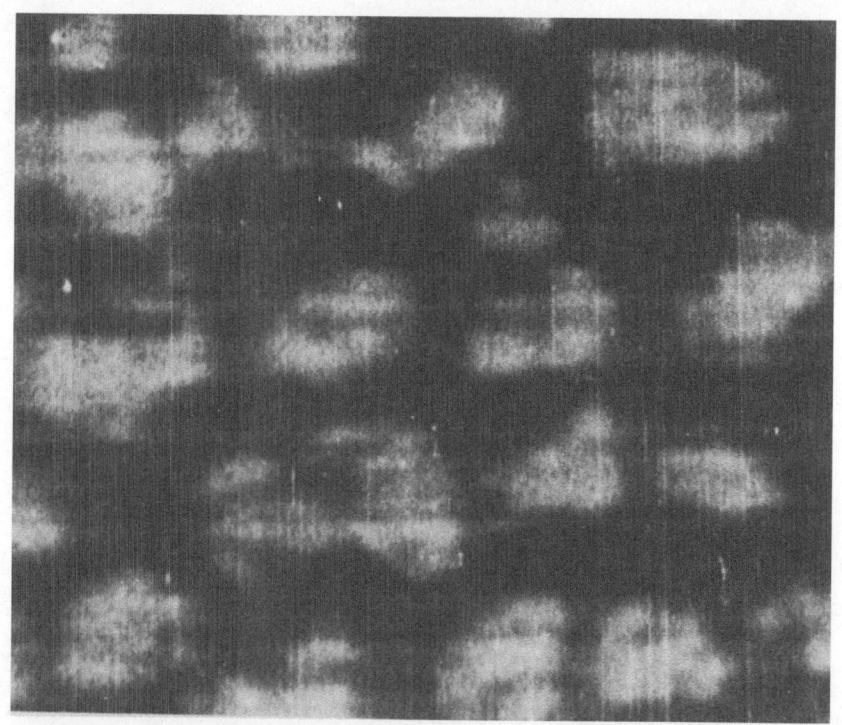

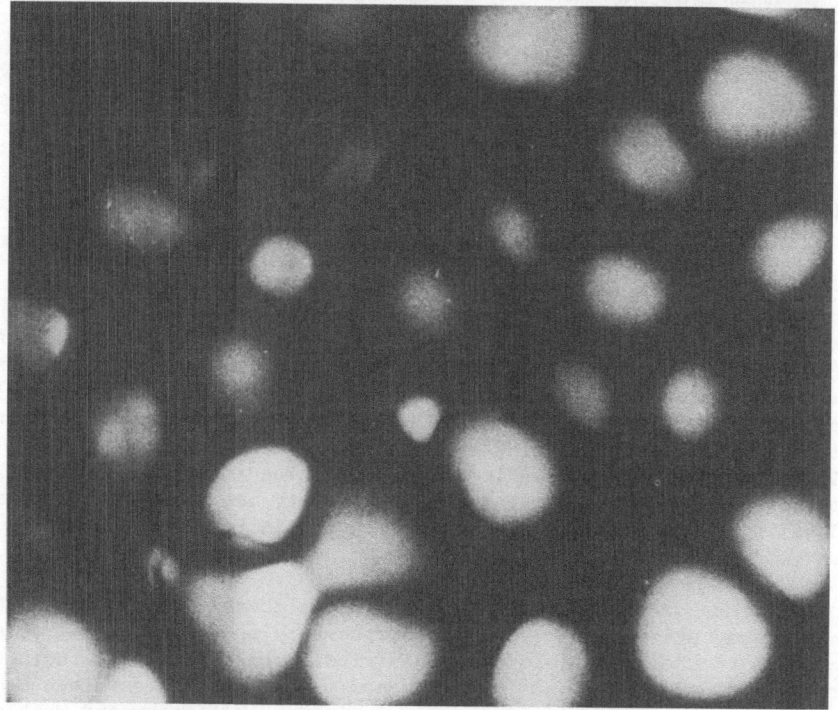

Fig. 1. Horizontal patterns of convection. *Upper*: Infrared
scanner image of sea surface; the typical white spot is
around 20 m across. *Lower*: Dye layer in laboratory
convection experiment; the typical white spot is around
0.02 m across.

in the mixed layer of the ocean due to the asymmetric way the ocean is heated and cooled. The principal heating process in the ocean is solar radiation, and while the heating is principally confined to the upper few metres, significant radiative heating occurs in about the upper 10 m in coastal waters and in about the upper 40 m in the open ocean. One asymmetry arises due to the shift of the wavelength of maximum intensity of the incoming radiation at about 0.5 μm (due to a solar blackbody temperature of order 6000°K) to about 10 μm (due to an ocean blackbody temperature of order 300°K) for backradiation. Since the attenuation of electromagnetic radiation in seawater is at a minimum near 0.5 μm but is several orders of magnitude greater near 10 μm (Jerlov, 1968), the incoming solar radiation penetrates to several metres while the backradiation from the ocean is effectively confined to a layer less than 0.1 mm thick at the sea surface. Other cooling processes, such as conduction of heat to the atmosphere, are also surficial processes. Since the thermal diffusivity of seawater is quite small (order 10^{-7} m^2 s^{-1}), the upper ocean would heat up considerably if heat were not removed mechanically by wind stirring or convection. In fact, very stably-stratified salt waters (solar ponds) have been heated to temperatures near boiling point by solar radiation alone (Zangrando, 1979).

The second asymmetry arises because solar heating has a diurnal cycle while cooling processes remain nearly constant day and night. Thus net heat is usually added to the ocean during the day resulting in a basically stable upper ocean while at night heat is removed, resulting in a basically unstable situation. A similar asymmetry also exists in the seasonal cycle of heating and cooling of the ocean with a general tendency for instability in the upper ocean in the fall and winter.

Haline convection in a nearly pure form occurs under the rapidly grow-ing sheets of sea ice that form in newly-opened leads in the polar regions due to the exclusion of salt atoms from the crystalline lattice of ice. Since the water in contact with sea ice is already at the freezing point, thermal convection does not occur except when there is supercooling. Even then the degree of thermal convection is minimal since the coefficient of thermal expansion is small near the freezing point. Because the diffusion of salt is governed by the same mathematical equations as those for the diffusion of heat, the phenomena of haline convection are similar to those of thermal convection except that the spatial scale is about an order of magnitude smaller (Foster, 1968).

Combined thermal and haline convection or thermohaline convection occurs whenever evaporation takes place at the sea surface. The theory of thermohaline convection has been only incompletely worked out although numerical modeling has been carried out over a narrow range of parameters (Piacsek, 1972). Experiments on evaporation of salt water over a wide range of salinities have shown that the spatial scale of the convection is governed by the thermal effects and that the salinity effects are secondary (Foster, 1971b).

Convection in the upper ocean can cause important mixing effects that can have a direct effect on the distribution of chemical species. In addition convection can have important effects on the flow pattern in the top few centimetres of the ocean, especially under low wind conditions. Thus convection can affect processes that take place in the air-sea bound-ary layer of the ocean, such as gas exchange and photochemical reactions.

CONVECTION BASICS

The basic parameter governing whether buoyancy-driven convection can occur in the ocean is the stability of the water column. The stability can

easily be calculated from the difference between the actual density gradient as determined by the temperature and salinity profiles and the
adiabatic gradient. Stability is usually expressed by the Brunt-Väisälä
frequency, N, (Eckart, 1960) where

$$N^2 = -\frac{g}{\rho}\left[\frac{d\rho}{dz} - \frac{d\rho}{dz}\Big|_{adiabatic}\right] = -\frac{g}{\rho}\left[\frac{d\rho}{dz} + \frac{\rho g}{c^2}\right] \simeq -\frac{g}{\rho}\left[\frac{d\rho}{dz}\,\theta\right] \tag{1}$$

and g = acceleration of gravity, ρ = density, z = height (-depth),
c = speed of sound and ρ_θ = potential density. When N^2 is negative, the
water column is said to be unstable. Except in the case of double-
diffusive convection the water column must be unstable for convection to
occur. However, that the water column is unstable is not a sufficient
condition for convection to occur.

Starting with Rayleigh (1916) and summarized by Chandrasekhar (1961) a
number of investigations have shown by stability analysis that the water
column must be more than unstable as defined by N due to the dissipation of
buoyancy by diffusion of momentum, heat and/or salt. They have shown that
thermal convection in a layer of fluid can be described by three dimensionless parameters: the Rayleigh number,

$$R_T = \frac{\alpha g \Delta T h^3}{\kappa \nu} \tag{2}$$

where α = coefficient of thermal expansion, ΔT = temperature difference
across the fluid layer, h = thickness of the layer, κ = thermal diffusivity
and ν = kinematic viscosity;

the Prandtl number,

$$P = \frac{\nu}{\kappa} \tag{3}$$

and the Nusselt number,

$$N_T = \frac{Q_T h}{\rho C \kappa \Delta T} \tag{4}$$

where Q_T = heat flux across the layer, and C = specific heat. Physically,
the Rayleigh number represents the ratio of buoyancy forces to dissipative
forces, the Prandtl number the ratio of viscous dissipation to thermal
dissipation, and the Nusselt number the ratio of heat flow through the
fluid layer to that which would flow by thermal conduction alone. Thus for
unconvecting fluids the Nusselt number would be 1.

For haline convection, the Rayleigh number becomes

$$R_S = \frac{\beta g \Delta S h^3}{D \nu} \tag{5}$$

where β = coefficient of expansion due to salinity changes, ΔS = salinity
difference across the layer, and D = salt diffusivity; the Prandtl number
is replaced by the Schmidt number

$$\sigma = \frac{\nu}{D} \tag{6}$$

and the Nusselt number becomes

$$N_S = \frac{Q_S h}{\rho D \Delta S} \tag{7}$$

where Q_S = salt flux across the layer.

For thermohaline convection all six dimensionless parameters are
relevant.

110

Stability analysis shows that the Rayleigh number must exceed a critical quantity which depends upon the boundary conditions of the fluid layer
(657.5 for free boundaries, 1707.8 for rigid boundaries). In the ocean the
Rayleigh number can be many orders of magnitude greater than the critical
Rayleigh number for very slightly unstable density gradients since the
mixed layer is commonly tens of metres thick. Thus the Rayleigh number
criterion for onset of convection would seem to have very limited applicability in the upper ocean. Moreover, conditions are not constant at the
sea surface, and a satisfactory stability analysis has not been developed
for time-dependent density gradients.

A different approach is to consider the fluid dynamics of convection
as an initial value problem. In this approach the time-dependence of the
density field is calculated using the diffusion equation and the momentum
equation, which are coupled through the buoyancy force in the momentum
equation and the advective terms in the diffusion equation. Simple analytic solutions have not been found so it is necessary to obtain numerical
solutions using high speed computers. In this way it is possible to
describe in mathematical terms the convection that results from cooling a
thick layer of fluid from above (Foster, 1965a).

The results show that the initial convection is independent of the
thickness of the fluid layer for large Rayleigh numbers (greater than about
10^6). For a constant heat flux the horizontal scale of the convection
cells depends upon the heat flux to the -1/4 power, and with a heat flux
typical of the sea surface (100 W m^{-2}) the horizontal scale is about 0.02
m, which has been confirmed by laboratory experiments (Foster, 1965b).
Similarly, the horizontal scale of haline convection induced by the
freezing of seawater depends upon the -1/4 power of the salt flux (Foster,
1968), and for freezing at about the same heat flux as above, the salt flux
will be of the order of 10^{-5} kg m^{-2} s^{-1} and the horizontal scale of the
initial convection about 0.002 m, which has also been confirmed by laboratory experiments (Foster, 1969). Thus the horizontal scale for haline
convection is roughly an order of magnitude smaller than for thermal
convection at about the same buoyancy flux. The physical reason for this
scale change is that the onset of convection in a time-dependent system
depends upon the thickness of the unstable boundary layer, δ, and this in
turn depends upon the square root of the diffusivity ($\delta \sim \sqrt{\kappa t}$ or $\delta \sim \sqrt{Dt}$,
where t = time). Since salt diffusivity is about two orders of magnitude
smaller than that for heat, the unstable boundary layer will be about an
order of magnitude smaller. At onset of convection the horizontal scale or
wavelength is usually approximately twice the boundary layer thickness.

Despite the success of the initial value approach to convection theory
in predicting cell size at onset of convection, there remains the problem
of defining "onset" and properly choosing initial conditions. The work of
Jhaveri and Homsy (1982), which uses stochastic methods in determining
initial conditions, provides a promising approach to the solution of this
problem.

CONVECTION IN THICK LAYERS OF FLUID

The form of convection when the Rayleigh number becomes very large has
been assumed to be turbulent and therefore characterized by random fluctuations of velocity and temperature. Thus it is common to assume that one
can use an analogy between the random behaviour of molecules in a gas and
the random behaviour of fluid elements in a turbulent fluid to derive eddy
coefficients of viscosity and diffusion corresponding to the molecular
coefficients derived from kinetic theory (see e.g., Neumann and Pierson,

1966). For convection at a given heat flux, Q_T, the proper form for the Rayleigh number becomes

$$R_F = \frac{\alpha g Q_T h^4}{\kappa^2 \nu \rho C} \qquad (8)$$

Thus for a heat flux of 100 W m^{-2} from a mixed layer 10 m thick the flux Rayleigh number becomes about 3×10^{16} or many orders of magnitude greater than values ordinarily assumed needed for turbulent convection ($\sim 10^8$). While eddy coefficients have been used with some success in modeling oceanic processes, they should be used with caution in dealing with convective processes which exhibit regular structures.

Howard (1966) has proposed a model for convection at high Rayleigh numbers in which the convection becomes intermittent and thus is quite different phenomenally from the steady secondary flow usually assumed. In Howard's model, convection follows a cyclic process: the formation of a thermal boundary layer by diffusion, the instability of this layer when it becomes sufficiently thick, the destruction of the layer by the convective flow, the dying down of the convection, and the reforming of the thermal boundary layer by diffusion. Foster (1971a) has quantified Howard's model using a pseudo two-dimensional, finite amplitude analysis of convection in a thick layer of fluid cooled at a constant rate from above. Figure 2 shows the time development of the horizontally averaged temperature profile through one cycle starting with the thermal boundary layer just before instability. Note the very rapid destruction of the boundary layer at top (frames 1-3) and the slow reformation (frames 4-12). For flux Rayleigh numbers above about 10^7 the analysis predicted that the intermittent convection would have a characteristic period

$$\tau \simeq 14 \left[\frac{\nu \rho C}{\alpha g Q_T} \right]^{\frac{1}{2}} \qquad (9)$$

and a characteristic horizontal wavelength,

$$\lambda \simeq 48 \left[\frac{\kappa^2 \nu \rho C}{\alpha g Q_T} \right]^{\frac{1}{4}} \qquad (10)$$

that are independent of the depth of the fluid layer.

Recent experiments by Foster and Waller (1985) have generally confirmed the validity of the Howard model and the predicted characteristic period and wavelength of intermittent convection. The spectra of the temperature fluctuations show maximum energy close to the calculated characteristic periods (Fig. 3) though the peaks are rather broad (Fig. 4). Direct measurement of the wavelength of the convection is difficult even using visualization techniques, such as schlieren optical systems, since there is not a steady pattern of convection cells, as is found at low Rayleigh numbers, and the convection is characterized by the intermittent peeling off of the boundary layer (Spangenberg and Rowland, 1961) or meandering plumes (Sparrow et al., 1970). Still these observations are consistent with the predicted wavelengths. Foster and Waller (1985) used the correlation between temperatures measured at varying distances to estimate the horizontal wavelength and found that the coherency became nearly zero at about half the calculated wavelengths. The $-\frac{1}{4}$ power dependence of wavelength on heat flux was consistent with these measurements. Thus it is expected that at the sea surface with a heat flux that is typical ($Q_T = 100$ W m^{-2}) the characteristic period of the intermittent convection would be about 60 seconds and the characteristic wavelength about 0.04 m.

That convection should have such small horizontal scales in the upper ocean seems to be at variance with the infrared scanner observations of the sea surface by McAlister and McLeish (1965), who observed features with a

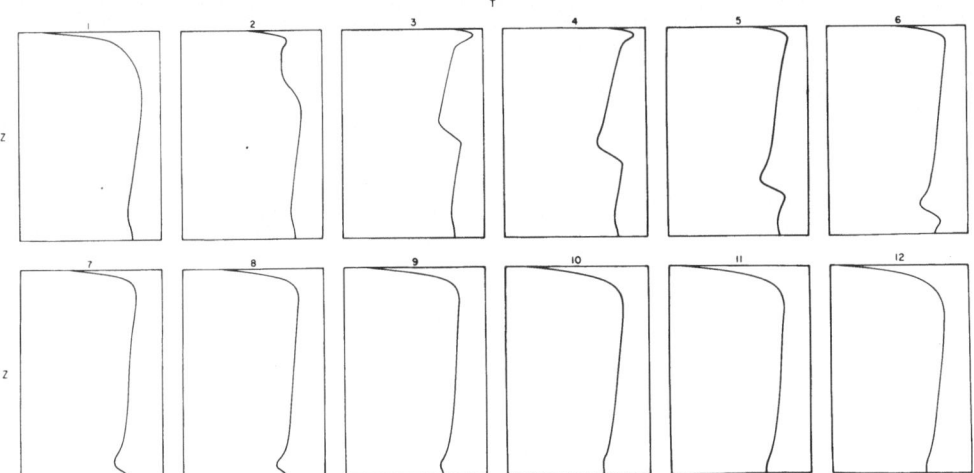

Fig. 2. Time development at equal time increments of the horizontally averaged temperature profile for a constant heat flux at the top with
a flux Rayleigh number of 10^7 (from Foster, 1971a).

horizontal scale of the order of 20 m, and with those of the MEDOC Group
(1970), who observed deep convective features in the Mediterranean with
horizontal scales of 1 to 5 km (perhaps the oceanic analog of a cumulonim-
bus cloud). Convective phenomena in the atmosphere have been extensively
studied owing to the striking presence of cumulus clouds and the easily
felt effects of thermals on aircraft. Priestly (1959), Townsend (1962),
Webb (1977) and many others have studied atmospheric convection from the
point of view of similarity theory or dimensional analysis. Similar studies
of the oceanic mixed layer have generally discounted buoyancy-driven
convection, and thus modeling has concentrated on relating wind stress to
mixed layer depth (see e.g., Pollard et al., 1973). The relation between
convection experiments in the laboratory and oceanic convection has been
tenuous at best.

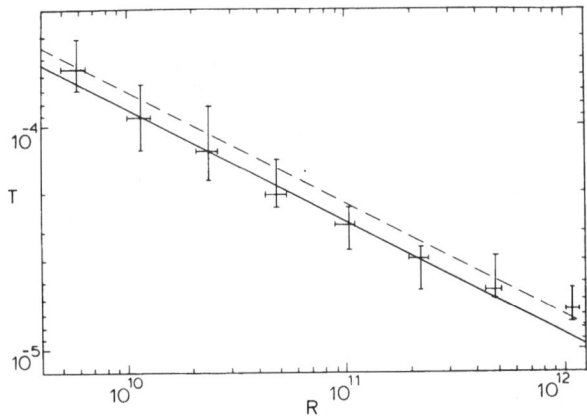

Fig. 3. Characteristic period of temperature
fluctuations as a function of flux
Rayleigh number. Solid line is the
theoretical prediction of Foster
(1971a). (After Foster and Waller,
1985.)

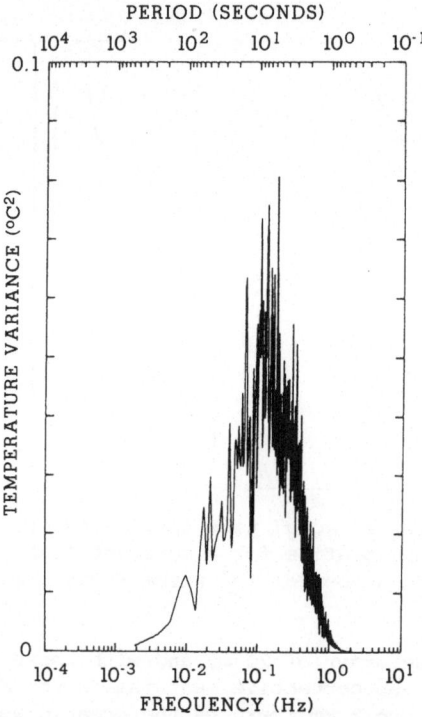

Fig. 4. Temperature spectrum for a
flux Rayleigh number of
8x10^{10} (after Foster and
Waller, 1985).

A somewhat different approach to explaining the large-scale features
of oceanic convection has been taken by Foster and his collaborators. The
original heuristic theory (Foster, 1974) proposed that the initial small-
scale convection resulting from the instability of the diffusive boundary
layer at the surface of the ocean could not penetrate in an organized
manner more than a depth of the order of one metre. Beyond this depth the
convection loses coherence and becomes diffusive in character due to the
disordered motion. A secondary boundary layer could then form that would
become unstable and induce an organized secondary convective flow with a
much larger scale in space and time. A hierarchy of scales of motion could
develop in this manner in which smaller scales would feed energy to larger
scales and thus exhibit what Starr (1968) has termed "negative viscosity
phenomena". The convective spatial and temporal scale could thus increase
until the spatial scale was of the same order as the depth of the mixed
layer. Green and Foster (1975) attempted to model the increase in spatial
scale using haline convection in a Hele Shaw cell. Laboratory and numerical experiments both showed that convective elements combined so that the
horizontal wave number of the convection increased with depth. For the
Hele Shaw cell the mechanism of scale increase turned out not to be a
boundary layer instability, but a pressure induced instability. Experiments with thermal convection in a 10 m deep tank cooled from above
(Foster and Waller, 1981) directly confirmed that convective motion is
coherent with increased spatial scales with increased depth. The characteristic period of the convection also increased with depth. In this case
the mechanism causing the scale changes could not be resolved. Experiments
designed to illuminate the mechanism of the scale increase with depth in
convection are now being planned.

REFERENCES

Chandrasekhar, S., 1961, "Hydrodynamic and Hydromagnetic Stability", Clarendon Press, Oxford.

Eckart, C., 1960, "Hydrodynamics of Oceans and Atmospheres", Pergamon Press, Oxford.

Foster, T.D., 1965a, Stability of a homogeneous fluid cooled uniformly from above, Phys. Fluids, 8:1249.

Foster, T.D., 1965b, Onset of convection in a layer of fluid cooled from above, Phys. Fluids, 8:1770.

Foster, T.D., 1968, Haline convection induced by the freezing of sea water, J. Geophys. Res., 73:1933.

Foster, T.D., 1969, Experiments on haline convection induced by the freezing of sea waters, J. Geophys. Res., 74:6967.

Foster, T.D., 1971a, Intermittent convection, Geophys. Fluid Dyn., 2:201.

Foster, T.D., 1971b, Experiments on convection induced by the evaporation of salt water (Abstract), Bull. Amer. Phys. Soc., 15:1533.

Foster, T.D., 1974, The hierarchy of convection, in: "Processus de Formation des Eaux Océaniques Profondes en Particulier en Méditerranée Occidentale", pp. 237-241, Centre National de la Recherche Scientifique, Paris.

Foster, T.D., and Waller, S., 1981, Free convection in a deep layer of fluid cooled from above (Abstract), Bull. Amer. Phys. Soc., 26:1286.

Foster, T.D., and Waller, S., 1985, Experiments on convection at very high Rayleigh numbers, Phys. Fluids, 28:455.

Green, L.L., and Foster, T.D., 1975, Secondary convection in a Hele Shaw cell, J. Fluid Mech., 71:675.

Howard, L.N., 1966, Convection at high Rayleigh numbers, in: "Proceedings of the Eleventh International Congress of Applied Mechanics, Munich (Germany) 1964", H. Görtler, ed., pp. 1109-1115, Springer-Verlag, Berlin.

Jerlov, N.G., 1968, "Optical Oceanography", Elsevier, Amsterdam.

Jhaveri, B.S., and Homsy, G.M., 1982, The onset of convection in fluid layers heated rapidly in a time-dependent manner, J. Fluid Mech., 114:251.

McAlister, E.D., and McLeish, W. L., 1965, Oceanic measurements with airborne infrared equipment and their limitations, in: "Oceanography from Space", G.C. Ewing, ed., pp.189-214, Woods Hole Oceanographic Institution.

MEDOC Group, 1970, Observations of the formation of deep water in the Mediterranean Sea, Nature, Lond., 227:1037.

Neumann, G., and Pierson, W.J., 1966, "Principles of Physical Oceanography", Prentice-Hall, Englewood Cliffs.

Piacsek S.A., 1972, Evaporation-driven convection currents at the ocean surface, in: "Proceedings of the Second Summer Simulation Conference", San Diego.

Pollard, R.T., Rhines, R.B., and Thompson, R.O.R.Y., 1973, The deepening of the wind-mixed layer, Geophys. Fluid Dyn., 4:381.

Priestly, C.H.B., 1959, "Turbulent Transfer in the Lower Atmosphere", University of Chicago Press.

Rayleigh, Lord, 1916, On convection currents in a horizontal layer of fluid when the higher temperature is on the under side, Phil. Mag., Ser. 6, 32:529.

Spangenberg, W.G., and Rowland, W.R., 1961, Convection circulation in water induced by evaporative cooling, Phys. Fluids, 4:743.

Sparrow, E.M., Husar, R.B., and Goldstein, R.J., 1970, Observations and other characteristics of thermals, J. Fluid Mech., 41:793.

Starr, V.P., 1968, "Physics of Negative Viscosity Phenomena", McGraw-Hill, New York.

Townsend, A.A., 1962, Natural convection in the earth's boundary layer, Q. J. Roy. Met. Soc., 88:51.

Webb, E.K., 1977, Convection mechanisms of atmospheric heat transfer from surface to global scales, in: "Second Australian Conference on Heat and Mass Transfer", R. W. Bilger, ed., pp.523–539, University of Sydney.

Williams, A.J., 1974, Salt fingers observed in the Mediterranean Outflow, Science, N.Y., 185:941.

Zangrando, F., 1979, Observations and analysis of the full-scale experimental, salt-gradient solar pond, Ph.D. Dissertation, University of New Mexico, Albuquerque.

FIELD MEASUREMENTS OF GAS EXCHANGE

Wolfgang Roether

Institut für Umweltphysik
University of Heidelberg
6900 Heidelberg F.R.G.

ABSTRACT

A transfer coefficient is required to convert a measured ocean—atmosphere gas solubility disequilibrium into a net interfacial gas flux density. For many gases, the transfer between the interface and the bulk liquid (transfer coefficient k_w) is rate controlling, whereas for other gases, notably reactive ones, a more complex situation has to be dealt with. The available oceanic field data for k_w are presented. The value of k_w increases considerably with wind velocity, and a total variation over more than one order of magnitude is probable. However, little field information exists for wind velocities lower than 5 m s^{-1} and, in particular, higher than about 10 m s^{-1}. Available field methods to measure k_w are described and their potential to fill the gaps in present knowledge is assessed.

INTRODUCTION

To understand the dynamics of a gaseous constituent in the upper ocean, knowledge of its ocean-atmosphere transfer is required. Commonly, the task will be to deduce gas transfer rates from measured concentrations. For a gas of low reactivity in both the water and air phases, the inter-facial flux density F (mol m^{-2} s^{-1}) can be factorized into the air-water solubility disequilibrium and a coefficient that characterizes the rate of transfer (Liss, 1973), i.e.,

$$F = k(c_w - \alpha c_a) \qquad (1)$$

where c_w, c_a = bulk gas concentrations (mol m^{-3}) in the water and air phases, respectively; α = Ostwald's solubility (α^{-1} = H = Henry's constant); k = bulk transfer coefficient (transfer velocity, piston velocity; m s^{-1}). In Equation (1), k apparently refers to the molar gas concentration in the water phase. The value of k must be known, in order to convert measured ocean-atmosphere solubility disequilibria into net fluxes according to Equation (1).

If, in addition to having low reactivity, the solubility of the gas is not exceedingly large (i.e., $\alpha \leq 10$) (Slinn et al., 1978), which is true for the majority of unreactive gases, such as the noble gases and oxygen, the resistance to interfacial transfer is concentrated in the laminar sub-layer at the *water* side of the air-water interface. A similarity exists (Deacon, 1977; Ledwell, 1984), in that the transfer depends on the small-scale mechanical stirring to which this layer is subjected, and on the molecular aqueous diffusivity of the gas. It is therefore possible to convert a transfer coefficient for one of these gases into that for any other (see below). Thus, in the limit of low reactivity and of no more than moderate solubility, in order to be able to use Equation (1), it is sufficient to determine k for any such gas.

However, for gases of very high solubility, additional transfer resistance in the air has to be taken into account. A similarity of the type just mentioned also holds here, and water vapour, for which the flux density, i.e., evaporation, is approximately known, can serve as a reference gas (Liss and Slater, 1974). For gases such as those involved in photochemical reactions (e.g., Garland et al., 1980), furthermore, the condition of low reactivity commonly breaks down. It is then necessary to take a more detailed view of the transfer process than would correspond to Equation (1). The subsequent steps of the transfer, i.e., by eddy diffusion in the air, molecular diffusion across minutely thin, quasi-stagnant layers to both sides of the interface, and eddy diffusion in the water, with parallel chemical reactions (source or sink), have to be considered. Zafiriou (this volume) presents a physical view of the situation. Liss and Slater (1974) describe a two-layer model which allows a quantification of such transfer in simple cases. In assessing any spec-ific situation, it has to be kept in mind that in the field the transfer time across the liquid laminar sublayer is typically one second or less. Therefore, moderate reactivity, such as that of CO_2 in seawater (reaction time scale ~100 s), is fairly easily accounted for (Emerson, 1975; Goldman and Dennett, 1983).

In summary it appears that, in order to convert measured concentrat-ions into net fluxes in the most general case, an adequate strategy will be to refine the model of Liss and Slater (1974) by an appropriate "micro-scopical" approach. Theoretical and laboratory work to this end is in progress (see Brutsaert and Jirka, 1984). However, in such a model, the liquid-phase transfer resistance (k_w^{-1}, i.e., the transfer resistance between the interface and the bulk liquid) will play an important, and for many gases even a predominant, part. The point here is that laboratory measurements of k_w exist, but that their applicability to the ocean-atmosphere interface is in doubt. Therefore, while other parts of the model may largely rely on laboratory results, a capability to predict ocean-atmosphere gas transfer coefficients, k_w, can only be developed on the basis of field measurements.

In the following, the liquid-phase gas transfer coefficient, k_w, for ocean-atmosphere gas transfer will be considered. The present knowledge of k_w will be assessed first. Subsequently, gaps will be identified and ways to fill them will be discussed.

AVAILABLE FIELD DATA

Most of the available open-ocean field measurements of the liquid-phase transfer coefficient, k_w, are summarized in Fig. 1. The data are converted to correspond to transfer of CO_2, at 20°C, and they are plotted *versus* wind velocity, which is supposed to be a dominant external variable in gas transfer. An increase with wind velocity is apparent, but the spread is large. Some of the data points carry error bars; for the others the uncertainty presumably is no smaller. An interpretation of the field results is attempted in the following material. To guide such interpretat-

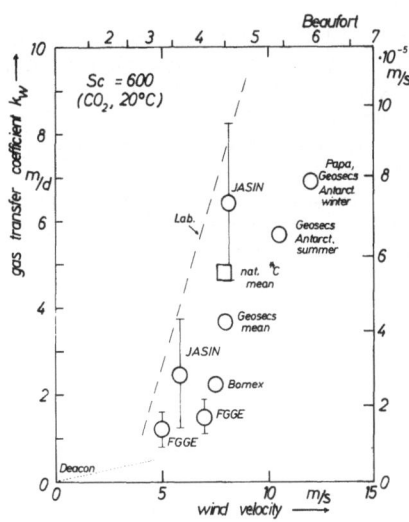

Fig. 1. Summary of measured open-ocean gas transfer coefficients. Circles: radon deficit
method; square: oceanic ^{14}C balance. Data
points: JASIN and FGGE, from Kromer and
Roether (1983); Papa, from Peng et al.
(1974); all others adapted from Figs 3-5
in Broecker and Peng (1982). Data points
denoted "mean" supposedly stand for
ocean-mean conditions. Dashed line: Upper
envelope of laboratory results after Jähne
et al. (1984a); dotted line: Deacon's
(1977) smooth rigid wall theory. All data
converted to Schmidt numbers, $\nu/D = Sc =
600$, according to $k_w \sim Sc^{-1/4}$ (see text for
details). A strong increase of k_w with
wind velocity is also supported by a large
^{3}He transfer that Weiss et al. (1980)
reported for observations in the
Mediterranean at $17 \ m \ s^{-1}$ wind velocity.

ion, a line (dashed) that represents the upper envelope of laboratory
results and a theoretical curve for low winds (dotted) are also shown in
Fig. 1.

The dashed curve is an upper limit also for the field data. The
considerable difference between the FGGE and JASIN data points in Fig. 1
has been ascribed to larger wind variability, i.e., a smaller-fetch
situation, for the JASIN observations (Kromer and Roether, 1983), which
effect would be in accordance with laboratory results (Jähne, 1980, 1982).
One might thus tentatively interpret Fig. 1 as indicating that the ocean-
atmosphere gas transfer coefficients as a function of wind velocity fall
into a certain range below the dashed curve, the position within this
range being determined by the fetch (i.e., the sea state), and that they
might converge towards the dotted line for very low wind. It is clear that
this interpretation can be no more than a preliminary guideline to estimate
a conceivable range of k_w for a certain situation. Such estimates are
presented in the next section.

Wind velocity and fetch (or sea state) are not the only governing
external parameters that must be considered (Hasse and Liss, 1980) but

temperature, air bubbles, and perhaps surface films will influence k_w as well. Temperature enters through the temperature dependence of both the aqueous molecular gas diffusivity, D, and the liquid-phase kinematic viscosity, ν, and one may use $k_w \sim (D/\nu)^{\frac{1}{2}} = Sc^{-\frac{1}{2}}$ ($\nu/D = Sc = $ Schmidt number) according to laboratory results (Ledwell, 1984), at least for higher winds. For low winds, $k_w \sim Sc^{-\frac{2}{3}}$ seems to apply (Jähne et al., 1984a). Table 1 lists Schmidt numbers for various gases and temperatures, which allow conversion between gases as well as between temperatures (however, the diffusivities for some gases are somewhat uncertain). According to Table 1 and the Schmidt number dependence, gas transfer coefficients at 0°C are almost a factor of two lower than those at 20°C, the reference temperature of Fig. 1. Under oceanic conditions, cold water tends to be positively correlated with high wind. Therefore, the temperature dependence will often to some extent counteract the effect of high wind.

From laboratory investigations (Merlivat and Memery, 1983; Broecker and Siems, 1984) bubbles have been estimated to enhance gas exchange above perhaps 10 m s^{-1} wind velocity, and to be considerably more effective for low-solubility gases. One should note that the data in Fig. 1 were obtained for either radon or CO_2 (i.e., $^{14}CO_2$). Both these gases have relatively large solubility, and extrapolations to higher wind velocities for low-solubility gases such as oxygen or helium should therefore tend on the high side. It is unclear whether surface films have a significant effect under common open-ocean conditions. Generally, one would expect an effect of films primarily for quite low winds, and under certain off-shore, as well as under polluted, conditions (Hühnerfuss et al., 1977).

In summary, the available open-ocean field data are restricted to a wind velocity between 5 and about 10 m s^{-1}, and even within this range, the uncertainties are large. It appears that for high winds, that are typical of higher-latitude winter situations where new subsurface water is formed, k_w must be expected to be substantially larger than its global average (4.8 m day^{-1}, based on natural ^{14}C; see Fig. 1), and *vice versa* for the lower winds that are common in low latitudes. Although measurements in continental lakes (see below) can to some extent serve as a substitute for oceanic measurements, more open-ocean field data are badly needed.

EXAMPLES OF GAS TRANSFER COEFFICIENT ESTIMATES

In order to illustrate both the conceivable range of k_w and the qualitative influence of the various external variables mentioned in the preceding section, a few specific situations are considered. It should be

Table 1. Schmidt numbers for some gases, 0 - 40°C

Temperature (°C)	Schmidt number					
	He	O$_2$	CO$_2$	Kr	Xe	Rn
0	510	1450	1860	2000	3000	3150
10		850	1010	1100	1400	1600
20	140	470	595	670	770	870
30			360	380	470	500
40	65	200	240	270	305	300

From Jähne (1980).

clear that an illustration, rather than a quantification, is aimed at, and primarily as a demonstration of the necessity of future work. The chosen examples are:

(i) CO_2 transfer, trade wind region, $\bar{u} = 7.5$ m s^{-1}, $T = 20°C$

(ii)   He transfer, Atlantic 40°N, winter, $\bar{u} = 10$ m s^{-1}, $T = 8°C$

(iii) O_2 transfer, Atlantic 60°N, winter, $\bar{u} = 12$ m s^{-1}, $T = 3°C$.

A possible procedure to estimate k_w for these situations is summarized in Table 2.

The only well-founded step in the procedure is the Schmidt number correction, all others are entirely tentative at the present stage of knowledge. The values of k_w range over one order of magnitude. The uncertainties presumably exceed a factor of two, except perhaps for the estimate for the trade wind region. It should be noted, however, that the situation for very low winds is again unclear: k_w may be larger than corresponds to the dotted curve in Fig. 1 because of the influence of the ubiquitous swell, or lower because of surface films.

FIELD METHODS

The preceding discussion should amply have demonstrated the necessity of future field measurements of the ocean-atmosphere gas transfer coefficient, k_w. Available methods (see also Liss (1983) and Broecker and Peng (1984)) are listed in Table 3.

[14]Carbon Balances

The use of ^{14}C balances (Broecker and Peng, 1974, 1984) can yield only global or basin-wide averages. They thus provide useful constraints, but

Table 2. Tentative gas transfer coefficient estimates for the three situations described in text (effects of surface films are not considered)

Situation	(i)	(ii)	(iii)
average wind velocity, $\bar{u}$ (m s^{-1})	7.5	10	12
average surface water temperature, T (°C)	20	8	3
(1) transfer coefficient estimated from Fig. 1 (m s^{-1})	4	6.5	8.5
(2) Schmidt number correction relative to Fig. 1	1	1.4	0.7
(3) correction for specific fetch (guess)	0.5	1.3	1.5
(4) bubble enhancement (guess)	1	1.5	1.8
resulting estimated k_w (m day^{-1})	2	17.5	16

will not give information for a specific situation such as those addressed in the preceding section. The other methods all determine a more or less instantaneous k_w.

Radon Deficit Method

This method, which has been developed by Broecker and co-workers (Broecker, 1965; Broecker and Peng, 1974), has recently been discussed in detail elsewhere (Kromer and Roether, 1983; Roether and Kromer, 1984). The basic features of the method are briefly as follows. Under stationary and horizontally homogeneous conditions, the ^{222}Rn deficit in the water column, I, integrated vertically through the disequilibrium surface layer (Fig. 2), is related to the interfacial ^{222}Rn flux density, F_{Rn}, by

$$F_{Rn} = \lambda \int_{\text{deficit layer}} [a_{Ra}(z) - a_{Rn}(z)]dz = \lambda I \qquad (2)$$

where, $\lambda = {}^{222}Rn$ radioactive decay constant, $2.10 \times 10^{-6}\ s^{-1} = 0.76\%\ h^{-1}$; $a_{Ra}(z)$, $a_{Rn}(z)$ = depth-dependent ^{226}Ra and ^{222}Rn radioactivity concentrations. The flux density at the same time is

$$F_{Rn} = k_{Rn}a_{Rn,z=0} \qquad (3)$$

where, k_{Rn} = gas transfer coefficient for radon. Therefore, under the said conditions,

$$k_{Rn} = \lambda I/a_{Rn,z=0} \qquad (4)$$

The use of Equation (4) is subject to two complications. Under conditions of changing gas transfer coefficient, k_w, the vertically integrated ^{222}Rn deficit, I, approaches a new equilibrium only with a few days relaxation time. In other words, the deficit I contains the gas transfer information in time-integrated form, while k_w as a rule will vary in time. The second point is that the oceanic surface layer through which the ^{222}Rn deficit extends is far from being static. Depth variations of the mixed layer base and of thermocline isopycnals are observed. Time scales range

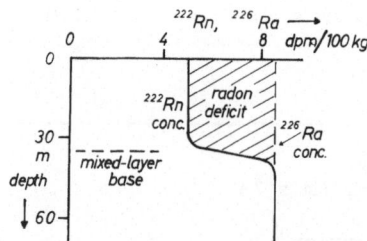

Fig. 2. Schematic illustration of the radon deficit method. Scales are approximate. In reality, the ^{226}Ra concentration also shows a certain increase with depth. (From Kromer and Roether, 1983).

Table 3. Field methods to determine ocean—atmosphere gas transfer
          coefficients

Method	Comment
(i) natural/bomb ^{14}C balances	input—inventory—decay; yield global or basin—wide averages
(ii) radon deficit method	uses $^{222}Rn/^{226}Ra$ disequilibrium in ocean surface layer; most generally used method
(iii) gas balance methods	evaluation of temporal evolution of an initial ocean—atmosphere solubility disequilibrium for an environmental gas (O_2, CO_2, ^3He,..)
(iv) added trace gas methods	similar to (iii), but employing purposefully added trace gas, in order to obtain signals that are easier to measure and to interpret
(v) containment gas balances	employ volume of air or water in contact with interface but separated from bulk air or water, time change of concentration of a gas in the containment being observed
(vi) atmospheric surface layer methods	vertical gas flux measurements above the air water interface (CO_2, $(CH_3)_2S$, CH_3I, ^{222}Rn,..)

It should be noted that all these methods use the same basic principle,
i.e., to measure simultaneously the solubility disequilibrium and the
interfacial flux for a selected gas, and to determine the transfer coefficient according to Equation (1).

between a fraction of an hour (buoyancy period) to days (mesoscale
adjustment time-scale). The depth variations are important because they
induce changes in I (via mixed-layer convergence/divergence), which
interfere with the gas transfer signal.

     Roether and Kromer (1984) conclude that the radon deficit method can
yield actual gas transfer coefficients at sea. To obtain a useful precision, averaging over 12 h or longer is necessary. Implementing the method
requires the following, quite elaborate procedure:

   (1) Selection of an observation area with sufficiently stable mixed—
        layer and thermocline and suitable wind conditions, as well as
        low mesoscale energy, and absence of strong currents and water—
        mass gradients.

   (2) A pre-survey in the selected area of the ^{226}Ra/nutrient/salinity
        relationships to allow real-time estimates of the ^{226}Ra distri-
        bution during the actual field phase.

   (3) An extended field phase comprising measurements to close the
        ^{222}Rn deficit balance (^{222}Rn, nutrients, salinity, temperature
        field, water samples for subsequent ^{226}Ra measurement), and to
        determine the required external variables (wind or friction

velocity, waves, etc.). These observations have to be carried out with the ship drifting with the mixed-layer flow, over periods of at least 2 days each, having assured oneself of the absence of fronts, and of excessive vertical shear of the horizontal flow, in the near vicinity of the observation point.

It is obvious that the radon deficit method in this form requires a major dedicated field effort. On the other hand, a "statistically" oriented approach, in which k_w values corresponding to individual isolated measurements of the radon deficit, I, are evaluated (Peng et al., 1979), is much less elaborate, but has to cope with large uncertainties of the individual values.

Gas Balance Methods

Results of gas balance methods have been reported in the literature (Johnson et al., 1979; Tsunogai and Tanaka, 1980). However, both the methods listed under (iii) and (iv) in Table 3 have to cope with the same problem that affects the radon deficit method, namely that of the complexity of the ever-transient oceanic surface layer, throughout which, and perhaps out of which (i.e., downwards), a gas becomes dispersed in a fashion that is difficult to quantify. So far, therefore, only qualitative results have been obtained. The trade-off between these methods and the radon deficit method is easier and faster measurement (e.g., O_2) *versus* a clearer signal (^{222}Rn). It appears that, in order to use these methods, again a major dedicated field effort would be required.

Containments

Use of containments, on the other hand, naturally avoids the said problems of the ocean surface layer. A recent example of such measurements, in which ^{222}Rn was collected into a containment reaching down to the interface from above, has been reported by Dueñas et al. (1983). It is doubtful, however, whether any containment measurement is really representative of the undisturbed water surface, as shielding of the wind or other surface disturbance effected by the containment may change k_w inside. On top of this uncertainty, containments naturally cannot be used at high wind speeds.

It may be instructive to consider that the methods (iii) and (iv) are accepted methods for continental lakes, streams, or water channels (see Brutsaert and Jirka (1984)). The explanation must be seen in the fact that these systems, contrary to the open ocean, because of their defined boundaries, which sometimes even allow a quasi one-dimensional treatment, confine any gas in a much more convenient fashion (the ^{14}C methods (i) likewise do not have a boundary problem, because the balance is set up for an entire ocean basin). As mentioned, measurements in lakes (Emerson, 1975; Torgersen et al., 1982; Jähne et al., 1984b) to some extent may serve as substitutes for oceanic ones, although lakes may be more strongly affected by surface films and are commonly more restricted in wind velocities. Fresh water, moreover, may generate bubble spectra that are different from those for sea water (Wu, 1981). Results for channels and streams certainly do not apply to the ocean because here, mostly, only a lesser part of the small-scale turbulence is wind induced (Hunt, 1984).

Atmospheric Surface Layer Methods

These methods, in which the gas flux is measured above the interface, open another way to avoid the problem with the oceanic surface layer, without, however, disturbing the water surface. The basic point is that in the lower atmospheric surface layer a quasi-stationary condition and lateral homogeneity are very much better approximations than in the oceanic surface layer, so that a localized, rather short-term measurement can give a representative gas flux. Carbon dioxide has been used so far, the flux being determined by an eddy correlation method (Weseley et al., 1982). However, extremely small variations in CO_2 concentration have to be determined, with the result that unambiguous results have not yet been obtained. Even if the formidable technical problems can be overcome (Leuning et al., 1982), minute advected variations in CO_2 concentration might mask the vertical flux signal.

Gases to replace CO_2 have recently been proposed (Roether, 1983). It was shown that those gases are suitable for which the ocean is a source and the atmosphere a sink with a destruction time scale between a few minutes and some days. Radon and methyliodide, CH_3I, were suggested as potentially useful gases. Vertical concentration signals for these gases are expected to amount to a few percent, as opposed to much less than one permille for CO_2. The practical feasibility of such radon and CH_3I measurements is yet to be demonstrated. Successful measurements of this type for dimethyl sulfide, $(CH_3)_2S$, have recently been reported (Nguyen et al., 1984). Atmospheric surface layer methods, in principle, appear to allow a fairly fast gas transfer determination, at the expense of a considerably lesser field effort than that required by the radon deficit and gas balance methods.

METHOD ASSESSMENT AND CONCLUSION

The potential of the various methods that were presented in the preceding section, in providing field data that would allow prediction of ocean-atmosphere gas transfer coefficients, k_w, is summarized as follows:

(1) The radon deficit and gas balance methods, if applied within a major dedicated field effort, appear suitable to provide adequate field data for k_w. For better economy, a combination of methods appears desirable in such an effort.

(2) Atmospheric surface layer methods (using gases such as CO_2, $(CH_3)_2S$, CH_3I, Rn, and perhaps others) show promise to provide data for k_w of similar quality, and presumably in larger quantity, at less expense of ship time. However, the methods have yet to be systematically developed and tested.

(3) Isolated radon deficit measurements and containment measurements can be regarded as a low-cost supplement that could provide useful information on ocean-atmosphere gas transfer. Containment bias would, however, have to be systematically explored.

Field data for k_w should ultimately be used to set up a parameterization of k_w in terms of the relevant external variables (wind velocity, wave parameters, etc.). Moreover, they should enter into the construction of "microscopical" models of the type mentioned above that would be suitable to describe ocean-atmosphere transfer for any gas whatever. These tasks will require a joint effort with laboratory and theoretical studies.

ACKNOWLEDGEMENT

Discussions at the Jouy—en—Josas meeting, particularly with P.S. Liss and H.C. Broecker, were extremely helpful to convert my original draft into the present manuscript.

REFERENCES

Broecker, H.C., and Siems, W., 1984, The role of bubbles for gas transfer from water to air at higher wind speeds. Experiments in the wind-wave facility in Hamburg, in: "Gas Transfer at Water Surfaces", W. Brutsaert and G.H. Jirka, eds, pp. 229–236, Reidel, Dordrecht.

Broecker, W.S., 1965, An application of natural radon to problems in ocean circulation, in: "Symposium on Diffusion in Ocean and Fresh Waters", pp. 116–145, Lamont-Doherty Geological Observatory, Palisades, N.Y.

Broecker, W.S., and Peng, T.-H., 1974, Gas exchange rates between air and sea, Tellus, 16:21.

Broecker, W.S., and Peng, T.-H., 1982, "Tracers in the Sea", Eldigio Press, Palisades, N.Y.

Broecker, W.S., and Peng, T.-H., 1984, Gas exchange measurements in natural systems, in: "Gas Transfer at Water Surfaces", W. Brutsaert and G.H. Jirka, eds, pp.479–494, Reidel, Dordrecht.

Brutsaert, W., and Jirka, G.H., eds, 1984, "Gas Transfer at Water Surfaces", Reidel, Dordrecht.

Deacon, E.L., 1977, Gas transfer to and across an air-water interface, Tellus, 29:363.

Dueñas, C., Fernandez, M.C., and Martinez, M.P., 1983, Radon 222 from the ocean surface, J. Geophys. Res., 88:8613.

Emerson, S., 1975, Gas exchange rates in small Canadian Shield lakes, Limnol. Oceanogr., 20:754.

Garland, J.A., Elzerman, A.W., and Penkett, S.A., 1980, The mechanism for dry deposition of ozone to seawater surfaces, J. Geophys. Res., 85:7488.

Goldman, J.C., and Dennett, M.R., 1983, Carbon dioxide exchange between air and seawater: No evidence for rate catalysis, Science, N.Y., 220:199.

Hasse, L., and Liss, P.S., 1980, Gas exchange across the air-sea interface, Tellus, 32:470.

Hühnerfuss, H., Walter, W., and Kruspe, G., 1977, On the variability of surface tension with mean wind speed, J. Phys. Oceanogr., 7:56.

Hunt, J.C.R., 1984, Turbulence structure and turbulent diffusion near gas-liquid interfaces, in: "Gas Transfer at Water Surfaces", W. Brutsaert and G.H. Jirka, eds, pp.67–82, Reidel, Dordrecht.

Jähne, B., 1980, Zur Parametrisierung des Gasaustausches mit Hilfe von Laborexperimenten, Doctoral dissertation, University of Heidelberg.

Jähne, B., 1982, Trockene Deposition von Gasen (Gasaustausch), in: "Exchange of Air Pollutants at the Air/Earth Interface", D. Flothmann, ed., Battelle Report BleV-R-64.284-2, pp.II–1–45, Frankfurt.

Jähne, B., Huber, W., Dutzi, A., Waiss, T., and Ilmberger, J., 1984a, Wind/wave tunnel experiment on the Schmidt number - and wave field dependence of air/water gas exchange, in: "Gas Transfer at Water Surfaces", W. Brutsaert and G.H. Jirka, eds, pp. 303–309, Reidel, Dordrecht.

Jähne, B., Fischer, K.H., Ilmberger, J., Libner, P., Weiss, W., Imboden, D., Lemnin, U., and Jaquet, J.M., 1984b, Parameterization of air/lake gas exchange, in: "Gas Transfer at Water Surfaces", W. Brutsaert and G.H. Jirka, eds, pp.459–466, Reidel, Dordrecht.

Johnson, K.S., Pytkowicz, R.M., and Wong, C.S., 1979, Biological production and the exchange of oxygen and carbon dioxide across the sea surface in Stuart Channel, British Columbia, Limnol. Oceanogr., 24:474.

Kromer, B., and Roether, W., 1983, Field measurements of air-sea gas exchange by the radon deficit method during JASIN 1978 and FGGE 1979, "Meteor"-Forsch-Ergebnisse A/B, 24:55.

Ledwell, J.J., 1984, The variation of the gas transfer coefficient with molecular diffusivity, in: "Gas Transfer at Water Surfaces", W. Brutsaert and G.H. Jirka, eds, pp. 293–302, Reidel, Dordrecht.

Leuning, R., Denmead, O.T., Lang, A.R.G., and Ohtaki, E., 1982, Effects of heat and water vapour transport on eddy covariance measurement of CO_2 fluxes, Boundary-Layer Meteorol., 23:209.

Liss, P.S., 1973, Processes of gas exchange across an air-water interface, Deep-Sea Res., 20:221.

Liss, P.S., 1983, Gas transfer : experiments and geochemical implications, in: "Air-Sea Exchange of Gases and Particles", P S. Liss and W.G.N. Slinn, eds, pp. 241–298, Reidel, Dordrecht.

Liss, P.S., and Slater, P.G., 1974, Flux of gases across the air-sea interface, Nature, Lond., 247:181.

Merlivat, L., and Memery, L., 1983, Gas exchange across an air-water interface: experimental results and modeling of bubble contribution to transfer, J. Geophys. Res., 88:707.

Nguyen, B.C., Bergeret, C., and Lambert, G., 1984, Exchange rates of dimethyl sulfide between ocean and atmosphere, in: "Gas Transfer at Water Surfaces", W. Brutsaert and G.H. Jirka, eds, pp. 539–545, Reidel, Dordrecht.

Peng, T.-H., Takahashi, T., and Broecker, W.S., 1974, Surface radon measurements in the North Pacific Ocean Station Papa, J. Geophys. Res., 79:1772.

Peng, T.-H., Broecker, W.S., Mathieu, G.G., and Li, Y.-H., 1979, Radon evasion rates in the Atlantic and Pacific Oceans as determined during the GEOSECS program, J. Geophys. Res., 84:2471.

Roether, W., 1983, Field measurement of air-sea gas transfer: a methodical search, Boundary-Layer Meteorol., 27:97.

Roether, W., and Kromer, B., 1984, Optimum application of the radon deficit method to obtain air-sea gas exchange rates, in: "Gas Transfer at Water Surfaces", W. Brutsaert and G.H. Jirka, eds, pp.447–457, Reidel, Dordrecht.

Slinn, W.G.N., Hasse, L., Hicks, B.B., Hogan, A.W., Lal, D., Liss, P.S., Munnich, K.O., Sehmel, G.A., and Vittori, O., 1978, Some aspects of the transfer of atmospheric trace constituents past the air-sea interface, Atmos. Environ., 12:2055.

Torgersen, T., Mathieu, G., Hesslein, R.H., and Broecker, W.S., 1982, Gas exchange dependency on diffusion coefficient: Direct ^{222}Rn and ^{3}He comparisons in a small lake, J. Geophys. Res., 87:546.

Tsunogai, S., and Tanaka, N., 1980, Flux of oxygen across the air-sea interface as determined by the analysis of dissolved components in sea water, Geochem.J., 14:227.

Weiss, W., Jenkins, W.J., and Fischer, K.H., 1980, Field determination of gas exchange with the tritium/helium-3 method, in: "Symposium on Capillary Waves and Gas Exchange", Berichte SFB 94, Universität Hamburg, Heft No. 17, pp. 94-97.

Weseley, M.L., Cook, D.R., Hart, R.L., and Williams, R.M., 1982, Air-sea exchange of CO_2 and evidence for enhanced upward fluxes, J. Geophys. Res., 87:8827.

Wu, J., 1981, Bubble populations and spectra in near-surface ocean: Summary and review of field measurements, J. Geophys. Res., 86:457.

PHOTOCHEMISTRY AND THE SEA-SURFACE MICROLAYER: NATURAL

PROCESSES AND POTENTIAL AS A TECHNIQUE

Oliver C. Zafiriou

Woods Hole Oceanographic Institution
Woods Hole, Massachusetts 02543 U.S.A.

ABSTRACT

The largest effects of photochemical processes in the Top Boundary
Layer probably involve reactions at the sea surface of species formed by
atmospheric processes. The ozone influx is probably the largest and most
significant one. Ozone reacts with dissolved iodide and with other unknown components. One consequence is the formation of volatile iodine
species, perhaps in sufficient amounts to balance the atmospheric iodine
budget. However, most of the ozone reaction products undergo other fates.

These reactions drive chemical fluxes that may be coupled to other Top
Boundary Layer processes in three ways: (1) modification of air-sea gas
exchange processes and fluxes, (2) chemical modification of trace components of the Top Boundary Layer, especially of species with long residence times there (indigenous biota, surface-active molecules, particles),
and (3) chemical modification of the physical properties of the interface
via reaction of surface active materials, leading to changes in surface
physical properties. These mechanisms may link the photochemically-
driven fluxes to trace element fluxes, surface biology, the fate of
particles at the interface, and the damping of capillary waves by surface
tension effects.

In order to substantiate some of these fluxes and effects, better
methods of studying and of modeling the air-sea interface on scales even
below 1 μm are needed. One possible tool in the required arsenal of new
techniques may be the use of photochemical processes to generate materials
at known rates and with known spatiotemporal distributions.

INTRODUCTION

In the present context we restrict the oceanic "Top Boundary Layer"
to the few centimetres on either side of the air-sea interface. This
region is considerably larger than the conventionally discussed "microlayer", or "stagnant boundary layer", but is smaller than the scale of
many processes with important effects on air-sea interaction, such as the
rising of bubbles generated by breaking of waves. This scale is chosen

Woods Hole Oceanographic Institution Contribution No. 5520

to be large enough to encompass *most* of the phenomena crucial to the exchange of momentum, materials, and energy across the interface. The associated timescale for diffusion/transport to the interface is <100 s.

Very little is known of this regime, despite its importance (Lion and Leckie, 1981). The difficulty of access and the small spatial and temporal scales associated with the boundary have undoubtedly contributed to the poor state of our knowledge. From the point of view of photochemical processes, there is as yet *no* direct and compelling evidence for the occurrence of easily detectable photoprocesses in the Top Boundary Layer. The author is aware of only one systematic, broad effort to detect such effects: Dr. P.M. Williams and co-workers at the Scripps Institution of Oceanography have studied the surface film on several cruises in coastal and oceanic waters of the southern California coast (P.M. Williams, personal communication). They measured sea surface potential and analyzed surface microlayer samples for inorganic nutrients, total organic carbon, particulate organic carbon, chlorophyll, ATP, total proteins, total and individual amino acids, total carbohydrates, and total lipids. Although they occupied about 15 stations during daylight and 5 stations at night, with one exception they found no distinct day-night contrast; the exception was apparently caused by the nocturnal feeding of pelagic crabs which released high levels of lipid at the sea surface.

Interest in photochemical processes at the sea surface thus derives from the a *priori* assumption that such processes might be important, not from evidence for specific effects of such reactions. However, given the difficulty of research in the Top Boundary Layer, the current lack of evidence can hardly be taken to indicate an absence of activity. Therefore, this chapter reviews the available evidence and ideas concerning photochemically driven reactions at the air-sea interface, emphasizing two questions:

(1) How does photochemistry affect the chemical composition of the air and water in the Top Boundary Layer?

(2) How might these chemical reactions be coupled to other processes of interest?

Finally, we close with a brief suggestion that photochemical processes and methods might also be useful tools for studying boundary layers in the laboratory and possibly even at sea.

NATURAL PHOTOCHEMISTRY OF THE TOP BOUNDARY LAYER

There are essentially three ways in which photochemical processes can influence the chemistry of the Top Boundary Layers:

(1) Photochemically generated species formed in the troposphere can be transported into and out of the boundary layer regime.

(2) Photochemically generated species formed in the upper ocean can be transported into the boundary layer.

(3) Processes occurring within the boundary layer itself can form and destroy chemical species.

Some relevant general reviews have been published concerning the photochemistry of the troposphere (Logan et al., 1981), and of seawater (Zika, 1981; Zepp, 1982; Zafiriou, 1983).

In the first two cases, the source of the material is distinct from its transport into the Boundary Layer, and the origin (photochemical or otherwise) of the material is irrelevant, since sources, transport processes, and sinks are well separated. For example, in the case of air-sea gas exchange, gases formed in the bulk water or in the troposphere by photochemical reactions can undergo exchange, behaving in an essentially conservative manner in the Top Boundary Layer. Examples include the flux of atmospherically formed CO from the air into the sea, or the counterflux of oceanic CO, formed photochemically, from sea to air (Conrad et al., 1982). Carbon monoxide has a turnover time much greater than 100 s in both bulk phases and probably is quasi-conservative in the Top Boundary Layer. Nevertheless, the direction and magnitude of the flux are ultimately determined by the tropospheric and oceanic photochemical reactions that set the concentrations in these two phases.

The third case is the most interesting, especially if we define "photochemical" processes to include secondary reactions of species derived from light-dependent reactions, as is usual. Then this case includes the set of materials which behave non-conservatively within the boundary layer region, either because of photochemical reactions taking place there, or because a species metastable in the phase in which it formed, is highly reactive in the other phase. The diffusion of ozone to the air-sea interface, where it reacts rapidly, exemplifies the second situation. The formation and reactions of free radicals within the air and the water of the Top Boundary Layer exemplifies the first case, since under almost all environmental conditions their turnover times with respect to chemical reactions are less than 100 s in both air and water.

Although there is no definitive quantitative evaluation of these possibilities, a preliminary attempt has been made to identify many of the major possibilities and to estimate the likely magnitudes of the fluxes. Thompson and Zafiriou (1983) utilized a conventional one-dimensional model of tropospheric photochemistry with parameters chosen to represent marine air. However, they specified as the lower (air-sea interface) boundary condition the stagnant-film model of the air-sea gas exchange process (Liss and Slater, 1974), including the term, α, for interfacial transport enhanced by reaction in the liquid phase. Free tropospheric photochemistry is simulated with the usual reaction set, but reactive species may diffuse to the surface and cross the interface. The rate of entry is influenced by the physical solubility and the chemical reactivity of the species in seawater.

This model has many deficiencies. For example, except for an efflux of NO (Zafiriou and McFarland, 1981), no allowance was made for photochemical reactions occurring in bulk seawater, as the requisite information is lacking. Also, the Top Boundary Layer is assumed to have the same chemical composition as bulk seawater, and no allowance is made for perturbations by aqueous phase interactions among the fluxes implied by the model. There is evidence that such photochemically important materials as chromophores (Carlson, 1982) and nitrite and iodide (Chapman and Liss, 1981) are present at higher than average concentrations near the air-sea interface.

Nevertheless, the results from this model have some interesting features. First, a number of photochemically generated species are predicted to have sizeable fluxes into the sea. Some of the influxes predicted by Thompson and Zafiriou (1983) are shown in Fig. 1A; for perspective, estimates of the fluxes of a variety of trace metals to the air-sea interface by atmospheric deposition and by transport on rising bubbles, as given by Wallace and Duce (1978), are shown in Fig. 1B. The similarities in flux magnitudes indicate that the reactive component fluxes

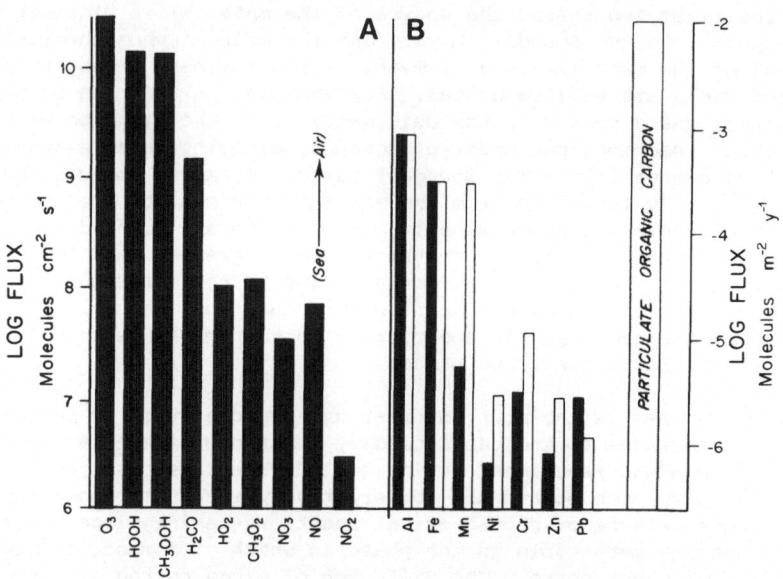

Fig. 1. A: Some estimated air-sea fluxes of reactive species
(from Thompson and Zafiriou, 1983); B: Estimated
transport fluxes of trace metals to the air-sea
interface by atmospheric dry deposition (solid bars)
and by transport by rising bubbles (open bars).
Data are from Wallace and Duce (1978). Particulate
organic carbon was calculated as molecules containing 20 carbon atoms.

are stoichiometrically sufficient, if they are coupled to reactions of the
metal species, to have a substantial effect on the trace metal speciations.
This comparison is not made to imply that such a coupling exists, but to
scale the atmospheric fluxes to some quantities more familiar to marine
scientists. However, some preliminary results suggest that such inter-
connections may exist (Lambert, 1981).

Flux magnitude is not the only criterion of significance for the
numbers indicated in Fig. 1A. The most interesting fluxes are, as stated
above, those which lead to reactions within the Top Boundary Layer. Again,
we lack direct evidence, but the available information strongly suggests
that HOOH, CH_3OOH, and CH_2O will have relatively long reaction turnover
times in the Top Boundary Layer (except that CH_2O hydrates to give
$HOCH_2OH$); they probably pass through quasi-conservatively. In contrast,
ozone and all the radicals should certainly react on a timescale much
shorter than 100 s. The reactions consuming these species also comprise
potentially significant source terms for the products.

If this model is even approximately correct, the atmospheric influx of
ozone on the top boundary layer exceeds by about a factor of 40 the fluxes
of the other identified substances that are expected to react rapidly, the
free radicals (largely CH_3O_2 and HO_2). Indeed, ozone reactions currently
constitute the strongest case illustrating the likelihood both of signif-
icant-sized flux and a coupling to processes of geochemical interest.
Thus, Garland et al. (1980) studied the reaction of ozone with seawater and
found chemical reaction rates fast enough to enhance the ozone influx,
which otherwise would be limited by the physical solubility of ozone in
seawater. They were unable to analyze this system completely, but did find

that reactivity of ozone with seawater seemed largely due to two components which they atttributed to: (1) reactions of unknown surface-active organic compounds, and (2) reactions of ozone with iodide ion. In a later paper, Garland and Curtis (1981) showed that radiolabelled I was liberated from the surface of seawater exposed in the laboratory to air/ozone. The volatile iodine species behaved operationally as I_2. They further estimated that ozone-iodide reactions could conceivably account for the sea-air fluxes required to balance the poorly understood but clearly anomalous atmospheric cycle of iodine.

These results, however exciting, also pose a chemical puzzle. The immediate product of the reaction of I^- and O_3 is HOI, not I_2 (Thompson and Zafiriou, 1983; J. Hoigné, personal communication). While the reaction of HOI to form I_2 can proceed to a small extent in the surface layer, the equilibrium is unfavourable and the reaction is slow on the 100 s time-scale. Hence, it is not clear how the volatile I species (which might not actually be I_2) are generated. Several alternatives were offered by Thompson and Zafiriou (1983).

However, the example of ozone fluxes and reactions serves not only to establish that the secondary reactions of photoproducts at the sea-air interface could be important, but also to point out that a much more sophisticated picture of the sea-air interface is needed to model such processes. For example, if one assumes a boundary layer at the air-sea interface which is truly stagnant (molecular diffusion), most of the incoming ozone reacts before penetrating beyond 2 or 3 μm in depth or only about one-tenth of the "conventional" stagnant film thickness invoked in gas exchange calculations. This creation of reaction products very close to the interface should give rise to very steep gradients near the interface, and hence to relatively large transport across the boundary.

Also, only a fraction of a percent of the products of O_3 reactions is required as volatile I to account for the I cycle deficiency. What happens to the remaining products? This question naturally leads to the second issue: how are photochemically-driven fluxes coupled to other processes in the Top Boundary Layer? Certain pathways seem sufficiently plausible to merit mention, if only to suggest future work:

(1) The air-sea exchange of gases may be influenced.

(2) The chemistry and biology of trace materials at the air-sea interface may be altered, especially in the case of substances having relatively long residence times in the Top Boundary Layer.

(3) The physical properties of surface-active molecules at the interface can be modified, giving rise to variations in surface tension that influence capillary wave damping (Broecker and Hasse, 1980). This effect could couple air-sea interaction in the meteorological sense to Top Boundary Layer photochemistry.

Of these possibilities, the first is illustrated by the example of ozone given above. The second effect, chemical reaction, has a relatively poor chance of causing major chemical changes for most solutes because their residence times in the Boundary Layer are short. However, surface-active molecules, particles, and the interfacial biota are three classes of important materials that should (at least under calm conditions) have much longer exposures to the reactive species. Hence, they are more likely to be affected in significant ways. If interfacial reactions change the subsequent behavior of particulate materials at the interface, the consequences could be important for atmospheric chemistry (Buat-Ménard, 1982), for the water column, and for transport of materials to the sediments. It also seems that near-surface organisms are subject to chemical stresses that do not exist in the water column. What are the effects? Finally, many of the most surface-active marine biomolecules are unsaturated lipids and other compounds (Liss, 1975; Lion and Leckie, 1981; Liss, this volume) that are expected to be highly reactive towards ozone and some free radicals. While their reaction products are expected to have different surfactant properties than the starting materials, more work is needed to know whether the products are more or less surface active and refractory than the starting materials.

PHOTOCHEMISTRY: A TOOL FOR STUDYING MICROLAYERS

It was noted above that there is very little useful information on the microphysics and chemistry of the boundary layer, despite qualitative indications that some geochemically important chemical transformations may be occurring in the upper few micrometres of the sea surface. Is this regime really stagnant? Can molecular exchange at the interface always keep up with other chemical events so that true equilibrium at the inter-face, as assumed by the Liss and Slater (1974) model, is always a viable assumption? How can these processes be studied? The answers to such questions are needed and yet are beyond the capabilities of currently utilized methods. One possible approach is through laboratory studies in which substances of known chemical reactivity react at an interface; for example, the reaction of ozone-containing gas phases with solutions of molecules with known, high rate constants for reaction with ozone (the I^- + O_3 rate, though fast, is not actually known).

However, another method of greater promise may be to use photolysis as a means of generating known distributions of material right at the gas-water interface. There are many variations on this general theme; it seems fitting to cite only a single example. By shining light from above on a solution of an appropriate chromophore, it is possible to destroy the chromophore and produce photoproducts at a precisely known rate and with known depth profile near the interface, on the molecular distance scale. The fraction of a volatile product escaping to the gas phase could be measured as a function of physical and chemical conditions, giving inform-ation about the microphysics at the interface. A particularly appealing aspect of such techniques is that it might ultimately be possible to apply them at sea to cross-calibrate field and laboratory results. For example, it may be feasible to create a detectable amount of NO, CO, or other exchangeable gas in the air by photolyzing the sea surface at night with an especially powerful pulse of light absorbed by chromophores naturally present in seawater, such as DOC, nitrite, or iodide.

ACKNOWLEDGEMENTS

This work was supported by National Science Foundation grants OCE-8111947 and OCE-8215614.

REFERENCES

Broecker, H.-Ch., and Hasse, L., 1980, "Symposium on Capillary Waves and Gas Exchange", Sonderforschungsbereich 94, Meeresforschung, Berichte Heft 17, Universität Hamburg.

Buat-Ménard, P., 1982, Particle geochemistry in the atmosphere and oceans, in: "Air-Sea Exchange of Gases and Particles", P.S. Liss and W.G.N. Slinn, eds, pp. 455-532, Reidel, Dordrecht.

Carlson, D.J., 1982, A field evaluation of plate and screen microlayer sampling techniques, Mar. Chem., 11:189.

Chapman, P., and Liss, P.S., 1981, The sea surface microlayer: Measurements of dissolved iodine species and nutrients in coastal waters, Limnol. Oceanogr., 26:387.

Conrad, R., Seiler, W., Bunse, G., and Giehl, H., 1982, Carbon monoxide in seawater (Atlantic Ocean), J. Geophys. Res., 87:8839.

Garland, J.A., and Curtis, H., 1981, Emission of iodine from the sea surface in the presence of ozone, J. Geophys. Res., 86:3183.

Garland, J.A., Elzerman, A., and Penkett, S.A., 1980, The mechanism for dry deposition of ozone to seawater surfaces, J. Geophys. Res., 85:7488.

Lambert, C.E., 1981, Le cycle interne du fer et du manganese et leurs interactions avec la matière organique dans l'océan, Ph.D. Thesis, Université de Picardie.

Lion, L.W., and Leckie, J.O., 1981, The biogeochemistry of the air-sea interface, Ann. Rev. Earth Planet. Sci., 9:449.

Liss, P.S., 1975, Chemistry of the seasurface microlayer, in: "Chemical Oceanography", Second Edition, Volume 2, J.P. Riley and G. Skirrow, eds, pp.193–243, Academic Press, London.

Liss, P.S., and Slater, P.G., 1974, Flux of gases across the air-sea interface, Nature, Lond., 247:181.

Logan, J.A., Prather, M.J., Wofsy, S.C., and McElroy, M.B., 1981, Tropospheric chemistry: A global perspective, J. Geophys. Res., 86:7210.

Thompson, A.M., and Zafiriou, O.C., 1983, Air-sea fluxes of transient atmospheric species, J. Geophys. Res., 88:6696.

Wallace, G.T., Jr., and Duce, R.A., 1978, Open-ocean transport of particulate trace metals by bubbles, Deep-Sea Res., 25:827.

Zafiriou, O.C., 1983, Natural water photochemistry, in: "Chemical Oceanography", Second Edition, Volume 8, J.P. Riley and R. Chester, eds, pp.339–379, Academic Press, London.

Zafiriou, O.C., and McFarland, M., 1981, Nitric oxide from nitrite photolysis in the central equatorial Pacific, J. Geophys. Res., 86:3173.

Zepp, R.G., 1982, Photochemical transformations induced by solar ultra violet radiation in marine ecosystems, in: "The Role of Solar Ultra-violet Radiation in Marine Ecosystems", J. Calkins, ed., pp.293–307, Plenum Press, New York.

Zika, R.G., 1981, Marine organic photochemistry, in: "Marine Organic Chemistry", E.K. Duursma and R. Dawson, eds, pp.299–325, Elsevier, Amsterdam.

ORGANIC CHEMICAL DYNAMICS OF THE MIXED LAYER: MEASUREMENT

OF DISSOLVED HYDROPHILIC ORGANICS AT SEA

Kenneth Mopper

University of Miami
Rosenstiel School of Marine and Atmospheric Science
Division of Marine and Atmospheric Chemistry
4600 Rickenbacker Causeway,
Miami, Florida 33149-1098 U.S.A.

ABSTRACT

Insights into important dynamic oceanographic processes can be gained
through analyses of natural dissolved organic compounds. These analyses
are being facilitated by the adaptation of sensitive and rapid measuring
techniques which can be used at sea to follow real-time variations in water
column processes. Particularly noteworthy is the analytical approach
employing reversed-phase high performance liquid chromatography coupled
with specific detection. In this review, this approach is discussed in
some detail. Actual and potential shipboard methods based on this approach
are described for amino acids, thiols, low molecular weight carbonyl
compounds, carboxylic acids, and metal-organic complexes. Emphasis is
placed on how the methods work, their limitations, and examples of field
applications.

INTRODUCTION

Dissolved organic matter (DOM) in seawater is present at concent—
rations of about 1-3 mg l^{-1}, or 1-3 ppm (MacKinnon, 1981). Specifically
identifiable components, such as amino acids, sugars and carboxylic acids,
may be present in only ppb or even lower concentrations. Despite these low
concentrations, DOM plays an important role in many dynamic geochemical and
biological oceanograpйgic processes, such as particle-organic interactions,
air—sea interactions, light absorption, trace metal complexation, and food
web dynamics (Morris and Eglinton, 1977; Pocklington, 1977; Wangersky,
1978; Mopper and Degens, 1979; Gagosian and Lee, 1981; Duursma and Dawson,
1981).

Most groups of dissolved organics are involved to a greater or lesser
extent in more than just one of these oceanic processes. For example
organoperoxides and organic free radicals should be present in the photic
zone as a result of photochemical processes (Zafiriou, 1983), and, due to
their high reactivity, they may be active in humification processes; thiols
originate chiefly from microbial sulfate reduction and protein degradation
in reducing environments but, if present in high concentrations in these
environments and at oxic—anoxic interfaces, these compounds may affect the

speciation and solubility of trace metals through their strong complexation abilities (Boulegue et al., 1982); and amino acids, which are present in seawater as a result of biological processes, are known to adsorb selectively to particles (Hedges, 1977), thereby affecting particle surface charge and dissolution rates.

Much of the past work on cycling, distribution, and transport of organic matter in the oceans has relied heavily on non-specific or bulk measurements, e.g., dissolved organic carbon (DOC). These measurements have been particularly useful for evaluating long-term geochemical processes, such as deep-sea nutrient regeneration and oxygen utilization (Menzel, 1974), fluxes of carbon from the land to the sea (Schlesinger and Melack, 1981), estuarine mixing (Mantoura and Woodward, 1983) and release of DOM from sediments (Williams et al., 1980). However, for the evaluation of short-term inter-related geochemical and biological processes, measurements of specific components within the pool of DOM in seawater are potentially much more useful.

CHARACTERISTICS OF SHIPBOARD METHODS

Marine organic chemistry is still a relatively new field, and therefore the data base for specific organic components in the sea is rather limited at present. A major stumbling block is the time required (usually hours to days) to extract and analyze specific trace organics in a seawater sample (Dawson and Liebezeit, 1981). In some cases, this stumbling block can be avoided by choosing analytical techniques which can be used at sea.

Unfortunately, analysis of trace organics at sea has some drawbacks: only a limited number of samples can be analyzed, as determined by the sample through-put time; the ship itself can be a major source of contamination during sample work-up, and therefore a chemical clean bench or van may be required; and the work schedule for shipboard studies, especially diurnal studies, is quite demanding. However, these disadvantages are far outweighed by the fact that field methods give real-time or near real-time data, and therefore the sampling program can be modified to follow water column processes while they are occurring. Also, contamination and chemical and biological alterations which may occur during sample storage are minimized. The latter point is particularly relevant to studies involving reactive or unstable compounds, such as thiols, carbonyl compounds, peroxides and free radicals, and compounds involved in rapid biological uptake and release, such as amino acids, monosaccharides and carboxylic acids.

Only a limited number of techniques are currently or potentially available for shipboard use; some examples are given in Table 1. The techniques listed fall into three general categories: (1) methods which rely on the specific detection of compounds directly in seawater with no extraction; these methods are typically based on reactions of selective reagents with organic functional groups, e.g., primary amines, monosaccharides (aldehyde group), and urea (guanidinium group), although, in the case of chlorophyll, its natural fluorescence enables one to detect it directly; (2) methods which are based on the selective, on-line extraction or stripping of compounds from seawater followed by chromatography and selective detection; these methods have been successfully applied to the analysis of volatile and hydrophobic compounds in seawater; and (3) methods which rely on the direct injection of seawater samples (derivatized or underivatized) onto reversed-phase high performance liquid chromatography (HPLC) columns followed by selective, on-line detection; these methods are particularly suited for the analysis of low molecular weight, moderately hydrophilic organic compounds in seawater.

Table 1. Examples of techniques applicable to shipboard determination of natural trace organic in seawater[8]

Compound	Method	Reference
Total free and combined sugars	Specific spectrophotometric technique (MBTH)	K.M.Johnson et al. (1981)
Glucose	Enzymatic	Hanson and Snyder (1979)
Total aldehydes	Spectrophotometric (MBTH)	Eberhardt and Sieburth (1985)
Urea	Specific spectrophotometric technique (diacetyl-monoxime)	Aminot and Kerouel (1982)
Total free primary amines	Specific spectrofluorometric technique	Sellner (1981); Sugimura and Suzuki (1983)
Organoperoxides/H_2O_2	Enzymatic	Zika and Saltzman (1982); Cooper and Zika (1983)
LMW hydrocarbons (C_1-C_4)	Purge and trap, GC/FID	Swinnerton and Linnenbom (1967)
Various volatile organics ($>C_6$)	Purge and trap, GC/MS	Schwarzenbach et al. (1978); Sauer (1980); Gschwend et al. (1982)
LMW halogenated hydrocarbons	Solvent extraction, GC/ECD; head space, GC/Hall detector or GC/ECD; purge and trap, GC	Lovelock et al.(1973); Helz and Hsu (1978); Fogelqvist et al. (1982); Weiss et al. (1985)
Fatty acid methyl esters and related lipophilic compounds	In situ concentration on XAD resin, GC/MS	Ehrhardt et al. (1980)
Methyltin species	Hydride generation, purge and trap, GC/FPD or atomic absorption	Byrd and Andreae (1982)
Methylantimony species	Hydride generation, purge and trap, atomic absorption	Andreae et al. (1981)
Methylarsenic species	Hydride generation, purge and trap, GC/MS; HPLC, hydride generation	Odanaka et al. (1983) Tye et al. (1985)

(continued)

Table 1. (continued)

Dimethyl sulfoxide, Dimethyl sulfide	Purge and trap, GC/FPD	Andreae (1980); Andreae and Barnard (1983)
Humic and fulvic substances	Extraction on XAD at sea	Harvey et al. (1983)
Flavins and pteridines	Extraction on cartridge, HPLC, fluorometric detection	Dunlap and Susic (1985); Mopper and Zika (1986)
Chlorophyll	Solvent extraction, fluorometry; *in situ* fluorometry	LeBouteiller and Herbland (1982); Derenbach et al. (1979)
Individual pigments (chlorophylls and carotenoids)	Solvent extraction, HPLC, UV-VIS and fluorometric detection	Mantoura and Llewellyn (1983); Falkowski and Sucher (1981); Repeta and Gagosian (1983); Gieskes and Kraay (1984)
Individual dissolved free amino acids	Fluorometric labelling, HPLC, fluorometric detection.	Mopper and Lindroth (1982); Mopper and Dawson (1986)
Adenosine nucleotides (ATP, ADP, AMP, c-AMP, adenosine, adenine)	Solvent extraction, fluorometric labelling, HPLC, fluorometric detection	Davis and White (1980) Preston (1983)
Individual phenolics	Fluorometric (or UV) labelling, extraction, HPLC, fluorometric (or UV) detection	Lawrence et al. (1976) Blo et al. (1983)
	Direct acetylation, extraction, GC/FID	Coutts et al. (1979)
Aldehydes and ketones	UV labelling (DNPH), HPLC, UV detection	van Hoof et al. (1985); Mopper and Stahovec (1986); Sikorski and Mopper (1986)
Alpha keto acids	Fluorometric or UV labelling, HPLC, fluorometric or UV detection	Kieber and Mopper (1983, 1986)
LMW carboxylic acids	Passive diffusion, UV labelling, HPLC, UV detection	Molongoski and Taylor (1985); Barcelona et al. (1980)

Table 1. (continued)

Thiols	Fluorometric labelling, fluorometric detection	Mopper and Delmas (1984)
Muramic acid	Hydrolysis, fluoro—metric labelling, HPLC, fluorometric detection	Moriarty (1983); Mimura and Delmas (1983)
Amino sugars	Hydrolysis, fluoro—metric labelling, HPLC, fluorometric detection	Hjerpe et al. (1980); Jones and Gilligan (1983)
Reducing sugars	Hydrolysis, fluoro—metric labelling, HPLC, fluorometric detection	Mopper and Johnson (1983); Eggert and Jones (1985)
Metal—organic complexes	Cartridge extraction, HPLC, metal detection	Mills et al. (1982); Mackey (1985); Osterroht et al. (1985)
Free radicals	Spin trapping, HPLC, UV, electrochemical, and ESR detection	Reich et al. (1981); Makino et al. (1984); Floyd et al. (1984)
Peroxides	Direct injection on HPLC column, amperometric detection	Funk and Baker (1985)

(a) Abbreviations used: MBTH = 3—methyl—2—benzothiazolinone hydrazone, GC = gas chromatography, FID = flame ionization detector, MS = mass spectrometry, ECD = electron capture detector, FPD = flame photometric detector, HPLC = high performance liquid chromatography, UV = ultraviolet, VIS = visible, ESR = electron spin resonance, LMW = low molecular weight.

Analytical capabilities of the methods within each category are constantly being improved and a number of them have received widespread acceptance (Wangersky and Zika, 1977; Dawson and Liebezeit, 1981). However, the methods in the latter category, those based on reversed-phase HPLC, are relatively new; therefore only a few of them have actually been tested in the field. Despite this, the results of these studies indicate that this analytical approach can potentially yield considerable information about dynamic biological, geochemical, and photochemical processes in the sea. The remainder of this paper deals with examples of actual and potential shipboard methods based on reversed-phase chromatographic techniques.

DERIVATIZATION AND REVERSED—PHASE LIQUID CHROMATOGRAPHY

The theory of reversed-phase liquid chromatography will not be reviewed here (see standard texts, such as Snyder and Kirkland, 1979). However, it is important to point out that as a consequence of the reversed phase mechanism the sample matrix (seawater) and the chromatographic mobile phase are completely compatible. Thus, aqueous samples such as seawater can be directly injected into the analytical system with little or no sample preparation, with the exception of a derivatization step. Pre-column derivatization serves two functions: it renders hydrophilic organic species such as sugars, amino acids and α-keto acids, more hydrophobic thereby making retention possible in a reversed-phase chromatographic system and, by converting the compounds into highly fluorescent or UV-absorbing or electrochemically active species, it makes specific detection possible. The separated components are thus detected on-line after the column and, since no post-column reaction system is necessary for locating eluted compounds, the analytical system is simple and flexible. In addition, by using pre-column derivatization, the compounds are detected against an almost dark background which generally results in extremely high sensitivity.

For seawater analysis, the ideal pre-column derivatization reaction is one that works in aqueous media and in pH and temperature ranges close to ambient. Further, the reaction must be quantitative and rapid, preferably instantaneous. The derivatization should also be specific for the compounds of interest and the derivatives should be sufficiently stable for chromatography. There should also be little or no interference from the excess reagent or from reaction by-products.

Of the numerous derivatization reactions that have been reported (Lawrence, 1982), very few fulfil all the requirements of the ideal, the most notable exceptions being those for the determination of dissolved free amino acids and thiols (see below). However, even reactions that fall somewhat short of the ideal may still be quite useful in shipboard studies if their limitations are recognized.

EXAMPLES

In the following examples emphasis is placed on how the methods work, their limitations, and actual marine applications, if available. A detailed discussion of the involvement of dissolved organics in various oceanographic processes and interactions will not be given here; the reader is referred to the reviews listed in the Introduction.

Dissolved Free Amino Acids (DFAA)

Amino acids have been one of the most frequently analyzed classes of organics in seawater and a wide variety of analytical methods have been employed in the past (as reviewed by Dawson and Pritchard, 1978; Dawson and Liebezeit, 1981). Most of these methods involved lengthy desalting and concentration steps prior to chromatography and the analyses were generally performed on preserved samples. As a result many of the older data on DFAA in seawater are questionable.

Recently, an HPLC method was introduced whereby DFAA are converted to highly fluorescent, moderately hydrophobic compounds, isoindoles (Fig. 1), by derivatization directly in seawater (Lindroth and Mopper, 1979). Reacted seawater samples (10-500 μl) are injected without further treatment onto a reversed-phase HPLC column and the eluted amino acid derivatives are detected fluorometrically (Fig. 1). This method was first tested at sea at a station in the Baltic Sea near the Gotland Deep (Mopper and Lindroth, 1982), where striking diel fluctuations in DFAA in the upper water column and pronounced depth variations in DFAA in the vicinity of the oxic-anoxic interface were observed.

Since its introduction, the method has been refined by other workers,

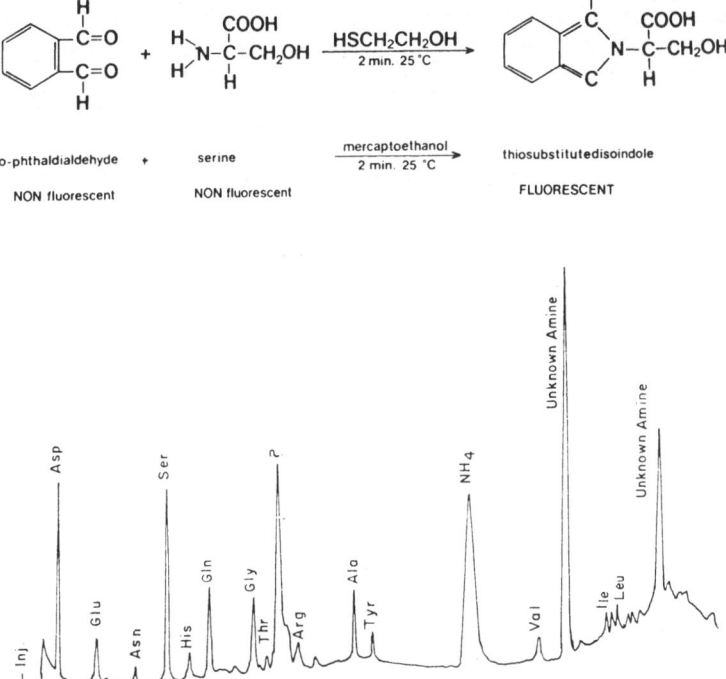

Fig. 1. *Upper*: Pre-column derivatization reaction of amino
acids performed directly in seawater (Mopper and
Lindroth, 1982). *Lower*: Chromatogram of fluor-
scently derivatized amino acids in a natural sea-
water sample (Delaware Bay); peak notation:
Asp-aspartic acid; Glu-glutamic acid;
Asn-asparagine; Ser-serine; His-histidine;
Gln-glutamine; Gly-glycine; Thr-threonine;
Arg-arginine; Ala-alanine; Tyr-tyrosine;
Val-valine; Ile-isoleucine; Leu-leucine.

as reviewed by Mopper and Dawson (1986). The chromatographic mobile phase
has been improved (Jones et al., 1981); a pH adjustment step may be added
to the derivatization procedure to protect the column (Evens et al., 1982);
the sample turn-around time can be reduced from 40 min to 15 min by using
short, high-efficiency HPLC columns (Price et al., 1984); the derivatization and injection steps can be automated (Cloete, 1984); and the detection of cysteine, which was previously undetectable, can be achieved through
the addition of a simple oxidation step to derivatization (Turnell and
Cooper, 1982). The latter two improvements have not yet been tested with
seawater samples. A major difficulty with the method is that proline and
hydroxyproline do not react with the reagent and therefore must be determined by other techniques.

The uptake and release of DFAA by organisms in seawater can be very
rapid, with turnover times in the range of minutes to hours (Fuhrman, 1983;
Lancelot and Billen, 1984). This implies that in order to obtain accurate
concentrations, analyses should be performed immediately after sampling.
In shipboard studies this is often not possible, especially for depth profiles, and therefore some form of short-term sample preservation is desirable. Retardation or elimination of biological activity may be effective,
if performed carefully. Figure 2 shows the effects of storage at 2°C,

addition of a preservative, and filtration, on the DFAA spectrum of a surface seawater sample. The results of this study indicate that the DFAA spectrum of a seawater sample may be preserved nearly intact up to a few hours if, upon retrieval, the sample is gently filtered and stored in the dark at 0-2°C. These results probably also apply to other groups of biologically utilizable organics, such as sugars and carboxylic acids. A thorough study of the effects of filtration on the determination of DFAA in seawater can be found in Fuhrman and Bell (1985).

Thiols

The same reagent used for the determination of DFAA, o-phthaldialdeyhde (Fig. 1), can also be used for the analysis of low molecular weight thiols in seawater and sediment porewater. In the presence of an excess primary amine, this reagent reacts with organic thiols, hydrogen sulfide and sulphite to form unique, chromatographable derivatives (Mopper and Delmas, 1984). Optimal reaction conditions are similar to those for the amino acid method and, like that method, the thiol method can be readily adapted to field use. But, because thiols oxidize rapidly, sample preparation and derivatization must be performed in an oxygen-free environment. The method should be of considerable use in the study of organosulfur species in reducing waters and sediments (Mopper and Taylor, 1986) and at oxic-anoxic interfaces. Typical chromatograms of a standard thiol mixture and a marine sediment porewater sample are depicted in Fig. 3. Organic disulfides cannot be directly determined by this method.

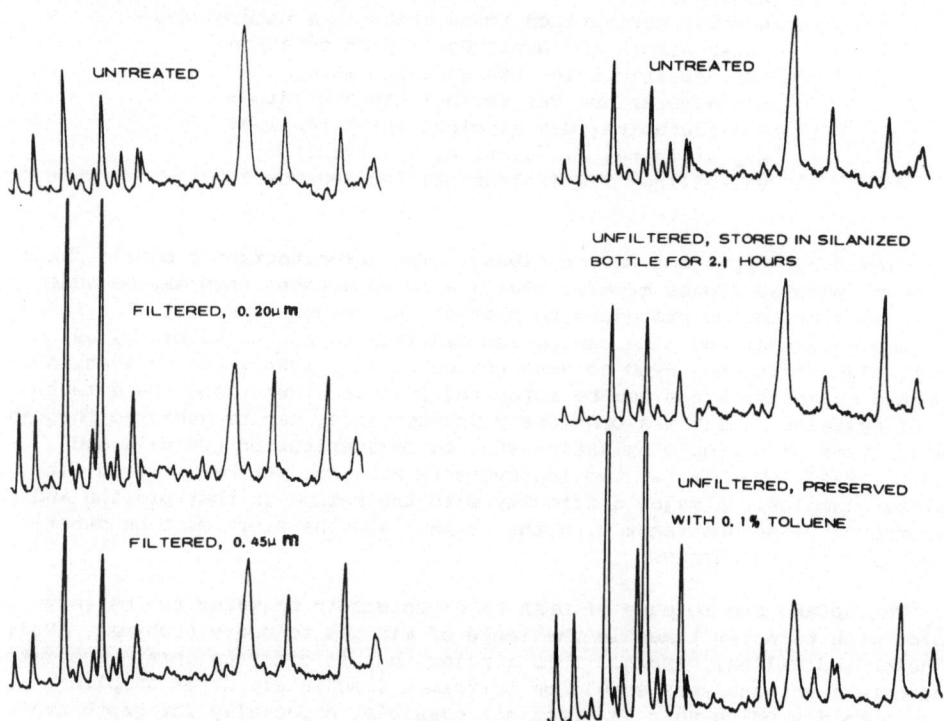

Fig. 2. Some effects of filtering and sample preservation on the spectra of dissolved free amino acids in aliquots of a surface seawater sample.

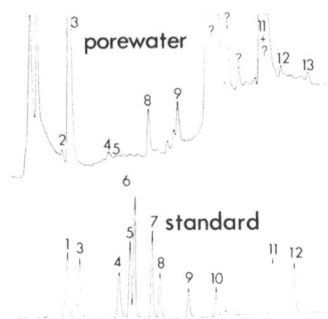

Fig. 3. *Upper*: Chromatogram of fluorescently derivatized thiols in reducing sediment porewater
(Biscayne Bay, Florida); the large unknown
peaks appear to be related to sulfide and may
be polysulfides (Mopper and Delmas, 1984).
Lower: Chromatogram of thiol standard; each
peak represents an injected amount of 10
pmol, except 3-mercaptosuccinic acid (1) and
sulfide (3) which are 1 nmol; peak notation:
(2) sulfite; (4) glutathione; (5) thioglyco-
late; (6) N-acetylcysteine; (7) β-mercapto-
ethanesulfonic acid; (8) 3-mercaptopropionic
acid; (9) monothioglycerol; (10) 2-mercapto-
ethanol; (11) methylmercaptan; (12) ethyl-
mercaptan; (13) propylmercaptan.

Low Molecular Weight Aldehydes and Ketones

Numerous photometric and chromatographic techniques can be found in
the literature for the determination of these compounds (Sawicki and
Sawicki, 1975). However, only two reagents have proven useful for pre-
column derivatization of these compounds in seawater: 2,4-dinitrophenyl-
hydrazine (DNPH) and dansylhydrazine (DnsH). The derivatization reactions
are shown in Fig. 4.

DNPH derivatives are detected spectrophotometrically after direct
injection of the reacted seawater sample onto a reversed-phase HPLC column
(Mopper and Stahovec, 1986). Detection limits are in the 2-50 nmol l^{-1}
range, but can be significantly lowered by a simple on-line preconcentration step (Sikorski and Mopper, 1984). The preconcentration step also
enhances the reactivity of the reagent towards ketones. Typical chromatograms of carbonyl compounds in seawater are depicted in Fig. 4. Using the
DNPH technique, shipboard studies in coastal waters revealed that these
compounds can undergo strong diurnal variations in the photic zone.
Laboratory experiments indicate that these variations are due to photochemical production and biological uptake (Mopper, 1985; Mopper and
Stahovec, 1986).

The DnsH reagent is potentially superior to the DNPH reagent because
the reaction is more rapid (10 min as opposed to 60 min); it occurs over a
pH range closer to that of seawater (pH 5-7 as opposed to pH 2); and, since
since the derivatives can be detected fluorometrically, the sensitivity may
be as much as a hundred-fold higher (L. Johnson et al., 1981). Unfortunately, difficulties arise when this reagent is used in conjunction with
HPLC for the determination of trace levels of carbonyl compounds in seawater because of interference from the excess DnsH reagent. This problem
may be overcome by using normal-phase HPLC after extraction of the derivatives into an apolar solvent (K. Mopper, unpublished results).

145

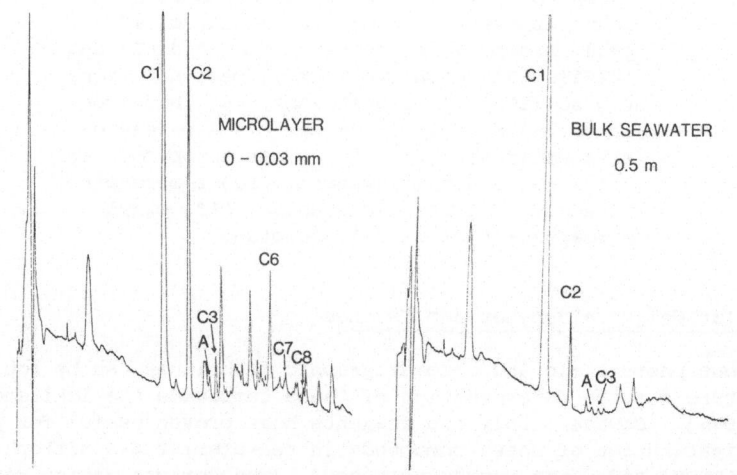

Fig. 4. *Upper*: Pre-column labelling reaction of low molecular
weight carbonyl compounds with dansylhydrazine to form
fluorescent derivatives. *Middle*: Pre-column labelling
reaction of low molecular weight carbonyls with
2,4-dinitrophenylhydrazine to form ultra-violet-
absorbing derivatives. *Lower*: Chromatogram of ultra-
violet-absorbing (DNP) derivatives of low molecular
weight carbonyls in surface seawater samples (Delaware
Bay); notation: C1-formaldehyde; C2-acetaldehyde;
A-acetone; C3-proponal; C6-C8-higher aldehydes (Mopper
and Stahovec, 1986).

A major difficulty encountered during trace analysis of carbonyl
compounds is the high levels of these compounds, especially formaldehyde
and acetone, in laboratory and ship air and in reagents and solvents.
Considerable effort must be expended to minimize these sources of contamination, such as using carbonyl-free solvents (Tuss et al., 1982) and performing sample manipulations in a carbonyl-free clean box or in unpolluted
outside air.

Low Molecular Weight Carboxylic Acids and α-Keto Acids

Like carbonyl compounds, numerous reagents are available for pre-

chromatographic derivatization of carboxylic acids (Lawrence, 1982).
Barcelona et al. (1980) used p-bromo-phenylacylbromide and Hordijk and
Cappenberg (1983) used 4-bromomethyl-7-methoxycoumarin as pre-column
labelling (esterification) reagents for the analysis of carboxylic acids
(e.g., formate and acetate) in natural water samples. Unfortunately, these
esterification reagents cannot be used directly in seawater because of
interference from chloride and divalent cations (Barcelona et al., 1980).
Furthermore, esterification reactions proceed only under anhydrous conditions and in aprotic solvents. Even if ion exchange and dehydration steps
are used, the manipulations involved introduce contamination resulting in a
detection limit ($\sim$0.5 - 5 μM) which would be too high for most seawater
studies (Billen et al., 1980). These problems may be overcome by extract-
ing organic acids from seawater by passive diffusion (Molongoski and
Taylor, 1985) or vacuum distillation techniques (Parkes and Taylor, 1983).
The desalted, concentrated samples can then be readily esterified and
analyzed by HPLC. Alternatively, the extracts can be analyzed by ion-
exclusion chromatography coupled with conductivity detection (Parkes and
Taylor, 1982; Keene et al., 1983; Jones and Simon, 1984) or by ion-exchange
chromatography coupled with a colorimetric reactor detection system (Wada
et al., 1984). In contrast to esterification methods, the latter system is
not affected by the major ions in seawater, thus it may be possible to
adapt it to the analysis of organic acids in seawater samples without ext--
raction or preconcentration (D.J. Kieber and K. Mopper, unpublished
results).

An important group of carboxylic acids, known as α-keto acids, can be
selectively determined in seawater by reaction with o-phenylenediamine.
The reaction results in the formation of stable, highly fluorescent
quinoxalinol derivatives (Fig. 5) which can be detected either after
solvent extraction from seawater (Steinberg and Bada, 1984) or by direct
injection of the derivatized seawater into the HPLC system (Kieber and
Mopper, 1983). The latter method is simpler and therefore more appropriate
to field studies. Low nanomolar levels of α-keto acids can be detected by
this method (Fig. 5). The main limitation is that the reaction requires
strongly acidic conditions and long reaction times at elevated temperatures. Since organisms are disrupted and hydrolyzed by these conditions,
samples must be sterile-filtered prior to reaction in order to obtain meaningful results. A considerably milder reaction procedure is based on derivatization of α-keto acids with 2,4-dinitrophenylhydrazine to form UV-
absorbing hydrazones (Fig. 4, middle), which can be readily separated by
reversed-phase HPLC (Kieber and Mopper, 1986). This technique has been
employed at sea to study the photochemical formation and spatial distribution of α-keto acids in the water column (Fig. 6) (Kieber and Mopper, 1985).

Flavins and Pteridines

Laboratory studies have indicated that flavins (e.g., riboflavin) and
pteridines may be important photosensitizers in surface seawater; these
compounds have been implicated in a number of redox processes ranging from
photogeneration of reactive species such as hydrogen peroxide to the photo-
oxidation of natural and anthropogenic compounds (Momzikoff et al., 1983;
Mopper et al., 1984; Mopper and Zika, 1986). Flavins and pteridines can be
readily determined in seawater by a simple extraction procedure followed by
reversed-phase HPLC coupled with fluorometric detection (Dunlap and Susic,
1985; Mopper and Zika, 1986). No pre-column derivatization steps are required. Shipboard studies using this method indicate that these compounds
undergo dynamic photochemical transformations in the sea (Vastano et al.,
1985; Mopper and Zika, 1986). For example, Fig. 7 shows that while riboflavin is present throughout the water column, its photochemical breakdown products, lumiflavin and lumichrome, occur dominantly in the upper
part of the water column (photic zone), as would be expected of photo-

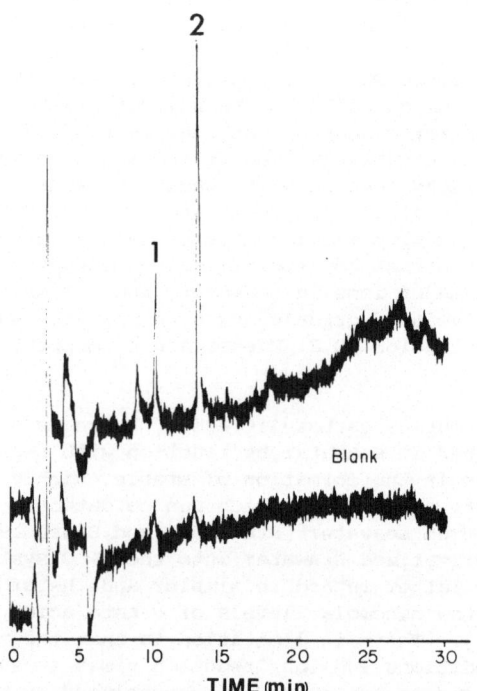

OPD   ∝-KETO ACID   2-QUINOXALINOL

Fig. 5. *Upper*: Pre–column labelling reaction between an α-keto acid
and o-phenylenediamine to form
fluorescent quinoxalinol derivative. *Lower*: Chromatogram of
fluorescently derivatized α–keto
acids in filtered natural seawater (Delaware Bay); peaks
approximately 1 pmol; notation:
(1) glyoxylic acid; (2) pyruvic
acid (Kieber and Mopper, 1983).

chemically produced species. In addition, riboflavin undergoes pronounced
diurnal fluctuations in the photic zone where it is presumably produced by
biological processes during the day and night, but photochemically degrad-
ed only during the day (Mopper and Zika, 1986).

Metal-organic Complexes

Perhaps one of the most exciting potential shipboard applications of
reversed-phase HPLC is the analysis of metal-organic complexes in seawater.
In the past, metal-organic interactions in seawater have been studied with
a variety of powerful, but *indirect* methods (e.g., Florence, 1982). In

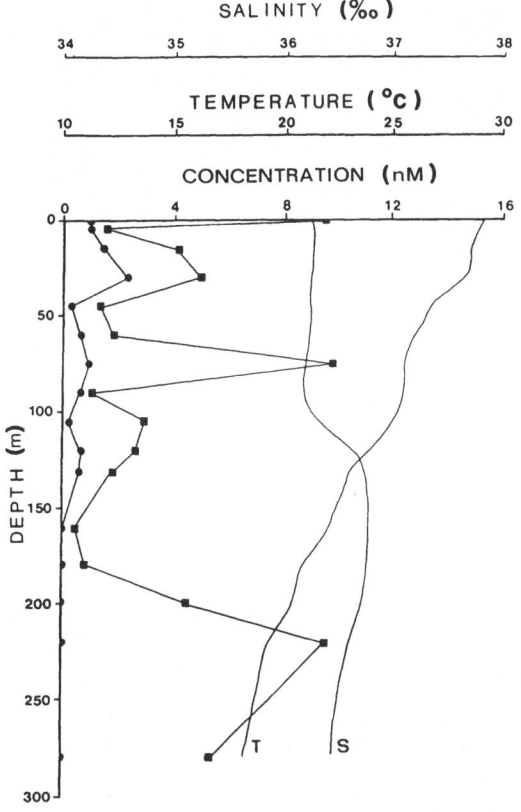

Fig. 6. Depth profiles of α-keto acids,
glyoxylate (circles) and pyr-
uvate (squares), temperature
(T) and salinity (S) in the Gulf
Stream (off Miami, Florida),
June 1985 (Kieber and Mopper,
1985, 1986).

contrast, reversed-phase techniques, which are based on the physical
isolation and chromatographic separation of complexes, provides a more
direct approach for their determination in seawater as well as their
eventual chemical characterization (Mills et al., 1982; Mackey, 1985;
Osterroht et al., 1985). Complexes are preconcentrated from several litres
of seawater on cartridges of reversed-phase material. After elution from
the cartridges the complexes are separated on analytical HPLC columns and
detected with metal-specific detectors, such as atomic fluorescence or
atomic absorption spectrometers. A major limitation of the method is that
it underestimates the extent of metal complexation by organics because
reversed-phase packings tend to discriminate against hydrophilic and weak
complexes.

CONCLUDING REMARKS

    Advances in organic chemical oceanography, like most other areas, have
hinged on the development of new analytical techniques. The approach
described in this review, the use of reversed-phase HPLC combined with
specific detection, opens up the possibility of measuring near real-time

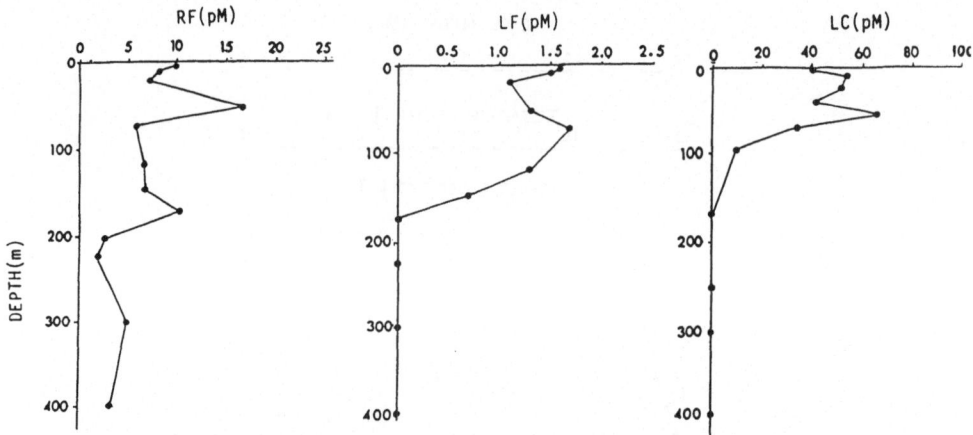

Fig. 7. Depth profiles of dissolved flavins, riboflavin (RF), lumiflavin
(LF) and lumichrome (LC) in the Tongue of the Ocean (Bahamas),
June, 1985 (Vastano et al., 1985; Mopper and Zika, 1986).

variations of a variety of trace organics commonly found in natural waters
(Table 1). Lengthy sample extraction and clean-up procedures can be
eliminated or greatly simplified, thereby reducing effects of contamination, adsorption, and variable extraction efficiencies. As a consequence,
results can be obtained rapidly and, more importantly, will reflect the
dynamic processes occurring at the time of sampling.

ACKNOWLEDGEMENTS

This work was supported by grants from the National Science Foundation, OCE 8207788, OCE 8411781, OCE 8516020, OCE 8517041, and from the
Office of Naval Research, NOOO 14-85-C-0020.

REFERENCES

Aminot, A., and Kerouel, R., 1982, Dosage automatique de l'uréé dans
     l'eau de mer: une méthode tres sensible a la diacétylmonoxime,
     Can. J. Fish. Aquatic Sci., 39:174.

Andreae, M.O., 1980, Dimethylsulfoxide in marine and freshwaters, Limnol.
     Oceanogr., 25:1054.

Andreae, M.O., and Barnard, W.R., 1983, Determination of trace quantities
     of dimethyl sulfide in aqueous solutions, Anal. Chem., 55:608.

Andreae, M.O., Asmodé, J.-F., Foster, P., and Van't dack, L., 1981,
     Determination of antimony (III), antimony (V), and methylantimony
     species in natural waters by atomic absorption spectrometry with
     hydride generation, Anal. Chem., 53:1766.

Barcelona, M.J., Liljestrand, H.M., and Morgan, J.J., 1980, Determination
     of low molecular weight volatile fatty acids in aqueous samples,
     Anal. Chem., 52:321.

Billen, G., Joiris, C., Wijnant, J., and Gillain, G., 1980, Concentration
     and microbiological utilization of small organic molecules in the
     Scheldt Estuary, the Belgian coastal zone of the North Sea and the
     English Channel, Estuarine Coastal Mar. Sci., 11:279.

Blo, G., Dondi, F., Betti, A., and Bighi, C., 1983, Determination of phenols in water samples as 4-aminoantipyrine derivatives by high-performance liquid chromatography, J. Chromatogr., 257:69.

Boulegue, J., Lord, C.J., III, and Church, T.M., 1982, Sulfur speciation and associated trace metals (Fe, Cu) in the pore waters of Great Marsh, Delaware, Geochim. Cosmochim. Acta, 46:453.

Byrd, J.T., and Andreae, M.O., 1982, Tin and methyltin species in seawater: Concentrations and fluxes, Science, N.Y., 218:565.

Cloete, C., 1984, Automated optimised high performance liquid chromatographic analysis of pre-column o-phthaldialdehyde-amino acid derivatives, J. Liquid Chromatogr., 7:1979.

Cooper, W.J., and Zika, R.G., 1983, Photochemical formation of hydrogen peroxide in surface and ground waters exposed to sunlight, Science, N.Y., 220:711.

Coutts, R.T., Hargesheimer, E.E., and Pasutto, F.M., 1979, Gas chromatographic analysis of trace phenols by direct acetylation in aqueous solution, J. Chromatogr., 179:291.

Davis, W.M., and White, D.C., 1980, Fluorometric determination of adenosine nucleotide derivatives as measures of the microfouling, detrital, and sedimentary microbial biomass and physiological status, Appl. Environ. Microbiol., 40:539.

Dawson, R., and Liebezeit, G., 1981, The analytical methods for the characterisation of organics in seawater, in: "Marine Organic Chemistry", E.K. Duursma and R. Dawson, eds, pp. 445-496, Elsevier, Amsterdam.

Dawson, R., and Pritchard, R.G., 1978, The determination of α-amino acids in seawater using a fluorimetric analyser, Mar. Chem., 6: 27.

Derenbach, J.B., Astheimer, H., Hansen, H.P., and Leach, H., 1979, Vertical microscale distribution of phytoplankton in relation to the thermocline, Mar. Ecol. Progr. Ser., 1:187.

Dunlap, W.C., and Susic, M., 1985, Determination of pteridines and flavins in seawater by reverse-phase, high-performance liquid chromatography with fluorometric detection, Mar. Chem., 17:185.

Duursma, E.K., and Dawson, R., eds, 1981, "Organic Marine Chemistry", Elsevier, Amsterdam.

Eberhardt, M.A., and Sieburth, J.McN., 1985, A colorimetric procedure for the determination of aldehydes in seawater in the cultures of methylotrophic bacteria, Mar. Chem., 17:199.

Eggert, F.M., and Jones, M., 1985, Measurement of neutral sugars in glycoproteins as dansyl derivatives by automated high-performance liquid chromatography, J. Chromatogr., 333:123.

Ehrhardt, M., Osterroht, Ch., and Petrick, G., 1980, Fatty-acid methyl esters dissolved in seawater and associated with suspended particulate material, Mar. Chem., 10:67.

Evens, R., Braven, J., and Brown, L., 1982, A high performance liquid chromatographic determination of free amino acids in natural water in picomolar ($M \times 10^{-12}$) range suitable for shipboard use, Chem. Ecol., 1:99.

Falkowski, P.G., and Sucher, J., 1981, Rapid, quantitative separation of chlorophylls and their degradation products by high-performance liquid chromatography, J. Chromatogr., 213:349.

Florence, T.M., 1982, The speciation of trace elements in waters, Talanta, 29:345.

Floyd, R.A., Watson, J.J., and Wong, P.K., 1984, Sensitive assay of hydroxyl free radical formation utilizing high pressure liquid chromatography with electrochemical detection of phenol and salicylate hydroxylation products, J. Biochem. Biophys. Methods, 10:221.

Fogelqvist, E., Josefsson, B., and Roos, C., 1982, Halocarbons as tracer substances in studies of the distribution patterns of chlorinated waters in coastal areas, Environ. Sci. Technol., 15:479.

Fuhrman, J., 1983, Close coupling between uptake and release of amino acids in seawater, Trans. Amer. Geophys. Un., 64:1095.

Fuhrman, J., and Bell, T.M., 1985, Biological considerations in the measurement of dissolved free amino acids in seawater and implications for chemical and microbiological studies, Mar. Ecol. Progr. Ser., 25:13.

Funk, M.O., Jr., and Baker, W.J., 1985, Determination of organic peroxides by high performance liquid chromatography with electrochemical detection, J. Liquid Chromatogr., 8:663.

Gagosian, R.B., and Lee, C., 1981, Processes controlling the distribution of biogenic organic compounds in seawater, in: "Marine Organic Chemistry", E.K. Duursma and R. Dawson, eds, pp. 91–123, Elsevier, Amsterdam.

Gieskes, W.W.C., and Kraay, G.W., 1984, Phytoplankton, its pigments, and primary production at a central North Sea station in May, July and September 1981, Netherlands J. Sea Res., 18–51.

Gschwend, P.M., Zafiriou, O.C., Mantoura, R.F.C., Schwarzenbach, R.P., and Gagosian, R.B., 1982, Volatile organic compounds at a coastal site. 1. Seasonal variations, Environ. Sci. Technol., 16:31.

Hanson, R.B., and Snyder, J., 1979, Enzymatic determination of glucose in marine environments: Improvement and note of caution, Mar. Chem., 7:353.

Harvey, G.R., Boran, D.A., Chesal, L.A., and Tokas, J.M., 1983, The structure of marine fulvic and humic acids, Mar. Chem., 12:119.

Hedges, J.I., 1977, The association of organic molecules with clay minerals in aqueous solutions, Geochim. Cosmochim. Acta, 41:1119.

Helz, G.R., and Hsu, R.Y., 1978, Volatile chloro- and bromocarbons in coastal waters, Limnol. Oceanogr., 23:858.

Hjerpe, A., Antonopoulus, C.A., Classon, B., and Engfeldt, B., 1980, Separation and quantitative determination of galactosamine and glucosamine at the nanogram level by sulphonyl chloride reaction and high-performance liquid chromatography, J. Chromatogr., 202:453.

Hordijik, K.A., and Cappenberg, T.E., 1983, Quantitative high-pressure liquid chromatography-fluorescence determination of some important lower fatty acids in lake sediments, Appl. Environ. Microbiol., 46:361.

Johnson, K.M., Burney, C.M., and Sieburth, J.McN., 1981, Doubling the production and precision of the MBTH spectrophotometric assay for dissolved carbohydrates in seawater, Mar. Chem., 10:467.

Johnson, L., Josefsson, B., Marstrop, P., and Eklund, G., 1981, Determination of carbonyl compounds in automobile exhausts and atmospheric samples, J. Environ. Anal. Chem., 9:7.

Jones, B.N., and Gilligan, J.P., 1983, o-Phthaldialdehyde precolumn derivatization and reversed-phase high-performance liquid chromatography of polypeptide hydrolysates and physiological fluids, J. Chromatogr., 266:471.

Jones, B.N., Paabo, S., and Stein, S., 1981, Amino acid analysis and enzymatic sequence determination of peptides by an improved o-phthaldialdehyde precolumn labeling procedure, J. Liquid Chromatogr., 4:565.

Jones, J.G., and Simon, B.M., 1984, Measure of microbial turnover of carbon in anoxic freshwater sediments: cautionary comments, J. Microbiol. Methods, 3:47.

Keene, W.C., Galloway, J.N., and Holden, J.D., Jr., 1983, Measurement of weak organic acidity in precipitation from remote areas of the world, J. Geophys. Res., 88:5122.

Kieber, D.J., and Mopper, K., 1983, Reversed-phase high-performance liquid chromatographic analysis of α-keto acid quinoxalinol derivatives. Optimization of technique and application to natural samples, J. Chromatogr., 281:135.

Kieber, D.J., and Mopper, K., 1985, Photochemical production of α-keto acids in seawater, Trans. Amer. Geophys. Un., 66:1266.

Kieber, D.J., and Mopper, K., 1986, Trace determination of α-keto acids in natural waters, Anal. Chim. Acta, in press.

Lancelot, C., and Billen, G., 1984, Activity of heterotrophic bacteria and its coupling to primary production during the spring phytoplankton bloom in the southern bight of the North Sea, Limnol. Oceanogr., 29:721.

Lawrence, J.F., 1982, Prechromatographic chemical derivatization in liquid chromatography, in: Chemical Derivatization in Analytical Chemistry", Volume 2, R.W. Frei and J.F. Lawrence, eds, pp. 191-242, Plenum Press, New York.

Lawrence, J.F., Renault, C., and Frei, R.W., 1976, Fluorogenic labeling of organophosphate pesticides with dansyl chloride. Application to residue analysis by high-pressure liquid chromatography and thin-layer chromatography, J. Chromatogr., 121:343.

Le Bouteiller, A., and Herbland, A., 1982, Diel variation of chlorophyll a as evidenced from a 13-day station in the equatorial Atlantic Ocean, Oceanol. Acta, 5:433.

Lindroth, P., and Mopper, K., 1979, High performance liquid chromatographic determination of subpicomole amounts of amino acids by precolumn fluorescence derivatization with o-phthaldialdehyde, Anal. Chem., 51:1667.

Lovelock, J.E., Maggs, R.J., and Wade, R.J., 1973, Halogenated hydrocarbons in and over the Atlantic, Nature, Lond., 241:194.

Mackey, D.J., 1985, HPLC analyses of metal—organics in seawater— interference effects attributed to stationary—phase free silanols, Mar. Chem., 16:105.

MacKinnon, M.D., 1981, The measurement of organic carbon in sea water, in: "Marine Organic Chemistry", E.K. Duursma and R. Dawson, eds, pp. 425—443, Elsevier, Amsterdam.

Makino, K., Moriya, F., and Hatano, H., 1984, Application of the spin—trap HPLC—ESR method to radiation chemistry of amino acids in aqueous solutions, Radiation Phys. Chem., 23:217.

Mantoura, R.F.C., and Llewellyn, C.A., 1983, The rapid determination of algal chlorophyll and carotenoid pigments and their breakdown products in natural waters by reverse-phase high-performance liquid chromatography, Anal. Chim. Acta, 151:297.

Mantoura, R.F.C., and Woodward, E.M.S., 1983, Conservative behaviour of riverine dissolved organic carbon in the Severn Estuary: chemical and geochemical implications, Geochim. Cosmochim. Acta, 47:1293.

Menzel, D.W., 1974, Primary productivity, dissolved and particulate organic matter, and the sites of oxidation of organic matter, in: "The Sea, Volume 5, Marine Chemistry", E.D. Goldberg, ed., pp. 659—678, Wiley—Interscience, New York.

Mills, G.L., Hanson, A.K., Jr., Quinn, J.G., Lammela, W.R., and Chasteen, N.D., 1982, Chemical studies of copper-organic complexes isolated from estuarine waters using C_{18} reverse-phase liquid chromatography, Mar. Chem., 11:355.

Mimura, T., and Delmas, D., 1983, Rapid and sensitive method for muramic acid determination by high-performance liquid chromatography with precolumn fluorescence derivatization, J. Chromatogr., 280:91.

Molongoski, J.J., and Taylor, C.D., 1985, Passive diffusion technique for concentration of short—chain volatile fatty acids from seawater, Appl. Environ. Microbiol., 50:1112.

Momzikoff, A., Santus, R., and Giraud, M., 1983, A study of the photosensitizing properties of seawater, Mar. Chem., 12:1

Mopper, K., 1985, Field evidence for the photochemical formation of low molecular weight carbonyl compounds in the sea, Trans. Amer. Geophys. Un., 66:1258.

Mopper, K., and Dawson, R., 1986, Determination of amino acids in sea water — recent chromatographic developments and future directions, Sci. Total Environ., in press.

Mopper, K., and Degens, E.T., 1979, Organic carbon in the ocean: nature and cycling, in: "The Global Carbon Cycle", B. Bolin, E.T. Degens, S. Kempe and P. Ketner, eds, pp. 293–316, Wiley, Chichester.

Mopper, K., and Delmas, D., 1984, Trace analysis of biological mercaptans by liquid chromatography and precolumn fluorometric labeling with o-phthaldialdehyde, Anal. Chem., 56:2557

Mopper, K., and Johnson, L., 1983, Reversed-phase liquid chromatographic analysis of Dns-sugars. Optimization of derivatization and chromatographic procedures and applications to natural samples, J. Chromatogr., 256:27.

Mopper, K., and Lindroth, P., 1982, Diel and depth variations in dissolved free amino acids and ammonium in the Baltic Sea determined by shipboard HPLC analysis, Limnol. Oceanogr., 27:336.

Mopper, K., and Stahovec, W.L., 1986, Photochemical production of low molecular weight organic carbonyl compounds in seawater, Mar. Chem., in press.

Mopper, K., and Taylor, B.F., 1986, Biogeochemical cycling of sulfur: Thiols in coastal marine sediments, in: "Organic Marine Geochemistry", M. Sohn, ed., in press, American Chemical Society, Washington, D.C.

Mopper, K., and Zika, R.G., 1986, Natural photosensitizers in seawater: riboflavin and its breakdown products, in: "Aquatic Photo-chemistry", W.J. Cooper and R.G. Zika, eds, in press, American Chemical Society, Washington, D.C.

Mopper, K., Stahovec, W., and Zika, R., 1984, Production of low molecular weight carbonyls by photochemical oxidation in seawater: possible role of photosensitizers, in: "Gas–Liquid Chemistry of Natural Waters", L. Newman, ed., Brookhaven National Laboratory Report No. 51751, Upton, New York.

Moriarty, D.J.W., 1983, Measurement of muramic acid in marine sediments by high performance liquid chromatography, J. Microbiol. Methods, 1:111.

Morris, R.J., and Eglinton, G., 1977, Fate and recycling of carbon compounds, Mar. Chem., 5:559.

Odanaka, Y., Tsuchlya, N., Matano, O., and Goto, S., 1983, Determination of inorganic arsenic and methylarsenic compounds by gas chromato-graphy and multiple ion detection mass spectrometry after hydride generation-heptane cold trap, Anal. Chem., 55:929.

Osterroht, C., Wenck, A., Kremling, K., and Gocke, K., 1985, Concentration of dissolved organic copper in relation to other chemical and biological parameters in coastal Baltic waters, Mar. Ecol. Progr. Ser., 22:273.

Parkes, R.J., and Taylor, J., 1983, Analysis of volatile fatty acids by ion–exclusion chromatography, with special reference to marine pore water, Mar. Biol., 77:113.

Pocklington, R., 1977, Chemical processes and interactions involving marine organic matter, Mar. Chem., 5:479.

Preston, M.R., 1983, Determination of adenine, adenosine and related nucleotides at the low picomole level by reversed-phase high-performance liquid chromatography with fluorescence detection, J. Chromatogr., 275:178.

Price, S.J., Palmer, T., and Griffin, M., 1984, High-speed assay of amino acids using reversed-phase liquid chromatography, Chromatographia, 18:62.

Reich, T.K., Chen, S., and Wan, J.K.S., 1981, Integral high-performance liquid chromatographic-electron spin resonance spectrometer for in situ studies of thermal and photochemical free radical reactions. Separation of phenoxy, hydrazyl, nitroxide and silyl-substituted semiquinone radicals, J. Chromatogr., 206:139.

Repeta, D.J., and Gagosian, R.B., 1983, Carotenoid transformation products in the upwelled waters off the Peruvian coast: suspended particulate matter, sediment trap material, and zooplankton fecal pellet analyses, in: "Advances in Organic Geochemistry 1981", M. Bjorøy, ed., pp. 380-388, Wiley, New York.

Sauer, T.C., Jr., 1980, Volatile liquid hydrocarbons in waters of the Gulf of Mexico and Caribbean Sea, Limnol. Oceanogr., 25:338.

Sawicki, E., and Sawicki, C.R., 1975, "Aldehydes - Photometric Analysis", Volume 1, Academic Press, New York.

Schlesinger, W.H., and Melack, J.M., 1981, Transport of organic carbon in the world's rivers, Tellus, 33:172.

Schwarzenbach, R.P., Bromund, R.H., Gschwend, P.M., and Zafiriou, O.C., 1978, Volatile organic compounds in coastal seawater, Org. Geochem., 1:93.

Sellner, K.G., 1981, Primary productivity and the flux of dissolved organic matter in several marine environments, Mar. Biol., 65:101.

Sikorski, R.J., and Mopper, K., 1986, Determination of trace amounts of carbonyl compounds in natural waters by liquid chromatography and on-line enrichment, Anal. Chem., submitted.

Snyder, L. R., and Kirkland, J. J., 1979, "Introduction to Modern Liquid Chromatography", Second Edition, Wiley, New York.

Steinberg, S.M., and Bada, J.L., 1984, Oxalic, glyoxalic and pyruvic acids in eastern Pacific Ocean waters, J. Mar. Res., 42:697.

Sugimura, Y., and Suzuki, Y., 1983, Amino acids dissolved in the western North Pacific waters, Pap. Meteorol. Geophys., 34:267.

Swinnerton, J.W., and Linnenbom, V.J., 1967, Gaseous hydrocarbons in seawater: determination, Science, N.Y., 156:1119.

Turnell, D.C., and Cooper, J.D.H., 1982, Rapid assay for amino acids in serum or urine by pre-column derivatization and reversed-phase liquid chromatography, Clin. Chem., 28:527.

Tuss, H., Neitzer, V., Seiler, W., and Neeb, R., 1982, Method for determination of formaldehyde in air in the pptv-range by HPLC after extraction as 2,4-dinitrophenyl-hydrazone, Fresenius Z. Anal. Chem., 312:613.

Tye, C.T., Haswell, S.J., O'Neill, P., and Bancroft, K.C.C., 1985, High-performance liquid chromatography with hydride generation/atomic absorption spectrometry for the determination of arsenic species with application to some water samples, Anal. Chim. Acta, 169:195.

Van Hoof, F., Wittocx, A., Van Buggenhout, E., and Janssens, J., 1985, Determination of aliphatic aldehydes in waters by high-performance liquid chromatography, Anal. Chim. Acta, 169:419.

Vastano, S., Milne, P., Mopper, K., and Zika, R., 1985, Photochemistry and distribution of flavins in the ocean, Trans. Amer. Geophys. Un., 66:1257.

Wada, A., Bonoshita, M., Tanaka, Y., and Hibi, K., 1984, A study of a reaction system for organic acid analysis using a pH indicator as post-column reagent, J. Chromatogr., 291:111.

Wangersky, P.J., 1978, Production of dissolved organic matter, Mar. Ecol., 4:115.

Wangersky, P.J., and Zika, R.G., 1978, "The Analysis of Organic Compounds in Seawater", Report 3, NRCC No. 16566, National Research Council of Canada, Halifax

Weiss, R.F., Bullister, J.L., Gammon, R.H., and Warner, M.J., 1985, Atmospheric chlorofluoromethanes in the deep equatorial Atlantic, Nature, Lond., 314:608.

Williams, P.M., Carlucci, A.F., and Olson, R., 1980, A deep profile of some biologically important properties in the central North Pacific gyre, Oceanol. Acta, 3:471.

Zafiriou, O.C., 1983, Natural water photochemistry, in: "Chemical Oceanography", Second Edition, Volume 8, J.P. Riley and R. Chester, eds, pp. 339-379, Academic Press, London.

Zika, R.G., and E.S. Saltzman, 1982, Interaction of ozone and hydrogen peroxide in water: implications for analysis of H_2O_2 in air, Geophys. Res. Lett., 9:231.

SURFACE WATER ^{234}Th/^{238}U DISEQUILIBRIA: SPATIAL AND TEMPORAL

VARIATIONS OF SCAVENGING RATES WITHIN THE PACIFIC OCEAN

Kenneth W. Bruland and Kenneth H. Coale

Institute of Marine Science
University of California
Santa Cruz, California 95064 U.S.A.

ABSTRACT

Dissolved and particulate ^{234}Th were determined on several surface transects within the Pacific Ocean. These transects, spanning major oceanographic regimes, ranged from oligotrophic subtropical gyres to eutrophic eastern boundary and equatorial divergence zones. Modeling of the disequilibria between ^{234}Th and ^{238}U within the surface waters provides estimates for the residence time of dissolved thorium with respect to particle scavenging, the residence time of particulate ^{234}Th, and the particulate ^{234}Th flux from the surface layer. The results provide a broad spatial view of the intensity and temporal variability of these scavenging processes. Within the oceanic surface waters dissolved ^{234}Th residence times vary from 6 to 220 days and the scavenging rate of dissolved ^{234}Th onto particles appears to be proportional to primary production. The scavenging intensity of dissolved ^{234}Th is strongly correlated with particulate organic carbon flux throughout a wide range of marine environments ranging from estuaries to the deep sea. Including these environments, which represent end-members for ^{234}Th scavenging intensity, the residence time for dissolved ^{234}Th is shown to vary from 0.3 days to 2 years depending on the particle flux. Particulate ^{234}Th residence times in oceanic surface waters are of the order of weeks and appear to be governed by the rate of zooplankton grazing. Model-derived particulate ^{234}Th fluxes can be used to constrain estimates based upon surface sediment trap collections.

INTRODUCTION

The removal of dissolved trace elements and radionuclides from seawater by adsorption onto sinking particles is an important mechanism controlling their oceanic concentrations. Goldberg (1954) was one of the first to discuss this process and referred to it as "scavenging". Subsequently, Turekian (1977) discussed this sequestering of trace elements via the "great particle conspiracy". This scavenging and removal process appears particularly important within the sunlit surface layer, the principal zone of particle production in the ocean. Trace elements and radionuclides can become associated with biogenic particles as a result of both active uptake (e.g., sequestering of required trace elements for enzyme systems) and passive adsorption processes (e.g., complexation of particle-reactive hydrolysed species such as $Th(OH)_4^\circ$ and $Fe(OH)_3^\circ$ with oxygen donor atoms in carboxylic and phenolic surface sites). Undigested residues from primary and secondary production in the surface zone can be packaged into fecal pellets or associate as amorphous aggregates. These large composite particles have sinking rates ranging from a few metres to a few thousand metres per day (Paffenhoffer and Knowles, 1979; Small et al., 1979; Shanks and Trent, 1979; Silver and Alldredge, 1981; Silver and Bruland, 1981) and are thought to be the principal means of particulate transport out of the surface layer. Recent studies using sediment traps and large volume *in situ* pumping techniques have demonstrated that these relatively large, rapidly sinking particles are responsible for most of the flux of material leaving the surface zone of oceanic regions (Bishop et al., 1977; Honjo, 1978; Wiebe et al., 1979; Knauer et al., 1979).

To understand better the surface water scavenging processes it would be useful to quantify: (1) the rates of transformation from dissolved to particulate form within the surface euphotic zone; (2) the resultant particulate elemental fluxes from the surface ocean to deep water; and (3) the dependence of scavenging rates and particulate elemental fluxes as a function of oceanographic regime and season.

There are a number of "particle-reactive daughter/inert parent" pairs of radionuclides in the uranium and thorium decay series that have proved useful for investigating scavenging rates within the surface euphotic zone. These include the pairs $^{234}Th/^{238}U$, $^{228}Th/^{228}Ra$ and $^{210}Pb/^{226}Ra$, with reactive daughter isotopes having half-lives ranging from 24 days to 22 years. Bhat et al. (1969) initiated the use of the short-lived nuclide ^{234}Th for studies of particulate removal, while Broecker et al., (1973) pioneered the work with ^{228}Th as a particle-reactive tracer of surface water scavenging processes. This initial work has been subsequently expanded by a number of investigators (^{234}Th: Matsumoto, 1975; Knauss et al., 1978; Kaufman et al., 1981; Li et al., 1981) (^{228}Th: Knauss et al., 1978; Li et al., 1979; Feely et al., 1980; Kaufman et al., 1981; Li et al., 1981). Rama et al. (1961) were the first to recognize the utility of ^{210}Pb to calculate scavenging rates of lead from oceanic surface waters, while Nozaki et al. (1976) and Bacon et al. (1976) have presented detailed investigations and models for the disequilibria existing between ^{226}Ra, ^{210}Pb, and ^{210}Po within oceanic surface waters.

Recently, Coale and Bruland (1985) demonstrated that ^{234}Th is an ideal particle reactive tracer for studying the scavenging of thorium from surface waters. Since the decay of conservative ^{238}U produces dissolved ^{234}Th at a constant rate, the $^{234}Th/^{238}U$ disequilibria in dissolved and particulate fractions yield information on both the rate of ^{234}Th transformation from dissolved to particulate forms and the rate of removal of particulate ^{234}Th from the surface photic zone. The 24.1 day half-life of ^{234}Th is optimal for elucidating these processes in the photic zone on time scales of one to hundreds of days; $^{234}Th/^{238}U$ disequilibria can also be used to monitor seasonal changes in the intensity of these removal processes.

The purpose of this investigation was to utilize $^{234}Th/^{238}U$ disequilibria to investigate the dependence of ^{234}Th scavenging rates and particulate thorium fluxes from oceanic surface waters as a function of oceanographic regime (spatial variability) and season (temporal variability). To accomplish this, we undertook surface water transects from the eutrophic coastal waters off central California across the California Current and out into the oligotrophic North Pacific central gyre. These transects were performed during both late summer, at the end of the summer insolation period, and early spring, at the end of the winter cooling and storm mixing. Additionally, a surface transect along 160°W, from 20°N to 20°S, was examined to contrast the oligotrophic central gyre regimes with the biologically productive equatorial divergence zone.

SAMPLING AND ANALYSIS

Samples for this investigation were collected on three research cruises, CEROP-III, VERTEX-I, and MC-80. Sample locations are presented in Fig. 1.

Seawater samples were collected with Teflon-coated, 30—litre Top-Drop or Go-Flo Samplers (General Oceanics). After collection, the sampler was pressurized with filtered nitrogen and the seawater passed via Teflon tubing to a Teflon filter sandwich supporting a tared 142 mm Nuclepore membrane filter (0.3 μm pore size). Dissolved samples were collected in 20—litre Cubitainers and acidified with 50 ml of concentrated hydrochloric acid. The membrane filters supporting the particulate samples (from 55-110 litres of seawater) were folded and stored frozen prior to analysis. Sample processing typically was completed within 30 minutes from the time the sampler reached the surface.

Due to the short half-life of ^{234}Th, rapid separation from its ^{238}U parent is necessary to avoid significant in-growth corrections. Thus, the spiking, equilibration, concentration and radiochemical separation steps were performed at sea. Details of these steps and the counting procedures are presented in Coale and Bruland (1985).

Coale and Bruland (1985) also presented results of a calibration study

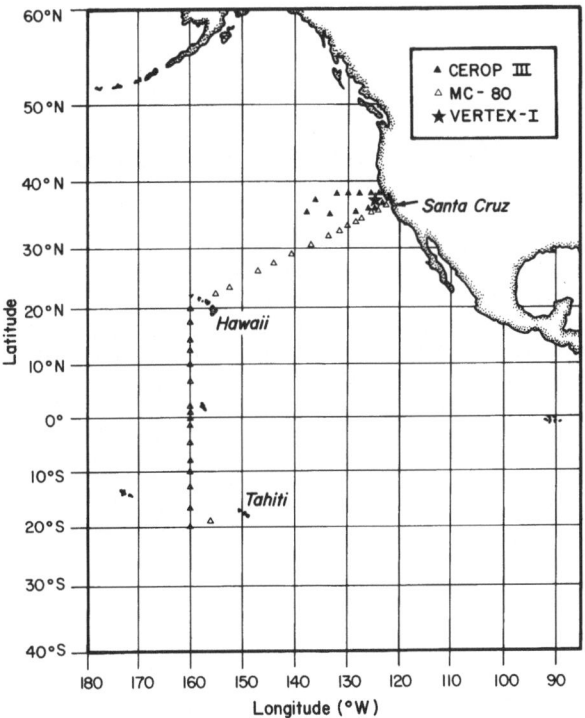

Fig. 1. Cruise track and station locations:
$\triangle$, MC-80; $\blacktriangle$, CEROP-III;
$\bigstar$, VERTEX-I.

yielding an estimate of the accuracy for determining the $^{234}Th/^{238}U$ activity ratio to within 1% of the true value, and estimated the precision of a normal sample determination to be ± 2.3% (one standard deviation). In addition, deep water total ^{234}Th results indicate equilibrium with ^{238}U, further supporting the accuracy of this technique.

RESULTS

Scavenging Model

The scavenging model applied to surface waters for the interpretation of the ^{234}Th disequilibria is similar to that presented by Coale and Bruland (1985). Assuming advection and diffusion to be negligible over the time scales relevant to these studies, the rate of change in dissolved ^{234}Th activity can be expressed by the equation

$$\frac{\delta A^d_{Th}}{\delta t} = A_U \lambda_{Th} - A^d_{Th} \lambda_{Th} - J_{Th} \tag{1}$$

where A_U is the activity of dissolved ^{238}U, λ_{Th} is the radioactive decay constant of ^{234}Th, and A^d_{Th} is the dissolved ^{234}Th activity; J_{Th} is the rate of removal of ^{234}Th from the dissolved to particulate form. If the removal of dissolved thorium is assumed to be first order with respect to the dissolved ^{234}Th activity, then $J_{Th} = A^d_{Th} \psi$, where ψ is a first order scavenging rate constant. The term $A_U \lambda_{Th}$ equals the rate of production of dissolved ^{234}Th (since ^{238}U exists essentially all in the dissolved form, its decay will, in turn, produce dissolved ^{234}Th). The term $A^d_{Th} \lambda_{Th}$ equals the loss rate by radioactive decay of dissolved ^{234}Th. Similarly, the rate of change of particulate ^{234}Th activity can be expressed by

$$\frac{\delta A^p_{Th}}{\delta t} = J_{Th} - A^p_{Th} \lambda_{Th} - P_{Th} \tag{2}$$

where A^p_{Th} is the activity of particulate ^{234}Th, and P_{Th} is the rate at which ^{234}Th is transported out of the surface layer by the particle flux. In this equation J_{Th} (equivalent to J_{Th} in equation 1) is the rate of production of particulate ^{234}Th while $A^p_{Th} \lambda_{Th}$ is the radioactive decay rate of particulate ^{234}Th.

If the system is close to steady state then equations (1) and (2) are simplified since $\delta A^d_{Th}/\delta t$ and $\delta A^p_{Th}/\delta t$ approach zero. The value of J_{Th} can be derived, using equation (1), from measurements of A^d_{Th} and estimates of A_U. Uranium has a conservative distribution in the ocean and the radioactivity of ^{238}U, in disintegrations per minute per litre (dpm l^{-1}), is calculated as 0.07081 x salinity (Ku et al., 1977). Then by measuring A^p_{Th}, values for P_{Th} can be calculated. When integrating over the surface mixed layer, values of A_U, A^d_{Th} and A^p_{Th} are in units of disintegrations per minute per m^2, while J_{Th} and P_{Th} are in disintegrations per minute per m^2 per day. The mean life of dissolved ^{234}Th with respect to removal onto particles is $\bar{t}_d = A^d_{Th}/J_{Th}$ while the mean life of particulate ^{234}Th with respect to removal from the surface layer is $\bar{t}_p = A^p_{Th}/P_{Th}$.

Vertical Profiles

Vertical profiles of A^d_{Th}/A_U and A^p_{Th}/A_U for the California Current, the central North Pacific, and the central South Pacific are presented in Fig. 2. The California Current station (VERTEX-I) exhibits a marked deficiency of dissolved and total ^{234}Th within the surface euphotic zone: $A^d_{Th}/A_U = 0.33 ± 0.06$, while $A^\Sigma_{Th}/A_U = 0.56 ± 0.10$, within the surface 50 m (A^Σ_{Th} = dissolved + particulate ^{234}Th activities). This deficiency of ^{234}Th with respect to its parent, ^{238}U, in California Current surface

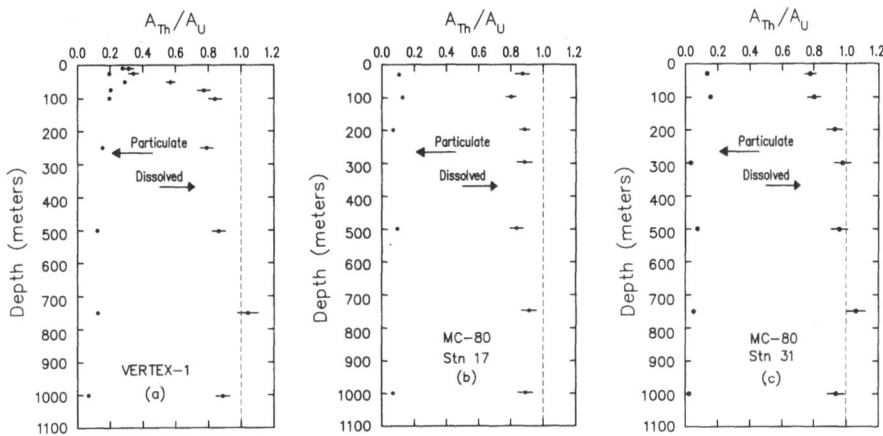

Fig. 2. Vertical profiles of the activity ratios, A_{Th}/A_U, for stations:
(a) VERTEX-I; (b) MC-80, station 17; (c) MC-80, station 31.
——•—, dissolved activity ratios (bar widths indicate 1σ standard
deviation error); • particulate activity ratios (errors less than
dissolved errors); vertical line at $A_{Th}/A_U = 1$ represents radioactive equilibrium.

water yields a mean life for dissolved ^{234}Th with respect to removal to
particulates of $\bar{t}_d$ = 17 days, and a mean life of particulate ^{234}Th prior to
sinking out of the photic zone of $\bar{t}_p$ = 18 days. In contrast, the oligotrophic stations show much smaller deficiencies of dissolved and total
^{234}Th within the surface layers in the central region of the North and
South Pacific: MC-80 station 17 (14°N, 160°W) has an average A^d_{Th}/A_U =
0.836 ± 0.05 and an average A^Σ_{Th}/A_U = 0.950 ± 0.035, while station 31
(20°S, 160°W) has an average A^d_{Th}/A_U = 0.792 ± 0.016 and A^Σ_{Th}/A_U = 0.938 ±
0.030 within the surface 100 m. The dissolved ^{234}Th activity ratios within
the surface 100 m at Stations 17 and 31 yield mean lives of dissolved Th
with respect to removal to particulate form of 177 and 132 days, respectively. The particle residence times in the surface waters of Stations 17
and 31 are both calculated to be 80 days.

Water deeper than 200 m at all three stations exhibits total ^{234}Th
values close to equilibrium with ^{238}U: VERTEX-I, A^Σ_{Th}/A_U = 1.003 ± 0.086;
MC-80 station 17, A^Σ_{Th}/A_U = 0.956 ± 0.040; station 31, A^Σ_{Th}/A_U = 1.005 ±
0.095. Thus, scavenging of total ^{234}Th from the deep waters of the Pacific
is not occurring on time scales of less than one year. There is, however,
measurable removal from dissolved to particulate form in the deep-sea as
evidenced by the average deep water A^d_{Th}/A_U = 0.94 ± 0.06.

Surface Transect: Monterey to Hawaii

Figure 3 presents surface mixed-layer observations of salinity together with dissolved and particulate ^{234}Th (presented as activity ratios with
respect to ^{238}U) from the relatively eutrophic California Current off central California, across the transition zone, and into the oligotrophic central North Pacific gyre. The MC-80 cruise (Fig. 3) took place in September
at the end of the summer insolation period. The surface mixed layer depth
was close to 30 m. Within the low salinity California Current, ^{234}Th was
markedly deficient: A^d_{Th}/A_U = 0.31 ± 0.07 and A^P_{Th}/A_U = 0.21 ± 0.07. These
activity ratios yield values of $\bar{t}_d$ = 16 days and $\bar{t}_p$ = 14 days, respectively. In contrast, ^{234}Th in central ocean surface water is much closer to
equilibrium. Between 130°W and 160°W, A^d_{Th}/A_U = 0.865 ± 0.039 and
A^P_{Th}/A_U = 0.095 ± 0.040 yielding an average $\bar{t}_d$ = 223 days and $\bar{t}_p$ = 83 days.

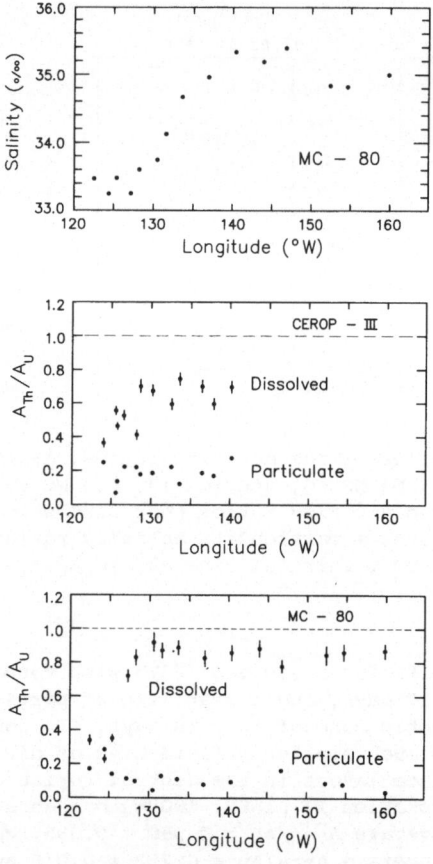

Fig. 3. Horizontal profiles of mixed
layer salinity and A_{Th}/A_U
ratios for MC-80 and CEROP—
III. Symbols as in Fig. 2.

The CEROP-III cruise (Fig. 3) took place in March at the end of the wintertime cooling and storms. As a result, the mixed layer depth approached 140 m. During this cruise, average dissolved and particulate ^{234}Th values within the surface waters of the California Current were $A^d_{Th} = 0.45 \pm 0.10$ and $A^P_{Th}/A_U = 0.18 \pm 0.10$, yielding an average $\bar{t}_d$ of 38 days and $\bar{t}_p$ of 17 days. The central ocean gyre water during March had average values of $A^d_{Th}/A_U = 0.676 \pm 0.058$ and $A^P_{Th}/A_U = 0.194 \pm 0.048$, yielding $\bar{t}_d$ = 73 days and $\bar{t}_p$ = 52 days. This latter $\bar{t}_d$ value was only one third that observed in September. Thus, there appears to be a substantial seasonal variation in the scavenging rate of dissolved thorium in the central gyre waters of the North Pacific, with the scavenging more intense in March following winter mixing, and less intense in September when maximum stratification occurs.

Surface Transect: 20°N to 20°S

Results of the surface transect along 160°W from the central North Pacific, across the equatorial divergence zone, and into the South Pacific central gyre are presented in Fig. 4. In the central North Pacific (15°N - 20°N) surface waters, the average $A^d_{Th}/A_U = 0.865 \pm 0.012$ and the average $A^P_{Th}/A_U = 0.07 \pm 0.03$ yielding $\bar{t}_d$ = 225 days and $\bar{t}_p$ = 35 days. In the equatorial divergence zone, the A^d_{Th}/A_U ranged from 0.42 to 0.59, yielding

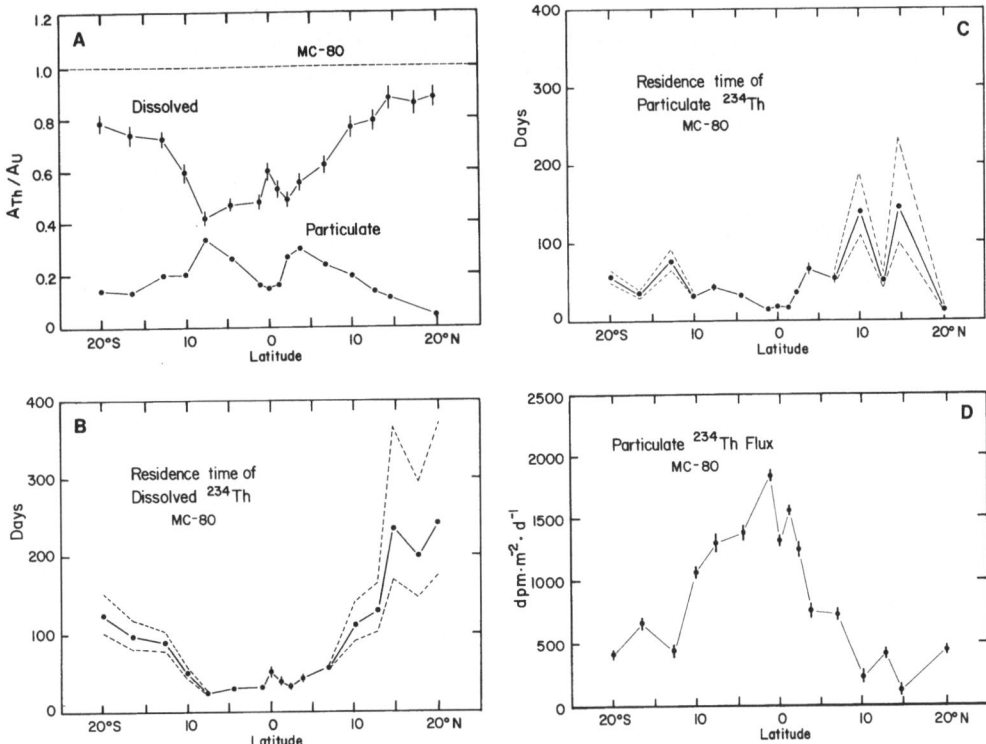

Fig. 4. Latitudinal variation in horizontal profiles of mixed layer (Cruise MC 80): (a) A_{Th}/A_U ratios for dissolved and particulate fractions, symbols as in Fig. 2; (b) Residence time (days) of dissolved ^{234}Th, +————+, calculated residence time *versus* latitude; +------+, one standard deviation in calculated residence time; (c) Residence time (days) of particulate ^{234}Th; (d) Model—calculated particulate ^{234}Th fluxes. Symbols as in Fig. 2.

values of $\bar{t}_d$ between 25 and 50 days. The A^P_{Th}/A_U in this region ranged from 0.142 to 0.325 yielding values of $\bar{t}_p$ from 14 to 40 days.

The surface waters of the central South Pacific (15°S - 20°S) had average $A^d_{Th}/A_U = 0.758 \pm 0.032$ yielding a $\bar{t}_d$ of 109 days. This residence time is shorter than observed in the central North Pacific gyre and is consistent with what one might anticipate for Southern Hemisphere late wintertime conditions based on the seasonal differences observed in the North Pacific. The average $A^P_{Th}/A_U = 0.132 \pm 0.006$, yielding $\bar{t}_p = 42$ days.

DISCUSSION

Variation of the ^{234}Th Scavenging Rate

Coale and Bruland (1985) have examined the scavenging rate of ^{234}Th within the surface waters of the California Current under various oceanographic conditions. They observed that the removal rate of thorium from dissolved to particulate form appeared to be proportional to primary productivity. Values of $\bar{t}_d$ ranged from 6 days, when coastal upwelling was intense, to 50 days under non-upwelling conditions. Kaufman et al. (1981) examined total ^{234}Th and ^{228}Th in waters of the New York Bight and found that the mean residence time of total thorium with respect to non-radioactive removal ranged from about 10 days near shore to 70 days at the shelf break. Residence times obtained using total thorium measurements on unfiltered samples, however, include both the time required for removal from dissolved to particulate form, and the time the thorium resides on suspended particles prior to leaving the surface layer.

In this study the mean life of dissolved ^{234}Th with respect to scavenging onto particulates, $\bar{t}_d$, for surface water of the central North Pacific varied temporally from an average of 227 days in September to 75 days in March. This corresponds to a three-fold seasonal difference in scavenging intensity, with removal being more intense in the spring at the end of wintertime mixing, and less intense in late summer when maximum stratification occurs. Presumably, this relationship is coupled to the higher nutrient input to surface waters by enhanced wintertime mixing and a corresponding higher productivity in the central gyre region during these seasonal cycles. The average mean life of dissolved ^{234}Th in the central South Pacific during October was 109 days. This value most probably reflects Southern Hemisphere springtime conditions, for it is comparable to the CEROP-III central North Pacific disequilibria observed during March.

We assumed a *priori* that the central North Pacific gyre would be temporally less variable than coastal upwelling systems. Our results indicate that the scavenging rate of thorium varies by a factor of three in these central gyres. This is, however, still less than the roughly eight-fold variation in scavenging intensity observed within the California Current.

The spatial variation of the dissolved ^{234}Th residence time (mean life of dissolved thorium with respect to removal to particulate forms) on the 20°N to 20°S transect is presented in Fig. 4b. High and low estimates associated with propagating the estimated error in the activity ratio are depicted. At latitudes between 10°N and 10°S, in the biologically productive equatorial divergence zone, markedly shorter dissolved thorium residence times were observed.

The results from this study can be used to extend the relationship between primary productivity and the scavenging rate of dissolved thorium previously observed within the California Current. It is difficult to make this comparison, however, because of the lack of oceanic primary productivity data and the controversy surrounding much of it. Figure 5 presents first-order scavenging rate constants and estimates of primary production for different regions including the California Current data from Coale and Bruland (1985). The scavenging rate of dissolved thorium onto particles in oceanic surface waters varies by a factor of 40, from 0.0044 d^{-1} in stratified oligotrophic central gyres, to 0.18 d^{-1} in high productivity coastal upwelling regimes. The variation in scavenging rate is consistent with the 15-fold variation in primary production between these two end-members. It should be noted that even if the central gyre estimates of primary production used are low, one would still observe a relationship between the scavenging intensity of thorium and primary production.

Perhaps a more likely parameter than total primary production for controlling scavenging intensity is new production or the particulate organic carbon (POC) flux out of the surface layer. In the central gyres, most of the primary production is recycled and only a small portion is thought to leave the surface layer as a particulate flux (Eppley and Peterson, 1979). In contrast, in the eutrophic boundaries a much greater percentage of the primary production is new production and leaves the system as sinking particles.

The mean life of dissolved ^{234}Th with respect to scavenging is extremely short in continental shelf and estuarine waters having high part-

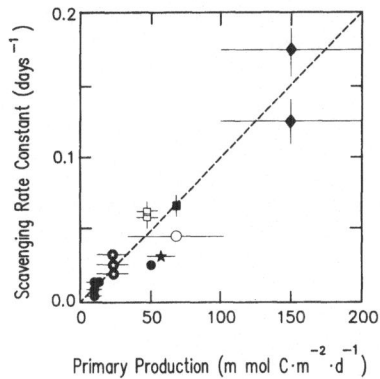

Fig. 5. Model derived pseudo first—order scavenging rate constant (day[-1])
versus primary productivity (mmol C m[-2] day[-1]). Scavenging rates:
our data. Primary production data from sources indicated: □ ,
measured, Martin et al. (1984); ⭐, estimated, based upon Ryther
(1969). Knauer et al. (1979) assigned this value to coastal non—
upwelling waters in this area; this is equivalent to the values
observed by Eppley et al. (1979) for this type of water off
Southern California; it is assigned a 50% error.
○ , estimated, based upon Ryther (1969). Knauer et al. (1979)
assigned this value to upwelling waters in this area; this is
mid-range for the values observed by Eppley et al. (1979) in
similar waters off Southern California; it is assigned a 50% error.
■, estimated, Knauer et al. (1979); this value has been assigned a
50% error. ★ , measured, Knauer and Martin (1981). ♦ , estimated,
this value represents the range of primary production for intense
coastal upwelling observed by Eppley et al. (1979) off Southern
California and is assigned a 50% error. ● , measured, Betzer
et al. (1984).

iculate loads. Santschi et al. (1979) found $\bar{t}_d$ values in Narragansett
Bay to be only 1.5 days. McKee et al. (1984) determined scavenging rates
for [234]Th from continental shelf waters near the mouth of the Yangtze
River. In waters with extremely high suspended loads, $\bar{t}_d$ ranged from 0.3
days nearshore to 4.0 days offshore. In our studies we deliberately
avoided coastal shelf waters where resuspended sediments can be the major
contributor to the particle flux. Instead, we have examined regions where
particles of biogenic origin provide the bulk of the particulate material
in order to evaluate more accurately the autochthonous removal processes.
We can compare our data with the estuarine data of Santschi et al. (1979)
and McKee et al. (1984) by using their particle residence times to estimate
a resuspension flux. In order to compare scavenging removal based on
resuspension fluxes to that based on oceanic particulate organic carbon
fluxes, we assumed that suspended particles in shallow estuaries and bays
contained 5% organic carbon.

Figure 6 displays a log-log plot of the various regimes. Within the
surface waters, the first-order scavenging rate constant varies by about
three orders of magnitude, from 0.0044 d[-1] in oligotrophic central gyre
regions to 3.0 d[-1] in estuarine or nearshore water where particle loads are
exceptionally high. This scavenging rate constant appears to be proportional to the POC flux, which varies by roughly four orders of magnitude.
Moreover, the scavenging rate of thorium in the deep sea also fits into
this general trend. This extends the range of values for the scavenging
rate constant of thorium to many types of marine environments yielding
greater confidence in suggesting a coupling between scavenging intensity

and POC flux. This relationship is consistent with that presented by Santschi (1984) who noted a simple inverse relationship between total thorium residence time and particle flux through the water column.

Nozaki and Horibe (1981) and Bacon and Anderson (1982) point out that the scavenging of thorium in the deep sea is best modeled (or described) as a reversible exchange process that occurs continuously throughout the deep water column. Bacon and Anderson (1982) have examined the kinetics of thorium exchange between dissolved and particulate forms. In the deep sea, the residence time of suspended particles in the water column is suffic- iently long (5 to 10 years) that particles reach a steady-state at which adsorption of ^{234}Th is balanced primarily by desorption and radioactive decay. An average value for the forward adsorption rate, K_1, is approx- imately 0.52 yr^{-1} (mean life of 1.9 years), while the desorption rate, K_{-1} is approximately 2.6 yr^{-1}. Within surface waters, the scavenging rate is much more intense, and the residence time of suspended particles is only of the order of weeks. Thus, in surface waters much of the particulate thorium can be packaged and sink out before it gets a chance to reversibly desorb. Rapid scavenging and particulate removal control the abundance of dissolved ^{234}Th in productive surface waters.

Particulate ^{234}Th Residence Times

Coale and Bruland (1985) concluded that within the California Current the residence time of suspended particulate ^{234}Th is a function of grazing pressure by herbivorous zooplankton. They observed residence times to vary from 2 days under intense grazing by gelatinous zooplankton, to 20 days where relatively sparse zooplankton populations occur. The average particulate ^{234}Th residence times observed within the California Current during the CEROP-III and MC–80 cruises were 17 and 14 days, respectively. The particulate ^{234}Th residence time is plotted against latitude for the 20°N to 20°S MC-80 surface transect, in Fig. 6. Note that the shortest residence times are observed in the regions of higher productivity near the equatorial divergence. This corresponds to the increased abundance of

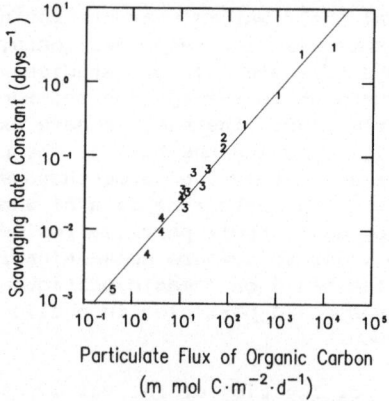

Fig. 6. Log-log plot of pseudo first–order scavenging rate constant for dissolved ^{234}Th (day^{-1}) *versus* flux of particulate organic carbon (mmol C m^{-2} day^{-1}): 1, Yangtze shelf region and Narragansett Bay, Santschi et al. (1979) and McKee et al. (1984); 2, Coastal up- welling stations, Coale and Bruland (1985); 3, California Current and equatorial divergence stations, this study; 4, Central gyre stations, this study; 5, Deep Pacific, Bacon and Anderson (1982) and this study. Dotted line represents best fit to all data; r = +0.988 on untransformed data.

zooplankton grazers in these higher productivity areas as observed by SCUBA divers during these studies. Unfortunately, we have no quantitative estimates of zooplankton grazing rates to compare with the particulate thorium residence times. Within the California Current and within a few degrees of the equatorial divergence, particulate ^{234}Th residence times are of the order of weeks. In contrast, within the central gyre regimes the particulate thorium residence times are of the order of 1 to 3 months.

These particulate ^{234}Th residence times agree well with the residence time of particulate organic carbon in the surface layer of the ocean (Eppley et al., 1983). Within the California Current, these workers observed the residence time of POC relative to new production (or the sinking flux of POC out of the photic zone) to vary from 3 days to greater than 100 days. The residence time of POC relative to total primary production ranged from 2 to 15 days. The difference between the two sets of residence times is a result of much of the POC being recycled numerous times prior to leaving the surface layer. The shortest POC residence times were associated with periods of high productivity. In the central North Pacific, the POC residence time with respect to new production was observed to be of the order of 1 year, while the residence time with respect to total production was close to 3 weeks. The ^{234}Th residence times are closer to the POC residence times relative to total primary production. This suggests that ^{234}Th is not regenerated with the more labile components of the POC. Instead, it appears more refractory and it is retained with the particulate phase and transported out of the surface layer with the flux of large particles.

Figure 4d presents the model-derived particulate ^{234}Th flux, P_{Th}, leaving a 70 m surface layer along the 20°N to 20°S surface transect. These predicted fluxes vary by roughly an order of magnitude between the central North Pacific gyre water and the zone of equatorial divergence. There is also an inter-hemispheric difference in P_{Th}, with the central South Pacific showing higher fluxes than the central North Pacific. We attribute this inter-hemisphere difference to the seasonal variability discussed earlier. Since $P_{Th} = J_{Th} - A^P_{Th}\lambda_{Th}$, an increase in dissolved ^{234}Th scavenging intensity (J_{Th}) in the spring can result in an increase in P_{Th} during this season given that the particulate ^{234}Th activities in the two hemispheres are roughly equivalent.

These model-derived values of P_{Th} can be used to compare and constrain flux data obtained with subsurface sediment trap deployments (Coale and Bruland, 1985). A comparison between model-derived particulate ^{234}Th fluxes and those measured in sediment trap collections at the base of the surface layer enables evaluation of these independent flux estimates. For CEROP-III, VERTEX-I, and MC-80 station 17, model-derived ^{234}Th fluxes agree to within 30% of the fluxes measured by sub-surface sediment traps.

CONCLUSIONS

The results from this study further support the conclusions of Coale and Bruland (1985) that in areas of predominantly biogenic particle production, the rate of scavenging of ^{234}Th from dissolved to particulate form varies as a function of primary production, whereas the residence time of particulate ^{234}Th varies as a function of zooplankton grazing pressure. In this study, we show that the scavenging rate of dissolved ^{234}Th is proportional to the POC flux over a wide range of environments. When the results from this study are combined with the estuarine and nearshore results of Santschi et al. (1979) and McKee et al. (1984) and the California Current results of Coale and Bruland (1985), a strong coupling is observed between the pseudo first-order scavenging rate constant for dissolved ^{234}Th (i.e., removal from dissolved to particulate form) and the flux of particulate organic carbon leaving the surface layer. This relationship appears to be linear over roughly five orders of magnitude in POC flux and four orders of magnitude in scavenging rate constant. These results further demonstrate the utility of $^{234}Th/^{238}U$ disequilibria in modeling ocean-wide mixed-layer scavenging and particulate removal processes

ACKNOWLEDGEMENTS

We wish to express our thanks and appreciation to the crew and officers of the research vessels T. G. Thompson and Wecoma and to Marcia Campbell for her assistance in shipboard processing of radionuclide samples during MC-80. We also thank Gary Gill and John Donat for reviewing the manuscript and providing valuable comments. This research was supported by NSF Grants OCE 79-19928 and OCE 79-23322.

REFERENCES

Bacon, M.P., and Anderson, F., 1982, Distribution of thorium isotopes between dissolved and particulate forms in the deep sea, J. Geophys. Res., 87:2045.

Bacon, M.P., Spencer, D.W., and Brewer, P.G., 1976, $^{210}Pb/^{226}Ra$ and $^{210}Po/^{210}Pb$ disequilibria in seawater and suspended particulate matter, Earth Planet. Sci. Lett., 32:277.

Betzer, P.R., Showers, W.J., Laws, E.A., Winn, C.D., DiTullio, G.R., and Kroopnick, P.M., 1984, Primary productivity and particle fluxes on a transect of the equator at 153°W in the Pacific Ocean, Deep-Sea Res., 31:1.

Bhat, S.G., Krishnaswami, S., Lal, D., Rama, and Moore, W.S., 1969, Th-234/U-238 ratios in the ocean, Earth Planet. Sci. Lett., 5:483.

Bishop, J.K.B., Edmond, J.M., Ketten, D.R., Bacon, M.P., and Silker, W.B., 1977, The chemistry, biology, and vertical flux of particulate matter from the upper 400 m of the equatorial Atlantic Ocean, Deep-Sea Res., 24:511.

Broecker, W.A., Kaufman, A., and Trier, R.M., 1973, The residence time of thorium in surface sea water and its implications regarding the fate of reactive pollutants, Earth Planet. Sci. Lett., 20:35.

Coale, K.H., and Bruland, K.W., 1985, $^{234}Th/^{238}U$ disequilibrium within the California Current, Limnol. Oceanogr., 30:22.

Eppley, R.W., and Peterson, B.J., 1979, Particulate organic matter flux and planktonic new production in the deep ocean, Nature, Lond., 282:677.

Eppley, R.W., Renger, E.H., and Harrison, W.G., 1979, Nitrate and phytoplankton production in the Southern California coastal waters and its role in the growth of phytoplankton, Limnol. Oceanogr., 24:483.

Eppley, R.W., Renger, E.H., and Betzer, P.R., 1983, The residence time of particulate organic carbon in the surface layer of the ocean, Deep-Sea Res., 30:311.

Feely, H.W., 1981, Natural radionuclides in waters of the New York Bight, Earth Planet. Sci. Lett., 55:217.

Feely, H.W., Kipphut, G.W., Trier, R.M., and Kent, C., 1980, ^{228}Ra and ^{228}Th in coastal waters, Estuarine Coastal Mar. Sci., 11:179.

Goldberg, E.D., 1954, Marine Geochemistry, 1, Chemical Scavengers of the Sea, J. Geol., 62:249.

Honjo, S., 1978, Sedimentation of materials in the Sargasso Sea at 5367 m deep station, J. Mar. Res., 36:469.

Kaufman, A., Li, Y.-H., and Turekian, K.K., 1981, The removal rates of ^{234}Th and ^{228}Th from waters of the New York Bight, Earth Planet. Sci. Lett., 54:385.

Knauer, G.A., and Martin, J.H., 1981, Primary production and carbon-nitrogen fluxes in the upper 1500 m of the northeast Pacific, Limnol. Oceanogr., 26:181.

Knauer, G.A., Martin, J.H., and Bruland, K.W., 1979, Fluxes of particulate carbon, nitrogen and phosphorus in the upper water column of the Northeast Pacific, Deep-Sea Res., 26:97.

Knauss, K.G., Ku, T., and Moore, W.S., 1978, Radium and thorium isotopes in the surface waters of the east Pacific and coastal southern California, Earth Planet. Sci. Lett., 39:235.

Ku, T.L., Knauss, K.G., and Mathieu, G.G., 1977, Uranium in the open ocean: Concentration and isotopic composition, Deep-Sea Res., 24:1005.

Li, Y.H., Feely, W.H., and Santschi, P.H., 1979, ^{228}Th/^{228}Ra radioactive disequilibrium in the New York Bight and its implications for coastal pollution, Earth Planet. Sci. Lett., 42:13.

Li, Y.H., Feely, W.H., and Toggweiler, J.R., 1980, ^{228}Ra and ^{228}Th concentrations in GEOSECS Atlantic surface waters, Deep-Sea Res., 27:545.

Li, Y.H., Santschi, P.H., Kaufman, A., Benninger, L.K., and Feely, H.W., 1981, Natural radionuclides in waters of the New York Bight, Earth Planet. Sci. Lett., 55:217.

Martin, J.H., Knauer, G.A., Broenkow, W.W., Bruland, K.W., Karl, D.L., Small, L.F., Silver, M.W., and Gowing, M.M., 1984, Vertical transport and exchange of materials in the upper waters of the oceans (VERTEX): Introduction to the program, hydrographic conditions and major component fluxes during VERTEX-I, Moss Landing Technical Publication 84-1.

Matsumoto, E., 1975, ^{234}Th-^{238}U radioactive disequilibrium in the surface layer of the ocean, Geochim. Cosmochim. Acta, 39:205.

McKee, B.A., DeMaster, D.J., and Nittrouer, C.A., 1984, The use of ^{234}Th/^{238}U disequilibrium to examine the fate of particle-reactive species on the Yangtze continental shelf, Earth Planet. Sci. Lett., 68:431.

Nozaki, Y., and Horibe, Y., 1981, The water column distributions of thorium isotopes in the western North Pacific, Earth Planet. Sci. Lett., 54:203.

Nozaki, W.M., Thompson, J., and Turekian, K.K., 1976, The distribution of Pb-210 and Po-210 in the surface waters of the Pacific Ocean, Earth Planet. Sci. Lett., 32:304.

Paffenhofer, G., and Knowles, S.C., 1979, Ecological implications of fecal pellet size, production and consumption by copepods, J. Mar. Res., 37:35.

Rama, Koide, M., and Goldberg, E.D., 1961, Lead-210 in natural waters, Science, N.Y., 134:98.

Ryther, J.H., 1969, Photosynthesis and fish production in the sea, Science, N.Y., 166:72.

Santschi, P.H., 1984, Particle flux and trace metal residence time in natural waters, Limnol. Oceanogr., 29:1100

Santschi, P.H., Li, Y.H., and Bell, J., 1979, Natural radionuclides in the water of Narragansett Bay, Earth Planet. Sci. Lett., 45:201.

Shanks, A.L., and Trent, J.D., 1979, Marine Snow: Microscale nutrient patches, Limnol. Oceanogr., 24:850.

Silver, M.W., and Alldredge, A.L., 1981, Bathypelagic marine snow: Vertical transport system and deep-sea algal and detrital community, J. Mar. Res., 39:227.

Silver, M.W., and Bruland, K.W., 1981, Differential feeding and fecal pellet composition of salps and pteropods, and the possible origin of the deep-water flora and olive-green "cells", Mar. Biol., 62:263.

Small, L.F., Fowler, S.W., and Ünlü, M.Y., 1979, Sinking rates of natural copepod fecal pellets, Mar. Biol., 51:233.

Turekian, K.K., 1977, The fate of metals in the oceans, Geochim. Cosmochim. Acta, 41:1139.

Wiebe, P.H., Madin, L.P., Haury, L.R., Harbison, G.R., and Philbin, L.M., 1979, Diel vertical migration by Salpa aspera and its potential for large-scale particulate organic matter transport to the deep-sea, Mar. Biol., 53:249.

DOMINANT MICROORGANISMS OF THE UPPER OCEAN: FORM AND FUNCTION,

SPATIAL DISTRIBUTION AND PHOTOREGULATION OF BIOCHEMICAL PROCESSES

John McNeill Sieburth

Graduate School of Oceanography
University of Rhode Island Bay Campus
Narragansett, Rhode Island 02882 U.S.A.

ABSTRACT

The microorganisms that account for 90% of the biomass in the upper layer of the open sea are so small they pass through a 10 μm porosity plankton net, but are so numerous that a 3 or 30 ml sample yields a statistically significant count of the bacteria and flagellates, respectively. The bacterial population contains not only the organotrophs responsible for organic matter decomposition and mineralization, but conspicuous and important populations of both phototrophic and chemotrophic autotrophs as well as an undetermined population of lysotrophic bacteria which prey upon other bacteria. These procaryotic cells lacking membrane-bound organelles interact with the eucaryotic (true) cells of the protists (unicellular microorganisms with organelles) that are equally divided between chloroplast-containing phototrophic flagellates and colorless phagotrophic (particle eating) flagellates.

These dominant microorganisms have two major distributions, as plankters free in the water and as epibionts associated with the aggregated seston that makes up marine snow. Their physical state as plankters or as epibionts has a major effect on their spatial distribution and, therefore, their activity and survival in clear oceanic waters during the photoperiod. Near ultra-violet and the shorter visible wavelengths of light which penetrate to depths exceeding 30 m apparently inhibit bacterial processes by photodenaturing bacterial enzymes at intensities far below that which kills the bacteria by denaturing DNA and RNA. Light sensitive bacteria include not only the organotrophs responsible for organic matter decay but the chemotrophs that oxidize the terminal products of decay, ammonia and methane. Therefore, the daily chemical cycles of CO_2, O_2, H_2, CH_4, and labile components of organic matter which are due, to a greater or lesser degree, to microbiological processes, are in turn controlled by the solar cycle directly through photoinhibition and phototoxicity and indirectly through vertical mixing caused by wind-drive turbulence and heat-driven convection that brings the less dense microbiota into the higher intensity light near the surface.

FORM AND FUNCTION

The forms that pass through a 10 μm net or Nuclepore filter include

microalgae, protozoa and bacteria. The first two groups have eucaryotic
cells with membrane bound organelles which can be seen when thin sections
of plastic embedded cells are examined by transmission electron microscopy
as shown in Fig. 1. The unique feature of the small flagellated microalgae
are their chloroplasts where photosynthesis takes place. These forms are
detected and enumerated in field samples by the autofluorescence of the
chlorphyll seen with the epifluorescence microscope (Davis and Sieburth,
1982; Caron, 1983). The unique feature of the small flagellated protozoa
(which lack chloroplasts) are the food vacuoles containing the ingested and
digesting prey. Procedures for their detection, differentiation from
microalgae, and enumeration by staining with fluorochromes are discussed in
the above references. These microflagellates, which number about 900 cells
ml^{-1}, are usually equally divided between phototrophs and phagotrophs.

The bacteria or procaryotes lack the membrane bound organelles of the
eucaryotic protists, yet many have distinctive morphologies and ultrastructure when viewed with electron microscopy (Sieburth, 1979). The
phototrophic bacteria are characterized by widely spaced cytomembranes
called thylakoids which function like chloroplasts. The ubiquity and
relatively numerous populations of 10^3 to 10^5 cells ml^{-1} of these cyanobacteria (blue green algae) in the open sea has only recently been recognized (Johnson and Sieburth, 1979; Waterbury et al., 1979). Similar sized
eucaryotic phototrophs, which can be as numerous (Johnson and Sieburth,
1982), together with the cyanobacteria occurring in the picoplankton size
fraction (0.2-2.0 μm) (Sieburth et al., 1978) have been reported to account
for 20-90% of both the biomass and productivity of oceanic waters (Li et
al., 1983; Platt et al., 1983). The ecology of different species of
cyanobacteria has recently been examined using fluorescent antibodies (FA)
to specific strains to study their occurrence in natural populations
(Campbell et al., 1983).

Equally distinctive bacteria with more closely spaced cytomembranes
are the chemotrophs responsible for the oxidation of ammonia and nitrite,
the nitrifying bacteria (Watson et al., 1981), and for the oxidation of
methane, the methanotrophs (Whittenbury and Dalton, 1981). Cells with
ultrastructures suggestive of these chemotrophic bacteria can be seen in
natural populations of oceanic picoplankton at concentrations exceeding the
phototrophs by three to eight-fold. The distribution of nitrifiers has
been described also using FA techniques (Ward and Perry, 1980). Recently
methanotrophs have been detected and isolated from the Sargasso Sea but not
from the Gulf Stream (Sieburth et al., 1983). Their distribution in
natural populations using the FA approach is in progress.

The autotrophs (phototrophs and chemotrophs) discussed above appear to
be larger but less numerous cells in the picoplankton. The larger population of smaller cells which lack distinctive cytomembranes is presumed to
consist of organotrophs. They are dominated by 0.2 μm wide cells in C and
S shapes up to 1.5 μm in length, which apparently have not been brought
into culture (Sieburth, 1979). Many less dominant cells are brought into
culture along with the autotrophs, as dependent satellites. An example is
Hyphomicrobium which accompanies both ammonia and methane oxidizing bacteria and is able to use their by-products, nitrate and methanol, respectively (Attwood and Harder, 1972). The sporadically occurring populations (Sieburth, 1971) of larger, easy to grow, non-distinctive bacteria
(Sieburth, 1979) are possibly epibacteria associated with aggregated
seston.

The large population of bacteria we can see in seawater with the
microscope is very poorly cultivatable on conventional media (Jannasch and
Jones, 1959) resulting in a two to five order of magnitude discrepancy for
nearshore and offshore waters respectively. This enigma, when coupled with

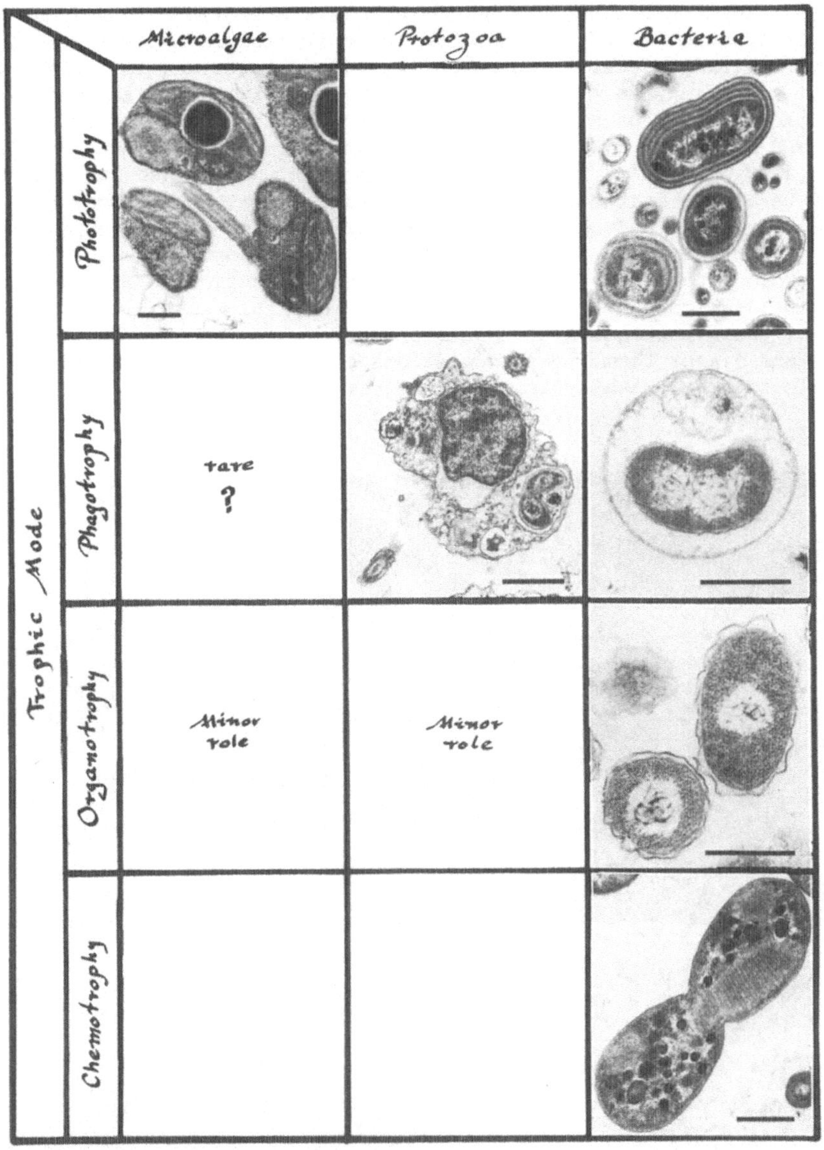

Fig. 1. A trophic mode-microorganism box that compares the great diversity of bacterial trophic forms with that of the microalgae and
protozoa. In the microalgae, the dominant flagellates are illustrated by *Micromonas pusilla*. Note that organotrophy is rare,
chemotrophy is absent and phagotrophy has been inadequately
observed. In the protozoa the dominant bacterivorous flagellates
are illustrated by a species of *Bodo* isolated from seawater. Note
that organotrophy is also minor and that phototrophy and chemotrophy are absent. The bacteria have representatives in each
trophic mode. The photographs are illustrated by the cyanobacterium *Synechococcus*, the phagotrophs (really lysotrophs) are
illustrated by the parasitic bacterium *Bdellovibrio* in its host
bacterium from Caribbean Sea water, the organotrophs show the
typical gram negative cell ultrastructure also seen in Fig. 2,
and chemotrophs are illustrated by the same methane oxidizing
bacterium, *Methylomonas*, from the Sargasso Sea, that was used to
obtain the data in Fig. 5. Marker bars, 0.5 μm.

an a *priori* assumption that all bacterial cells are organotrophs, has led
to much frustration in trying to prove the existence of hypothetical low-
nutrient requiring bacteria (Jannasch, 1968) and in trying to isolate them
(Carlucci and Shimp, 1974). Today we are beginning to realize that organotrophic bacteria may be multiphasic (Azam and Hodson, 1981) or eurytrophic
(Baxter and Sieburth, 1984) being able to utilize substrates ranging from
10 μg to 3000 mg C 1^{-1}. The discrepancy between what we see and what we
grow may be due to inattention to trophic mode (i.e., phototrophs and
chemotrophs) and to unique substrate requirements such as C_1 compounds
(Hanson, 1980). A small but potentially important population is that of
the lysotrophic bacteria. They consist of the much studied *Bdellovibrio*
(Stolp, 1981) which penetrate host cells and multiply within them, and
members of the cytophagales such as *Saprospira* which coil around their prey
bacteria and digest them. The populations of these lytic bacteria and
their role together with lytic bacterial viruses (bacteriophage), is
unknown.

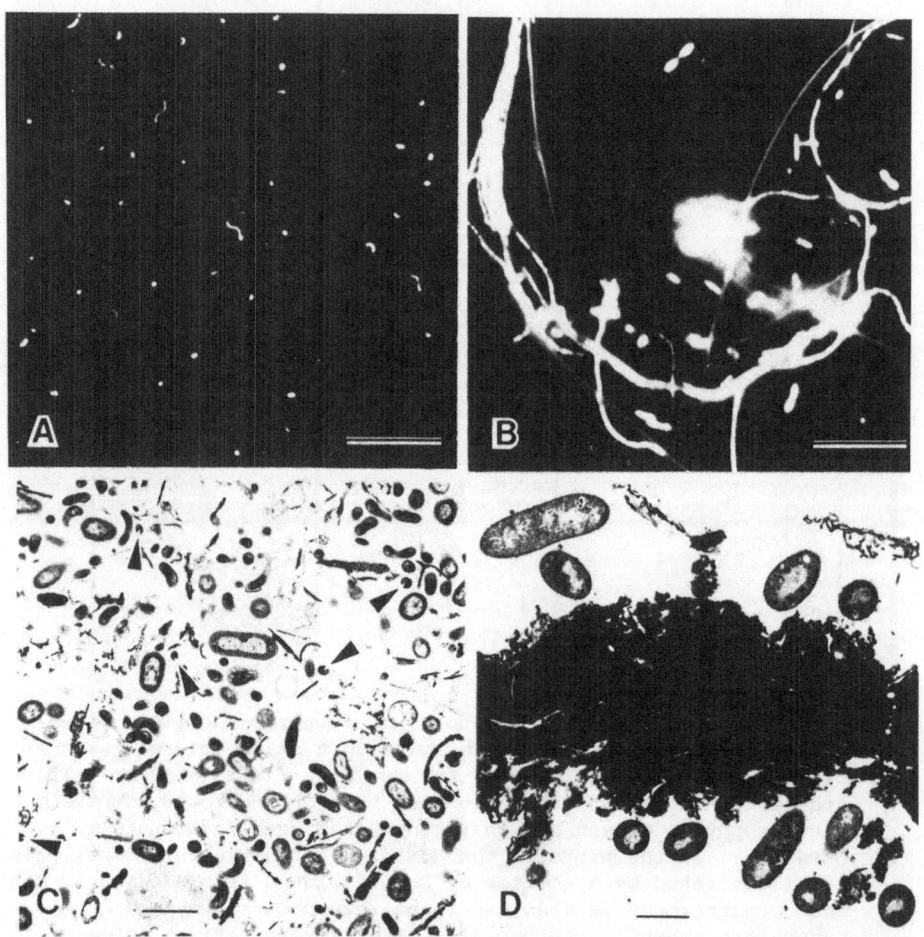

Fig. 2. Morphology and ultrastructure of free (A, C) and particle associated (B, D) organotrophic bacteria representative of those in the
mixing layer. The cells free in the water (A, C) are much smaller
and poorer in ribosomes than those associated with marine snow (B,
D) which are larger and ribosome rich. The micrographs in A and B
are DAPI stained cells viewed with epifluorescence microscopy while
C and D are thin sections, as seen with transmission electron
microscopy. Marker bars A & B, 10 μm; C & D, 1 μm.

SPATIAL DISTRIBUTION

    The spatial distribution of the smaller more dominant micro-organisms
is complicated by the fact that by evolving together they have learned to
live together in tight associations. This is due in large measure to their
small size and the highly viscous nature of water at low Reynolds numbers
which impedes their mobility (Purcell, 1977; Vogel, 1981). The major difference in habitat is between those microorganisms that live in "planktonic
constellations" and those that live as epibionts on "sestonic aggregations". This is shown for bacteria in Fig. 2. The free cells in the
plankton are shown with epifluorescence microscopy in A and transmission
electron microscopy (TEM) in C. These can be seen to be small cells while
those associated with fragments of seston are seen to be much larger in
both epifluorescence (B) and in TEM (D). The differences in state of
nutrition (Maaløe and Kjeldgaard, 1966) and rate of sedimentation (McCave,
1975) for the microorganisms in the two habitats is very marked. The
microhabitat, therefore, influences both growth rates and spatial distribution. A third microbial habitat is as ecto- and endo-symbionts associated
with the larger microorganisms visible to the naked eye. These "microbial
consortia" are like miniature coral reefs or "oases" intermingled with
aggregations of marine snow against a sub-visible background of the more
numerous free cells in the planktonic constellations.

    A conceptual model of how the microorganisms in these three micro-
habitats may be distributed spatially is shown in Fig. 3. The microflocs
of marine snow coalesce due to their downward sedimentation against the
force of vertically rising water in Langmuir cells to form macroflocs of
marine snow which are fragile and easily fragment when disturbed (Kranck
and Milligan, 1980). These macroflocs appear to concentrate as clouds in
an approximately 30 m band across the thermocline as viewed from the Harbor
Branch Foundation's manned submersible Sea-Link (Marsh Youngbluth, personal
communication) and shown in Fig. 3. Above the thermocline, the water
column is occupied by microscopic constellations of the microbial plankton

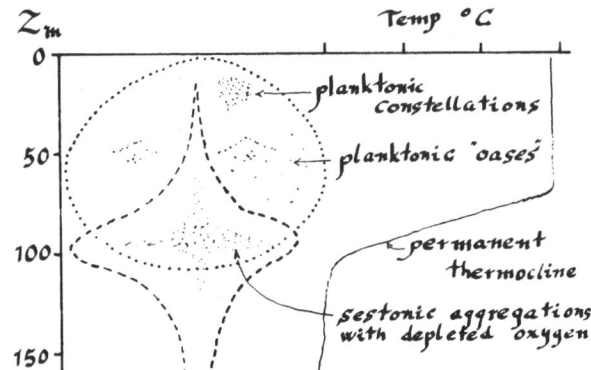

Fig. 3. Assumed distribution of the three major
        microbial assemblages in the upper ocean.
        These assemblages, the "planktonic con-
        stellations", the microbial associations
        which function as "reefs" or "oases", and
        the "sestonic aggregations", are described
        in Tables 1-3. All three co-exist in the
        mixing layer but the latter concentrate as
        clouds of marine snow in the thermocline
        which may form anaerobic microsites that
        function as a false benthos.

Table 1. Planktonic constellations occurring in one microlitre of seawater

Trophic compartment	Number per μl	Biovolume (μm^3) per cell	per μl^{-1}
Primary producers			
phototrophic flagellates	0.4	24.0	9.6
phototrophic bacteria	5.0	0.2	12 { 1.0
chemotrophic bacteria	7.0	0.2	1.4
			} 23
Organotrophic bacterial satellites	520.0	0.04	20.8
Bacterivorous flagellates	0.5	24.0	10

that are repeated over and over again, while much more infrequently, associations of large host cells containing symbionts are visible to the naked eye. Let us consider the nature and spatial distribution of these three microbial habitats.

The multitrophic constellation that occurs in 1 μl of Caribbean seawater is shown in Table 1. By multiplying these values by two to yield whole numbers, each 2 μl is seen to contain 1 phototrophic flagellate, 10 phototrophic bacteria, 14 chemotrophic bacteria, 1040 organotrophic bacteria and one bacterivorous flagellate. Therefore, if evenly distributed, all these trophic forms could be contained in just one five hundreth of a ml of seawater. But biovolume relationships are probably more important Some 12 μm^3 μl^{-1} of autotrophs appear to support some 21 μm^3 μl^{-1} of organotrophs.

This apparent paradox of an imbalance between standing stocks of autotrophs and organotrophs may have several causes. First, the autotrophs may be growing faster than the organotrophs, e.g., 2.5 h *versus* 8 h generation times, respectively. Secondly, autotrophs produce exocellular substances that feed the organotrophs without the cells themselves being consumed. This feeding on "soluble" or colloidal substances where the by-product of one bacterial process is consumed in a subsequent bacterial process occurs between different trophic forms of bacteria, and explains their success as well as the difficulty in assessing their trophic contribution. A third explanation for the paradox is that the feeding efficiency at this trophic level approaches 60 to 70% (Calow, 1977). Where the bacteria are presumably being consumed by phagotrophic flagellates, some 2 μm^3 μl^{-1} of bacteria appear to support 10 μm^3 μl^{-1} of bacterivorous flagellates. Assuming approximately equal growth rates, this would indicate a 43% feeding efficiency, far above the 10% values (Slobodkin, 1962) expected at the higher trophic levels of macroorganisms.

Host microorganisms and their endo- and ecto-symbionts which form microbial associations are shown in Table 2. These visible assemblages occur much less frequently than the planktonic constellations but must be important to the specialized harpacticoid copepods that prey upon them. Although the populations and biomass of the zooplankters are very small compared with the smaller microbial plankton, they are voracious grazers and filterers and are very important in consuming the microbial plankton and in forming seston.

Table 2. Microbial associations serving as planktonic "oases"

Host microorganism	Associated microorganisms	
	Endosymbionts	Ectosymbionts
Phagotrophs		
Zoothamnium	—	organotrophic bacteria
colonial radiolaria	zooxanthellae	—
foraminifera	zooxanthellae	—
Phototrophs		
Rhizosolenia		
solitary	N$_2$—fixing cyanobacteria	—
matts	N$_2$—fixing organo—trophic bacteria	—
Trichodesmium	—	dinoflagellates organotrophic bacteria bacterivorous ciliates and flagellates

Fascinating microbiology and chemistry appears to be associated with the aggregations of seston known as marine snow. The possible nature or cause of some of these flocs and slimes is shown in Table 3. These micro-sites are apparently heavily colonized (Caron et al., 1982) and may be ser-iously depleted in oxygen by the facultative organotrophs. They could be the site of methanogenesis, since methane concentrations peak at the thermocline (Scranton and Brewer, 1977). The very recent discovery that methanogens can use molecular sulfur in the presence of oxygen to form H$_2$S and, therefore, their own anoxic microenvironment (Stettar and Gaag, 1983) may explain why methanogenesis can occur in an otherwise oxygen saturated environment. Whatever the mechanism, methanogenesis appears to be suffic-ient to support an enrichable population of methane oxidizing bacteria in the Sargasso Sea (Sieburth et al., 1983). Other chemotrophic bacteria which appear to be present in conspicuous numbers in thin sections of pelleted picoplankton are bacteria with an ultrastructure very suggestive of ammonia and methane oxidizers (see Type II and III cells, respectively, in Johnson and Sieburth, 1979). Although such nitrifiers are somewhat evenly distributed throughout the mixing layer (Ward and Perry, 1980), the occurrence of nitrite maxima near 50 m has led to speculation con-cerning selective photosensitivity between ammonia and nitrite oxidizing bacteria (Olson, 1981a,b).

PHOTOREGULATION OF BIOCHEMICAL PROCESSES

On the basis of published ^{14}C primary productivity measurements of nearshore and offshore waters, Ryther (1969) characterized the open ocean as a desert. From a bacteriological point of view, both culture estimates (Sieburth, 1971) and glucose uptake kinetics (Vaccaro et al., 1968) would tend to confirm this. But perhaps these accepted procedures for nearshore waters are ill suited for detecting in oligotrophic oceans, either viable organotrophic bacterial populations or their activities. Alternative procedures are to use ATP determinations (Holm-Hansen and Booth, 1966) of the <3 μm particles as an estimate of biomass and to measure variations in

Table 3.   Sestonic aggregations which function like sewage flocs
----

Amorphous marine snow:
    fragile detritus with bacteria, flagellates and ciliates

Gelatinous zooplankton feeding nets:
    fecal—pellets, cyanobacteria, sulfate reducing bacteria, flagellates
    and ciliates

Fecal pellets:
    organotrophic bacteria, cyanobacteria, flagellates and ciliates

Bacterial aggregations:
    cysts and zoogloea of methane and ammonia oxidizing bacteria
----

major and labile components of the DOM pool such as total monosaccharides
(Johnson and Sieburth, 1977; Johnson et al., 1981a), or total carbohydrate
and polysaccharide by difference (Burney and Sieburth, 1977) as an index of
bacterial activity.

    Such procedures, which avoid bottle incubations and determine temporal
changes that occur in the water itself, were used by my laboratory during a
crossing of the North Atlantic (R.V. Trident cruise 170-1975) in which
vertical profiles were obtained at dusk and the following dawn. Negligible
^{14}C—glucose uptake rates obtained on the same samples by a Richard Wright
protégé, Peter Yorgey, would appear to confirm Ryther's conclusions.
However, high carbohydrate values at dusk were consistently followed by
lower dawn values and by significant changes in the <3 μm particulate ATP
which accounted for 30-40% of total biomass (Burney et al., 1979). The
high apparent rates of productivity suggested by these observations
(Sieburth et al., 1977) prompted Sieburth (1977) to question the ^{14}C data
base upon which Ryther made his calculations. Measuring temporal differences in chemical constituents such as concentration of organic matter, CO_2
and O_2, as well as the populations of the major microorganisms that occur
in bulk water (Burney et al., 1979, 1981, 1982; Johnson et al., 1981b,
1983) as an alternative to ^{14}C bottle assays (Peterson, 1980) is not without problems. But the problems of sampling the same water mass may not
be any worse than the multitude of problems associated with the deck incub-
ation of ^{14}C bottle assays (Wright, 1973; Venrick et al., 1977).

    Ferguson and Rublee (1976) observed a temporal change in total bact-
erial population off Sandy Hook, New Jersey, but due to problems in sampling different water masses, did not consider the results unequivocal.
Our carbohydrate studies show a dominant pattern of a diurnal increase in
carbohydrate roughly accompanied by a decrease in total bacteria and a
nocturnal decrease in carbohydrate roughly accompanied by an increase in
the bacterial population (Burney et al., 1979, 1982). The enigmatic
differences we have found between TCO_2 and O_2 (Johnson et al., 1981b)
prompted us to look more closely at the possible microbiological causes
(Johnson et al., 1983). At four Caribbean stations away from the influence
of coral reefs, the total bacterial numbers peaked at 09.00 and underwent a
marked reduction or inhibition during the rest of the photoperiod. One
might expect ultra—violet destruction of nucleic acids in the bacteria of
the neuston such as nitrifying bacteria (Horrigan et al., 1981) but a more
definite example of bacterial photoinhibition at depth is more difficult to
demonstrate.

    Ward et al. (1982) conducted a much needed comparison of the vertical
distribution of nitrifying bacteria and the rates of nitrification with

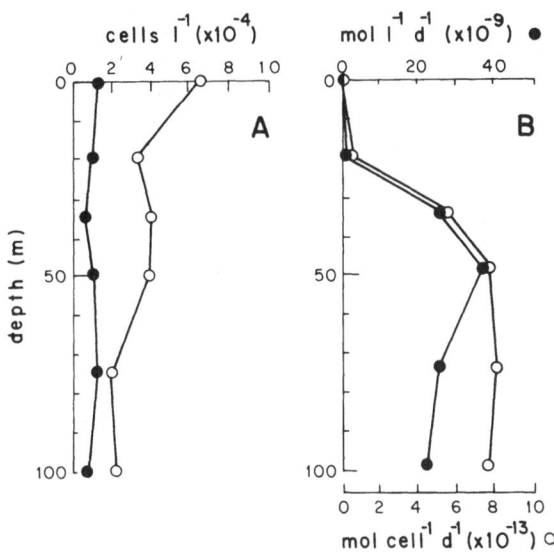

Fig. 4. (A) Abundance of ammonium-oxidizing
bacteria (O, *Nitrosomonas marinus*; ●,
Nitrosococcus oceanus), and (B)
^{15}N-ammonium oxidation rates (O,
rates per bacterial cell; ●, total
rates) in California coastal waters,
showing approximately constant
populations with depth, but activity
only below the photic zone (adapted
from Ward et al., 1982).

depth. The distribution of two species of nitrifying bacteria as determined by fluorescent antibody (FA) was somewhat constant in the upper 100 m
(Fig. 4A), but the rate of ^{15}N nitrification expressed either totally or
per cell was markedly inhibited above a depth of 50 m (Fig. 4B). One
might infer that such an inhibition to depths of 30 m might be due to the
penetration of near ultra-violet light to such depths (Smith and Baker,
1979). Temporal effects of light may give a more convincing explanation.

    In seeking the cause of the greater diel variation that we have obs-
erved in TCO$_2$ than in O$_2$, we started to explore the possibility of methane cycling (Sieburth, 1983; Johnson et al., 1983). The enrichment and
isolation of methane oxidizing bacteria from the Sargasso Sea has given us
the first cultures of methanotrophs from upper ocean waters. These have
permitted us to study photoinhibition with another group of chemotrophs.
As can be seen in Fig. 5, the dark oxidation of methane is quite marked,
while full sunlight virtually stops the process. One important feature of
these data is that the end-point of this inhibition occurs between 0.7 and
7.0% light, intensities that occur between 30 and 50 m. This was also the
depth to which nitrification inhibition occurred (Fig. 4). Another similarity in the two data sets is that the number of bacterial cells is a
poor index of microbiological activity.

    Can solar radiation penetrating to depths between 30 and 50 m really
cause this great a photoinhibition? The photooxidative changes discussed
by Zafiriou (this volume) would have little influence past the upper metre.
Smith and Baker (1979) among others point out that near ultra-violet penetration can approach depths of 30 to 50 m in oceanic seawater. But is

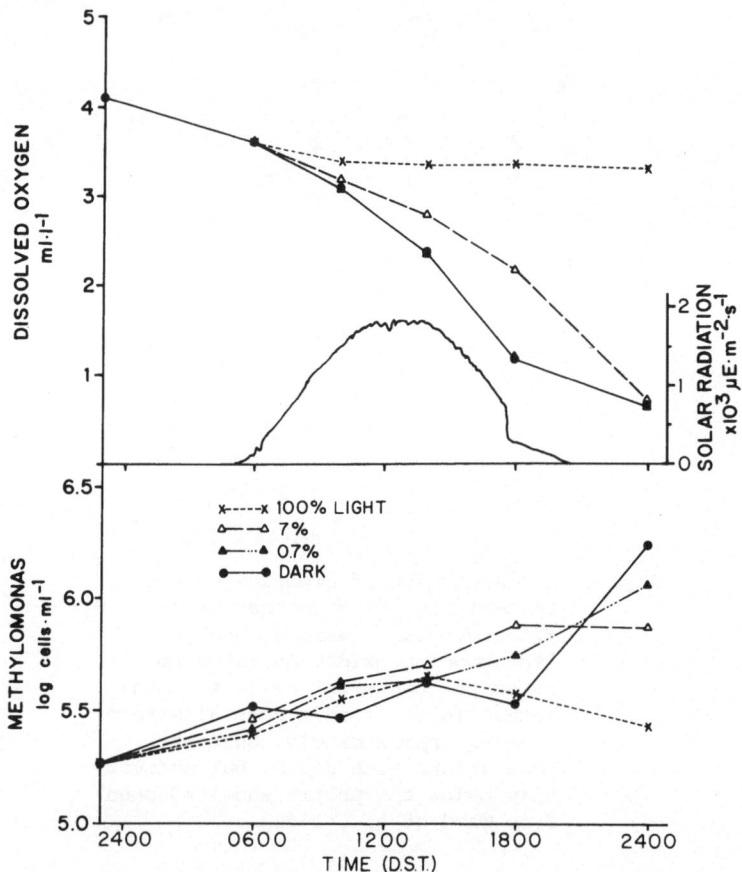

Fig. 5. The influence of light of differing intensities on
the rate of methane oxidation (as shown by oxygen
consumption) and on population changes in a
Sargasso Sea isolate of *Methylomonas*. Note that
the end point of photoinhibition is between 0.7
and 7.0% light intensity.

there really an effect of near ultra—violet or ultra-violet-B on bacterial
RNA and DNA? In our methanotrophs we can see apparent nucleic acid damage
in 4',6-diamidino-2-phenylindole stained bacteria from the 100% light
bottles, which pass >400 nm wavelengths, but not at the 7% light level.
The answer again may be coming from the ammonia oxidizers. Hooper and
Terry (1974) found that the apparent light sensitivity of *Nitrosomonas* was
due to the photodenaturation of the enzymes responsible for the oxidation
of hydroxylamine, which had an absorption shoulder at 420 nm, but not the
enzyme responsible for the oxidation of ammonia to hydroxylamine, which
lacked this shoulder.

For a long time we have known that the Gulf Stream waters are imp-
overished in regard to the adjacent slope and Sargasso Sea waters. In
this regard, it is curious that the light sensitive methane oxidizer,
Methylomonas, was obtained from 72% of the enrichments from the Sargasso
Sea (n = 32) but not from the Gulf Stream (n = 72) (Sieburth et al., 1984).
Could it be that at this time of year (February-March) the deep winter
thermocline above which methane oxidation must occur was below 100 m while
that for the Gulf Stream is usually <30 m and might preclude the survival

of light sensitive bacteria in the mixing layer? An understanding of upper
ocean chemistry must go hand in hand with an understanding of both the
microbiological processes and the physical oceanographic processes that
control them.

REFERENCES

Attwood, M.M., and Harder, W., 1972, A rapid and specific enrichment
        procedure for Hypomicrobium spp. Antonie von Leeuwenhoek,
        J. Microbiol. Serol., 38:369.

Azam, F., and Hodson, R.E., 1981, Multiphasic kinetics for D-glucose
        uptake for assemblages of natural marine bacteria, Mar. Ecol.
        Progr. Ser., 6:213.

Baxter, M., and Sieburth, J.McN., 1984, Metabolic and ultrastructural
        response to glucose of two eurytrophic bacteria isolated from
        seawater at different enriching concentrations, Appl. Environ.
        Microbiol., 47:31.

Burney, C.M., and Sieburth, J.McN., 1977, Dissolved carbohydrates in
        seawater. II. A spectrophotometric procedure for total carbo-
        hydrate analysis and polysaccharide estimation, Mar. Chem., 5:15.

Burney, C.M., Johnson, K.M., Lavoie, D.M., and Sieburth, J.McN., 1979,
        Dissolved carbohydrate and microbial ATP in the North Atlantic:
        concentrations and interactions, Deep-Sea Res., 26:1267.

Burney, C.M., Davis, P.G., Johnson, K.M., and Sieburth, J.McN., 1981,
        Dependence of dissolved carbohydrate concentrations upon small
        scale nanoplankton and bacterioplankton distributions in the
        Western Sargasso Sea, Mar. Biol., 65:289.

Burney, C.M., Davis, P.G., Johnson, K.M., and Sieburth, J.McN., 1982,
        Diel relationships of microbial trophic groups and in situ
        dissolved carbohydrate dynamics in the Caribbean Sea, Mar. Biol.,
        67:311.

Calow, P., 1977, Conversion efficiencies in heterotrophic organisms,
        Biol. Rev., 52:385.

Campbell, L., Carpenter, E.J., and Iacono, V.J., 1983, Identification and
        enumeration of marine chroococcoid cyanobacteria by immunofluor-
        escence, Appl. Environ. Microbiol., 46:553.

Carlucci, A.F., and Shimp, S.L., 1974, Isolation and growth of a marine
        bacterium in low concentrations of substrate, in: "Effect of the
        Ocean Environment on Marine Bacteria", R.R. Colwell and R.Y.
        Morita, eds, pp. 363-367, Baltimore University Press.

Caron, D.A., 1983, Technique for enumeration of heterotrophic and photo-
        trophic nanoplankton, using epifluorescence microscopy, and
        comparison with other procedures, Appl. Environ. Microbiol.,
        46:491.

Caron, D.A., Davis, P.G., Madin, L.P., and Sieburth, J.McN., 1982,
        Heterotrophic bacteria and bacterivorous protozoa in oceanic
        macroaggregates, Science, N.Y., 218:795.

Davis, P.G., and Sieburth, J.McN., 1982, Differentiation of the phototrophic and heterotrophic populations of nanoplankton populations
in marine waters by epifluorescence microscopy, Ann. Inst.
Oceanogr., Paris, suppl., 58:249.

Ferguson, R.L., and Rublee, P., 1976, Contribution of bacteria to standing
crop of coastal plankton, Limnol Oceanogr., 21:141.

Hanson, R.S., 1980, Ecology and diversity of methylotrophic organisms,
Adv. Appl. Microbiol., 26:3.

Holm-Hansen, O., and Booth, C.R., 1966, The measurement of adenosine
triphosphate in the ocean and its ecological significance,
Limnol. Oceanogr., 11:510.

Hooper, A.B., and Terry, K.T., 1974, Photoinactivation of ammonia
oxidation in Nitrosomonas, J. Bacteriol., 119:899.

Horrigan, S.G., Carlucci, A.F., and Williams, P.M., 1981, Light inhibition of nitrification in sea-surface films, J. Mar.Res. 39:557.

Jannasch, H.W., 1968, Growth characteristics of heterotrophic bacteria in
seawater, J. Bacteriol., 95:722.

Jannasch, H.W., and Jones, G.E., 1959, Bacterial populations in sea
water as determined by different methods of enumeration,
Limnol. Oceanogr., 4:128.

Johnson, K.M., and Sieburth, J.McN., 1977, Dissolved carbohydrates in
seawater. I. A precise spectrophotometric analysis for monosaccharides, Mar. Chem., 5:1.

Johnson, K.M., Burney, C.M., and Sieburth, J.McN., 1981a, Doubling the
production and precision of the MBTH spectrophotometric assay for
dissolved carbohydrates in seawater, Mar. Chem., 10:467.

Johnson, K.M., Burney, C.M., and Sieburth, J.McN., 1981b, Enigmatic
marine ecosystem metabolism measured by direct diel ΣCO_2 and O_2
flux in conjunction with DOC release and uptake, Mar. Biol.,
65:49.

Johnson, K.M., Davis, P.G., and Sieburth, J.McN., 1983, Diel variation of
TCO_2 in the upper layer of oceanic waters reflects microbial
composition, variation and possibly methane cycling, Mar. Biol.,
74:1.

Johnson, P.W., and Sieburth, J.McN., 1979, Chroococcoid cyanobacteria in
the sea: A ubiquitous and diverse phototrophic biomass, Limnol.
Oceanogr., 24:928.

Johnson, P.W., and Sieburth, J.McN., 1982, In-situ morphology and
occurrence of eucaryotic phototrophs of bacterial size in the
picoplankton of estuarine and oceanic waters, J. Phycol., 18:318.

Kranck, K., and Milligan, T., 1980, Macroflocs: Production of marine snow
in the laboratory, Mar. Ecol. Progr. Ser., 3:19.

Li, W.K.W., Subba Rao, D.V., Harrison, W.G., Smith, J.C., Cullen, J.J.,
Irwin, B., and Platt, T., 1983, Autotrophic picoplankton in the
tropical ocean, Science, N.Y., 219:292.

Maaløe, O., and Kjeldgaard, N.O., 1966, "Control of Macromolecular Synthesis", Benjamin, New York.

McCave, I.M., 1975, Vertical flux of particles in the ocean, Deep-Sea Res., 22:491.

Olson, R.J., 1981a, ^{15}N tracer studies of the primary nitrite maximum, J. Mar. Res., 39:203.

Olson, R.J., 1981b, Differential photoinhibition of marine nitrifying bacteria: a possible mechanism for the formation of the primary nitrite maximum, J. Mar. Res., 39:227.

Peterson, B.J., 1980, Aquatic primary productivity and the ^{14}C-CO_2 method: a history of the productivity problem, Ann. Rev. Ecol. Systematics, 11:359.

Platt, T., Subba Rao, D.V., and Irwin, B., 1983, Photosynthesis of picoplankton in the oligotrophic ocean, Nature, Lond., 301:701.

Purcell, E.M., 1977, Life at low Reynolds number, Amer. J. Physics., 45:242.

Ryther, J.G., 1969, Photosynthesis and fish production in the sea, Science, N.Y., 166:72.

Scranton, M.I., and Brewer, P.G., 1977, Occurrence of methane in the near surface waters of the western subtropical North-Atlantic, Deep-Sea Res., 24:127.

Sieburth, J.McN., 1971, Distribution and activity of oceanic bacteria, Deep-Sea Res., 18:1111.

Sieburth, J.McN., 1977, International Helgoland Symposium: Convener's report on the informal session on biomass and productivity of microorganisms in plankton ecosystems, Helgolander Wiss. Meeresunters., 30:697.

Sieburth, J.McN., 1979, "Sea Microbes", Oxford University Press, New York.

Sieburth, J.McN., 1983, Microbiological and organic-chemical processes in the surface and mixed layers, in: "Air-Sea Exchange of Gases and Particles", P.S. Liss and W.G.N. Slinn, eds, pp. 121-172, Reidel, Dordrecht.

Sieburth, J.McN., Johnson, K.M., Burney, C.M., and Lavoie, D.M., 1977, Estimation of in situ rates of heterotrophy using diurnal changes in dissolved organic matter and growth rates of picoplankton in diffusion culture, Helgolander Wiss. Meeresunters., 30:565.

Sieburth, J.McN., Smetacek, V., and Lenz, J., 1978, Pelagic ecosystem structure: heterotrophic compartments of the plankton and their relationship to plankton size fractions, Limnol Oceanogr., 23:1256.

Sieburth, J.McN., Johnson, P.W., Eberhardt, M.A., and Sieracki, M.E., 1984, Methane-oxidizing bacteria from the mixing layer of the Sargasso Sea and their photosensitivity, Abstract, Trans. Amer. Geophys. Un., 64:1054.

Slobodkin, L.B., 1962, "Growth and Regulation of Animal Populations", Holt, Rinehart and Winston, New York.

Smith, R.C., and Baker, D.S., 1979, Penetration of UV-B and biologically effective dose-rates in natural waters, Photochem. Photobiol., 29:311.

Stetter, K.O., and Gaag, G., 1983, Reduction of molecular sulphur by methanogenic bacteria, Nature, Lond., 305:309.

Stolp, H., 1981, The genus Bdellovibrio, in: "The Prokaryotes", M.P. Starr, H. Stolp, H.G. Trüper, A. Balows and H.G. Schlegel, eds, pp. 618-629, Springer-Verlag, New York.

Vaccaro, R.F., Hicks, S., Jannasch, H.W., and Carey, F.G., 1968, The occurrence and role of glucose in seawater, Limnol. Oceanogr., 13:356.

Venrick, E.L., Beers, J.R., and Heinbokel, J.F., 1977, Possible consequences of containing microplankton for physiological rate measurements, J. Exp. Mar. Biol. Ecol., 36:55.

Vogel, S., 1981, "Life in Moving Fluids", Willard Grant Press, Boston.

Ward, B.B., and Perry, M.J., 1980, Immunofluorescent assay for the marine ammonium-oxidizing bacterium Nitrosococcus oceanus, Appl. Environ. Microbiol., 39:913.

Ward, B.B., Olson, R.J., and Perry, M.J., 1982, Microbial nitrification rates in the primary nitrite maximum off southern California, Deep-Sea Res., 29:247.

Waterbury, J.B., Watson, S.W., Guillard, R.R.L., and Brand, L.E., 1979, Widespread occurrence of a unicellular, marine, planktonic cyanobacterium, Nature, Lond., 277:293.

Watson, S.W., Valois, F.W., and Waterbury, J.B., 1981, The family Nitrobacteraceae, in: "The Prokaryotes", M.P. Starr, H. Stolp, H. G. Trüper, A. Balows and H.G. Schlegel, eds, pp. 1003-1022, Springer-Verlag, New York.

Whittenbury, R., and Dalton, H., 1981, The methylotrophic bacteria, in: "The Prokaryotes", M.P. Starr, H. Stolp, H.G. Trüper, A. Balows and H.G. Schlegel, eds, pp. 894-902, Springer-Verlag, New York.

Wright, R.T., 1973, Some difficulties in using ^{14}C-organic solutes to measure heterotrophic bacterial activity, in: "Estuarine Microbial Ecology", H. Stevenson and R.R. Colwell, eds, pp. 199-217, University of South Carolina Press, Columbia.

ACKNOWLEDGEMENTS

I would like to acknowledge David A. Caron and Paul G. Davis for their observations on marine snow and Michael E. Sieracki who provided the data for Figure 5. This work was supported by the Biological Oceanography Program of the National Science Foundation through grant No. OCE-8121881.

SHORT TERM VARIATIONS IN PRIMARY PRODUCTIVITY

Richard E. Eppley

Institute of Marine Resources, A-018
Scripps Institution of Oceanography
University of California, San Diego
La Jolla, California 92093 U.S.A.

ABSTRACT

   This review briefly considers two topics in primary production. In
both cases time scales appear to be short (hours to days) and both have
significance for ocean chemistry and physics. The first concerns implications, for studies of mixing in the surface layer, of the fact that phytoplankton "remember" their past light history. Here time scales of minutes
to hours are important. The second concerns primary production as the
driving force for the sinking flux of biogenic organic particles, and significant time and space scales of variability of the production of sinking
particles in the surface layer. The minimum time scale of interest here
appears to be 12-24 hours.

EFFECTS OF LIGHT PREHISTORY ON PHYTOPLANKTON PHOTOSYNTHESIS; OR, CAN
PHOTOSYNTHESIS *VERSUS* LIGHT CURVES PROVIDE A TIME SCALE FOR VERTICAL
MIXING?

   Steemann Nielsen and Hansen (1959) took water samples from different
depths in the euphotic zone in the North Atlantic, spiked them with ^{14}C-
labelled bicarbonate, and incubated subsamples at each of several light
intensities. After collecting the particulate matter on filters and
determining its radiocarbon activity these authors were able to express the
rate of photosynthesis (P) as a function of light intensity (I): to construct P *versus* I, or PI, curves. The P *versus* I curves varied over the
sampling depth if the water was stratified but were essentially identical
if the water was well mixed (isothermal).

   Similar observations have been made by many others since then in both
oceans and lakes. Further, laboratory cultures grown under different light
intensity have different P *versus* I curves, and by 1960 there was already
an extensive literature on the P *versus* I response of terrestrial plants
and *Chlorella* cultures. Distinct sun and shade species of higher plants
exist. But most phytoplankton species seem very plastic and able to adjust
their response to the ambient illumination, i.e., to "photoadapt".

   Examples of the P *versus* I curves are shown in Fig. 1. The characteristic features of these curves can be described by: (1) the initial slope,

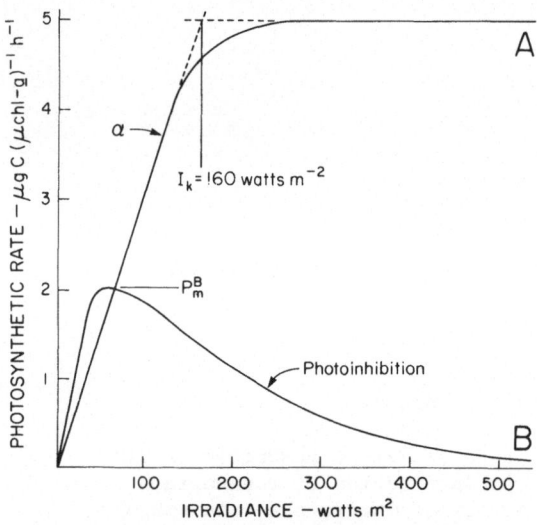

Fig. 1. Hypothetical P *versus* I curves for
near surface phytoplankton (A)
compared with phytoplankton from
near the bottom of the euphotic
zone (B). Noteworthy differences
between the two curves are in P_m^B,
the maximum rate; the initial
slopes of the lines; and I_k, a
measure of the saturating irradiance for photo—synthesis.
Photoinhibition is apparent only
in the deep sample. Based upon
stratified, sub-tropical waters
off Southern California.

α, where photosynthetic rate is a linear function of irradiance; (2) the
maximum observed rate, P_m^B ; (3) the ratio P_m^B /α, a measure of light saturation termed I_K (Talling, 1957); (4) the irradiance at P_m^B, termed I_m;
(5) some measure of the irradiance at the onset of any decline in rate at
supra-optimal irradiance, if such a decline is observed (Steele, 1962;
Jasby and Platt, 1976); and (6) the slope of the decline, a measure of
photoinhibition (Platt et al., 1980). When photosynthesis is measured as
oxygen exchange there will be a consumption of oxygen, R, in the dark due
to cellular respiration. At some (low) irradiance, called $I_{compensation}$,
the respiratory oxygen consumption and photosynthetic oxygen production
will be equal.

The rates are usually normalized to some measure of photosynthetic
biomass, B, such as chlorophyll a, giving rise to expressions such as P_m^B
(Jassby and Platt, 1976).

Phytoplankton grown at low irradiance differ from the same species
grown at higher irradiance in several respects. Both P_m^B and I_K are less
in the low-light plankton than in the high-light plankton; I_m, α and the
irradiances at the onset of photoinhibition are often measurably less as
well (Fig. 1). In addition, with laboratory cultures the low light cells
contain more chlorophyll per unit biomass than the high light cells. In
the literature on higher plants such changes are referred to as sun-shade
adaptation (Yentsch and Lee, 1966).

Differences in the parameters of PI curves are seen over the seasons as well as over depth in stratified surface water (Steemann Nielsen, 1975). The maximum photosynthetic rate, P_m^B, is thought to represent the enzymatic dark reactions of photosynthesis and is thus a function of the ambient temperature. On the other hand the initial slope, α, reflects photochemical reactions and would be expected to be less dependent on temperature. In actual field data α and P_m^B are often correlated, perhaps because it is difficult to measure with high precision (Platt and Jassby, 1976). To use P *versus* I curves in the study of mixing processes requires knowledge of the time rate of change of P *versus* I curve parameters with changing ambient illumination and temperature. Originally, the sun-shade adaptation was studied by aquatic biologists with respect to the depth variability of photosynthesis and the problem of integrating photosynthesis over depth in order to express productivity per unit area of surface. Nevertheless, there were a few measurements in the early 1960s of the time rate of change of P *versus* I curves, i.e., of the rate of photoadaptation (Steemann Nielsen et al., 1962; Steemann Nielsen and Park, 1964; Brooks, 1964). The idea was already extant that P *versus* I curves might be used to study mixing in a quantitative way. But there were more important matters to attend to and the rebirth of interest in this prospect for ocean waters is fairly recent.

Several kinds of study have supported this rekindling of interest: (1) investigations of the changes in structure of the photosynthetic machinery in phytoplankton during sun-shade adaptation (Falkowski, 1980, 1981; Perry et al., 1981; Prezelin, 1981, and references therein); (2) new mathematical descriptions of the photosynthetic light curve, with attention to the statistical variability of its parameters and the ranges of parameter variability (Platt and Jassby, 1976; Platt et al., 1980); (3) the discovery in sev-

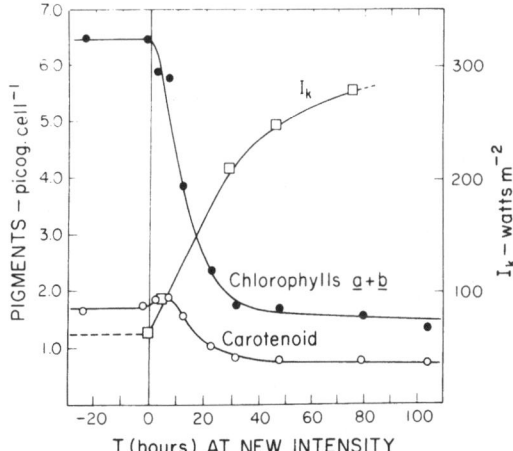

Fig. 2. Time course of change in pigment content and the saturating irradiance for photosynthesis, I_k, of *Dunaliella tertiolecta*. At time zero the culture was transferred from an irradiance of 76 W m^{-2} to 250 W m^{-2} at 20°C. Photosynthetic P *versus* I curves were determined at each sampling point, using an oxygen electrode mounted in a special reaction vessel, to evaluate I_K. (From Brooks, 1964.)

eral laboratories that the *in vivo* fluorescence of chlorophyll in plankton
provides information on previous light exposure and hence on mixing
(Harris, 1978); and (4) the continuing awareness that mixing is important
for photosynthesis in the sea (Marra, 1980).

Changes in P *versus* I curve parameters on changing light exposure of
the phytoplankton appear to follow first-order kinetics (Falkowski, 1980)
where examined, as do changes in chlorophyll fluorescence (Neale et al.,
1982). Early studies of the P *versus* I curve parameter, I_K (a measure of
the irradiance necessary to saturate, just, the rate of photosynthesis)
suggested a time scale for photoadaptation of several hours (Steemann
Nielsen et al., 1962; Steemann Nielsen and Park, 1964; Brooks, 1964).

Figure 2 shows changes in I_K of the green flagellate, *Dunaliella
tertiolecta,* on shifting the culture flask from a low irradiance to a
higher one at 20°C (Brooks, 1964). Also shown are changes in the chlorophyll and carotenoid pigments on a per cell basis. The time course suggests smooth transition over time. At least sometimes, the time courses
imply complex transients and steady values of the P *versus* I parameters are
not reached for several days; this was found by Prezelin and Matlick (1980)
with the dinoflagellate, *Glenodinium* sp.

An exciting development for studies of mixing in the euphotic zone is
that the several parameters of the P *versus* I curve, such as its initial
slope (α), maximum rate (p_m^β), and irradiance at the onset of photoinhibition, have different photoadaptive time scales (Horne et al., 1982; Lewis
et al., 1984). Farmer and Takahashi (1982) provide a description of how
the physical and biological information might be used in a field study.

Understanding of the time course features of photoadaptation remains
limited and some caution is appropriate in applying existing knowledge to
specific studies of mixing. To date several caveats have emerged; they
include:

(1) The complex time course of photoadaptation observed in
     Glenodinium (Prezelin and Matlick, 1980) may not be unique
     to that alga;

(2) P *versus* I curve parameters often show diurnal periodicity in
     algae growing with day-night light cycles (reviewed by Harris,
     1978) and vary with incubation time in natural seawater samples
     (Platt et al., 1980);

(3) Some phytoplankton can control their buoyancy, others can migrate
     vertically by directed swimming; the resulting vertical movements
     and photoadaptation will be independent of water motion in both
     cases (see Harris, 1981, for review);

(4) The P *versus* I curve parameters may be a function of the time
     taken to make the measurements (Marra, 1980).

Nevertheless the topic of photoadaptation of phytoplankton in natural
waters is relatively well studied and understood and problems such as those
above can be recognized and largely overcome under favourable circumstances. There is much additional information on the time scales of photo-
responses of the photosynthetic apparatus. The topic is too rich to cover
in a brief report. For example, chloroplast movements within diatom cells
and chloroplast shrinking and swelling in diatoms, in response to changes
in irradiance, take place in seconds or minutes (Kiefer, 1973). Changes
in fluorescence of photosynthetic pigments in response to environmental
changes have been reviewed by Harris (1981) and Prezelin (1981).

PRIMARY PRODUCTION AND THE SINKING FLUX: TIME SCALES OF VARIATION

There is little doubt that the sinking flux of biogenic organic particles in the deep ocean is driven by plankton photosynthesis (e.g., Suess, 1980). It is clear also that this flux plays an important role in the distribution of many elements, not only elements incorporated into the living material and participating in metabolic functions but also elements scavenged by organic particles. Photosynthesis is thought to be regulated by light and stratification in high latituides ($\geqslant 40°$), and in subtropical and tropical oceans by the rate of nutrient input to the euphotic zone. Indeed in the 'typical tropical situation' described by Herbland and Voituriez (1979), primary production is a function of the depth of the top of the nitracline (where nitrate is depleted at the surface, the depth at which it begins to increase more or less monotonically is called the top of the nitracline or just the 'nitracline depth').

Dugdale and Goering's (1967) concepts of 'new' and 'regenerated' primary production can be used to estimate the mean lifetime of carbon or nitrogen in biogenic organic particles in the surface layer of the deep ocean. Rates of regenerated production and the standing stocks of organic particulates suggest that the mean time for biogenic particles to decompose in the surface layer, i.e., to be remineralized in place, is uniformly about two weeks. The temperature dependence of decomposition results, of course, in shorter lifetimes in tropical waters and longer ones in polar seas (Eppley et al., 1983). On the other hand rates of new production suggest that export from the surface layer by sinking results in quite variable residence times of biogenic organic particles in the surface layer, depending upon the overall rates of primary production. Such residence times appear to range from 2-3 days, during an upwelling period in Southern California coastal waters, up to about 100 days in the subtropical gyre of the North Pacific (Eppley et al., 1983). Thorium isotope data suggest similar particle residence times in the surface waters (Bruland and Coale, this volume).

New production is the primary production available for export and is, in principle, quantitatively equivalent to the sinking flux of biogenic particles from the euphotic zone (Dugdale and Goering, 1967; Eppley and Peterson, 1979). Table 1 lists a sequence of events in the formation of sinking biogenic organic particles. It is steps 2 and 3 of this sequence that are included in the measurement of new production. Solute-particle and particle-particle interactions take place over steps 2-6 of the sequence, as well as during particle decomposition during and after this sequence. If new production (steps 2 and 3) is to be taken as a measure of step 6, then we must know something of the spatial and temporal relations among the steps in the sequence. We know that steps 1 and 2 are closely coupled in time in the nutrient-poor surface waters of the subtropical oceanic gyres and that they can be separated by months in temperate and polar waters where plants cannot grow in winter, yet surface and deep waters stir and mix in that season.

Nutrient injection into nutrient-depleted surface layers no doubt involves different physical mechanisms at different times and places in the oceans. It is not clear if it is useful to imagine such inputs, and the new production that results, as quasi-continuous or more or less quantized or pulsed. Billow turbulence and the capsizing of internal waves on the thermocline has been visualized (Woods and Fosberry, 1966-67; Haury et al., 1979) and the size of salt fingers can be determined (Williams, 1975). But the spatial scale of the biological response to nutrient inputs is not well defined and would be expected to vary with the nature, size and ambit of the organisms involved.

The photosynthetic new production of carbon can take place only in the light and is to that extent uncoupled from any nutrient uptake that takes place at night. Similarly, it is not the new production itself which sinks (i.e., phytoplankton) but largely products of its utilization by hetero-trophic organisms. Thus, new production and sinking cannot be simultaneous. In the case of herbivore grazing, followed by fecal pellet production and sinking of the pellets, the time between new production of carbon and the sinking of an equal amount of carbon may be quite close. The gut clearance time of herbivores can be as brief as a few minutes, and the ambit of a copepod or salp may be centimetres to metres in this time. But clearly the temporal and spatial scales must increase as we proceed from nutrient input, to phytoplankton new production, to the sinking of fecal material.

The corresponding time scales for these processes in food webs involving macroscopic aggregates (marine snow), instead of macrozooplankton herbivores, may be somewhat greater. But since we do not usually know the origin and nature of that material we can only speculate on such matters. The mechanisms and pathways of carbon flow clearly determine the time scales of importance for biogeochemical cycles that involve the sinking flux of biogenic organic matter. Time scales less than 12-24 hours, the daily light-dark cycle, may not be important. The corresponding minimal spatial scales of interest would not be the translational advective motion of the water over 12-24 hours but the ambits of the organisms contributing to production and the formation of sinking particles.

Two of the sediment trap studies provide information on the time sequence of particle formation and sinking from step 2 through step 6 (Table 1). In the Sargasso Sea the seasonal (spring) pulse of primary production off Bermuda is observed in sediment traps at >3000 m depth, and apparently within about a month (Deuser and Ross, 1980; Deuser et al., 1981). Since one month is the approximate transit time for a fecal pellet sinking at 100 m d^{-1} to reach 3000 m the steps may be nearly simultaneous when observed on such a coarse time scale. In Dabob Bay of Puget Sound, Washington, C. J. Lorensen (personal communication) and colleagues have collected sinking material in the upper 200 m on a time scale short enough to observe diurnal changes in the fecal pellet flux. The changes apparently result from diel vertical migrations of large zooplankton that spend evenings in the surface water where the primary production takes place. This result implies coupling between steps 2 and 6 on a 12-24 hour time scale, as inferred earlier (Welschmeyer et al., 1984).

The important geochemical significance of the sinking and decomposition of organic particles promises to lead to productive interaction between marine chemistry and biological oceanography. The history of ocean production, of the sinking flux, and of the distribution of the elements

Table 1. Sequence of events in the biogenic particle flux from surface waters

1	Nutrient input
2	Nutrient uptake by phytoplankton
3	Photosynthetic growth of phytoplankton
4	Utilization of phytoplankton and its products by animals and bacteria
5	Formation of fecal material and marine snow aggregates
6	Sinking of these particles out of the surface layer

over geologic time offers fascinating subject matter in yet another
dimension. These are indeed exciting times to be studying the oceans.

ACKNOWLEDGEMENT

This work was supported by U.S. Department of Energy Contract
DE-AM03-76SF00010. I thank C. J. Lorenzen, G. A. Knauer, J. E. Brooks, and
T. Platt for helpful discussion and access to unpublished results.

REFERENCES

Brooks, J.E., 1964, Acclimation to light intensity in two species of
     marine phytoplankton Dunaliella tertiolecta Butcher and
     Skeletonema costatum (Grev.) Cleve, M.S. Thesis, University of
     Southern California, Los Angeles.

Deuser, W.G., and Ross, E.H., 1980, Seasonal change in the flux of
     organic carbon to the deep Sargasso Sea, Nature, Lond., 283:364.

Deuser, W.G., Ross, E.H., and Anderson, R.F., 1981, Seasonality in the
     supply of sediment to the deep Sargasso Sea, and implications for
     the rapid transfer of matter to the deep ocean, Deep-Sea Res.,
     28:495.

Dugdale, R.C., and Goering, J.J., 1967, Uptake of new and regenerated
     forms of nitrogen in primary production, Limnol. Oceanogr.,
     12:196.

Eppley, R.W., and Peterson, B.J., 1979, Particulate organic matter flux
     and planktonic new production in the deep ocean, Nature, Lond.,
     282:677.

Eppley, R.W., Renger, E.H., and Betzer, P.R., 1983, The residence time
     of particulate organic carbon in the surface layer of the ocean,
     Deep-Sea Res., 30:311.

Falkowski, P.G., 1980, Light-shade adaptation in marine phytoplankton,
     in: "Primary Production in the Sea", P.G. Falkowski, ed., pp.
     99-119, Plenum Press, New York.

Farmer, D.M., and Takahashi, M., 1982, Effects of vertical mixing on
     photosynthetic responses, Japan. J. Limnol., 43:173.

Harris, G.P., 1978, Photosynthesis, productivity and growth: the
     physiological ecology of phytoplankton, Ergebnisse Limnologie,
     10:1.

Harris, G.P., 1981, The measurement of photosynthesis in natural
     populations of phytoplankton, in: "The Physiological Ecology
     of Phytoplankton", I. Morris, ed., pp. 129-187, University of
     California Press, Berkeley.

Haury, L.R., Briscoe, M.G. and Orr, M.H., 1979, Tidally generated internal
     wave packets in Massachusetts Bay, Nature, Lond., 278:312.

Herbland , A., and Voituriez, B., 1979, Hydrologic structure analysis
     for estimating the primary production in the tropical Atlantic
     Ocean, J. Mar. Res., 37:87.

Horne, E.P., Lewis, M.R., Cullen, J.J., Oakey, N.S., and Platt, T., 1982, Simultaneous measurements of algal photoadaptation and turbulence during the spring bloom, Trans. Amer. Geophys. Un., 63:962.

Jassby, A.D., and Platt, T., 1976, Mathematical formulation of the relationship between photosynthesis and light for phytoplankton, Limnol. Oceanogr., 21:540.

Kiefer, D.A., 1973, Chlorophyll a fluorescence in marine centric diatoms: responses of chloroplasts to light and nutrient stress, Mar. Biol., 29:39.

Lewis, M.R., Cullen, J.J., and Platt, T., 1984, Relationship between vertical mixing and photoadaptation by phytoplankton. Similarity criteria, Mar. Ecol. Progr. Ser., 15:141.

Marra, J., 1980, Vertical mixing and primary production, in: "Primary Productivity in the Sea", P.G. Falkowski, ed., pp. 121-137, Plenum Press, New York.

Neale, P.J., Vincent, W.F., and Richerson, P.J., 1982, Diel variation of photosynthesis in Lake Titicaca: adaptation to an extreme irradiance environment, Trans. Amer. Geophys. Un., 63:962.

Perry, M.J., Talbot, M.C., and Alberte, R.S., 1981, Photoadaptation in marine phytoplankton: response of the photosynthetic unit, Mar. Biol., 62:91.

Platt, T., and Jassby, A.D., 1976, The relationship between photosynthesis and light for natural assemblages of coastal marine phytoplankton, J. Phycol., 12:421.

Platt, T., Gallegos, C.L., and Harrison, W.G., 1980, Photoinhibition of photosynthesis in natural assemblages of marine phytoplankton, J. Mar. Res., 38:687.

Prezelin, B.B., 1981, Light reactions in photosynthesis, Can. Bull. Fish. Aquatic Sci., 210:1.

Prezelin, B.B., and Matlick, H.A., 1980, Time-course of photoadaptation in the photosynthesis-irradiance relationship of a dinoflagellate exhibiting photosynthetic periodicity, Mar. Biol., 58:85.

Steele, J.H., 1962, Environmental control of photosynthesis in the sea, Limnol. Oceanogr., 7:137.

Steemann Nielsen, E., 1975, "Marine Photosynthesis", Elsevier, Amsterdam.

Steemann Nielsen, E., and Hansen, V.K., 1959, Light adaptation in marine phytoplankton populations and its interrelation with temperature, Physiol. Plant., 12:353.

Steemann Nielsen, E., and Park, S.T., 1964, On the time course in adapting to low light intensities in marine phytoplankton, J. Cons. Int. Perm. Explor. Mer., 29:19.

Steemann Nielsen, E., Hansen, V.K., and Jorgensen, E.G., 1962, The adaptation to different light intensities in Chlorella vulgaris and the time dependence on transfer to a new light intensity, Physiol. Plant., 15:505.

Suess, E., 1980, Particulate organic carbon flux in the oceans-surface
    productivity and oxygen utilization, Nature, Lond., 288:260.

Talling, J., 1957, The phytoplankton population as a compound photo-
    synthetic system, New Phytol., 56:133.

Welschmeyer, N.A., Copping, A.E., Vernet, M., and Lorenzen, C.J., 1984,
    Diel fluctuation in grazing rate of zooplankton as determined from
    sinking of phaeopigments in feces, Mar. Biol., 83:263

Williams, A.J., 1975, Images of ocean microstructure, Deep-Sea Res.,
    22:811.

Woods, J.D., and Fosberry, G.G., 1966-67, The structure of the thermo-
    cline, Rep. Underwater Assoc., Lond., pp. 5-18.

Yentsch, C.S., and Lee, R.W., 1966, A study of photosynthetic light
    reactions and a new interpretation of sun and shade phytoplankton,
    J. Mar. Res., 24:319.

SOME PERSPECTIVES ON ECOLOGICAL MODELING

FOCUSED ON UPPER OCEAN PROCESSES

Daniel Kamykowski

Department of Marine, Earth and Atmospheric Sciences
Box 8208, North Carolina State University
Raleigh, North Carolina 27695-8208   U.S.A.

ABSTRACT

Biological processes clearly affect the often complex distributions of
many chemical elements in the ocean. The biological processes, in turn,
are inherently intricate and are variable in time and space due to environmental dependencies. Modeling provides a mechanism by which complex
biological-environmental interactions can be studied. This potential is
evident in the progress exhibited in the construction of biological-
physical subsystem models in the last decade. Future progress in these
models depends on an increased understanding of how the vertical motion
affects the environmental exposure of plankton and on improved capabilities
of real time and space biological parameterization. From the present
perspective, the latter includes not only small scale responses due to
planktonic physiology and behavior but also geographic differences in
oceanographic conditions. Biologically oriented subsystem models provide a
powerful tool when closely coupled with laboratory and field efforts that
can contribute significantly to the elucidation of upper ocean chemistry.

INTRODUCTION

Several recent reviews have dealt with aspects of upper ocean
modeling. The most general include Goldberg et al. (1977) and Nihoul
(1977) in which the various allied oceanographic sciences are represented.
Longhurst (1981) and Platt et al. (1981) have provided a more specific
biological treatment. Benefitting from these overviews and responding to
the suggestion of Radford et al. (1981) for increased attention to
biological subsystem models, the present paper focuses on certain aspects
of upper ocean modeling dealing with the physiological and behavioral
responses of plankton (especially phytoplankton) to the environment.
Subsystems are here defined as specific physical or biological processes
isolated from the complex of natural processes. As detailed biological
insight is gained into individual planktonic processes, chemical patterns
related to the biology will become more interpretable. An index of the
scope of this potential benefit is presented in Table 1 which lists the
elements that exhibit distributional relationships with the classical
nutrient elements (carbon, nitrogen, silicon, phosphorus). Since it is
unlikely that dynamics for correlated elements will be exactly parallel
(see Steele, this volume) increased information on the classical nutrient

Table 1. Summary of the chemical elements that have been reported to have distributional relationships with carbon, nitrogen, silicon or phosphorus

Nutrient elements (symbol)	Nutrient relationships
Carbon (Alk)	Calcium (Alk)
Nitrogen (N)	Chromium (Si, N, P)
Silicon (Si)	Manganese
Phosphorus (P)	Iron
	Cobalt
	Nickel (Si, P)
	Copper
	Zinc (Si)
	Germanium (Si)
	Arsenic (P)
	Selenium (Si, P)
	Strontium (P)
	Cadmium (P)
	Iodine (N, P)
	Barium (Si, Alk)
	Lanthanides
	Mercury (Si)
	Radium

The symbols in parentheses following elements in the right hand column indicate the nutrient elements, as listed in the left hand column, with which there is the most close correlation. Where no relationship is shown the elements are uncorrelated with nutrients or are related to all of them. Information derived from Broecker and Peng (1982) and Quinby-Hunt and Turekian (1983).

elements may help to explain both the reasons for the correlations and the exact dynamics of the other elements.

A logical starting point for this discussion is the compilation of biologically active, upper ocean physical processes provided by Haury et al. (1978) within a time-space framework. Monin et al. (1978) and Huthnance (1981) have provided informative corroborating surveys of general physical processes. Though zooplankton are emphasized by Haury et al. (1978), a similar treatment is possible for phytoplankton with somewhat altered time relationships. The conceptual approach offered by Haury et al. (1978) emphasizes the continuum of physical forcing that affects biological patterns. Such a reminder is useful since the study of mesoscale events, defined as those with time scales of 100 days and length scales of 100-1000 km, like upwelling, rings or eddies, is of substantial current interest due to ease of satellite detection and reasonable compatibility with shipboard capabilities. The present orientation will also emphasize the continuum through concern for the internal details of small scale variability that occur within mesoscale events and the external details of the environment within which mesoscale events occur.

Table 2 lists some of the physical processes within the time-space spectrum that have been modeled for biological effects. All references are less than 10 years old and several emphasize the vertical component due to the greater effectiveness of the limited swimming capability of plankton in this dimension. This list reveals something of the state-of-the-art. Many studies consider either an isolated periodic source of advection or eddy diffusion. Larger scale studies often consider both advection and eddy diffusion simultaneously but generally emphasize mean flow. The

Table 2. Representative physical–biological models

Gyres	Vinogradov & Menshutkin (1977)
Currents	Walsh (1977)
Eddies	Hoffman et al. (1980)
Rings	Wroblewski (personal communication)
Upwelling	Wroblewski (1977, 1980); Barber & Smith (1981)
River Plumes	DiToro et al. (1977)
Internal Tides	Kamykowski (1974, 1976, 1979b)
Seiches	Kamykowski (1978, 1979a)
Oceanic Fronts	Papers in Bowman & Esaias (1978)
Vertical Shears	Evans (1978); Riley (1976)
Horizontal Mixing	Okubo (1978) among many others
Internal Waves	Kamykowski (1981a)
Convective Cells	Ledbetter (1979); Evans & Taylor (1980); Woods and Onken (1982)
Vertical Mixing	Jamart et al. (1977); Falkowski & Wirick (1981)

Adapted from Haury et al. (1978).

separation and simplification of these components is obviously artificial, and one major advance in this type of biological modeling is the integration of more realistic time-dependent advection and diffusion in models of all scales. Woods and Onken (1982) provide an example of this approach. The improved ability of physical oceanographers to measure vertical water motion through new current meter technology (T. B. Curtin, personal communication) may further catalyze such multi-process models. Biological oceanographers may then be able to construct organism trajectories and therefore organism environmental exposure in the upper ocean as determined by the biologically active portion of the time-varying physical motion spectrum at a given site.

In spite of future progress in integrating physical oceanography into biological modeling, the problem of more realistic biological parameterization of these models will remain. Harris (1980) has provided a thoughtprovoking discussion of this problem. The environment offers a spectrum of variables each with its own spectrum of variability to which the available plankton species respond in different ways depending on their individual inherent periodicities. The increasing temporal or spatial sequence of phytoplankton processes that correspond to the physical motion spectrum include variations in fluorescence, photosynthesis, nutrient uptake, chlorophyll synthesis, behavior, growth rate, seasonal community changes and yearto-year community and productivity changes.

Four examples of possible improvement in the biological parameterization of future interactive models will be discussed. These examples are mostly drawn from personal experience because of their familiarity. Other choices are certainly possible as exemplified by the discussion by Radford et al. (1981) of light dependence of phytoplankton photosynthesis. At small scales, some physiological aspects of competitive nutrient uptake under conditions of low nutrient concentration and the behavioral aspects of dinoflagellate velocity are considered. At large scales, the physiological aspects of the latitudinal relationships among temperature and plant nutrients and the geographic patterns of the nitrate-silicic acid ratio are considered. These examples demonstrate the beneficial feedback with field and laboratory studies that can be anticipated in future modeling efforts.

Example 1: Competitive Nutrient Uptake at Low Nutrient Concentration

Ammonium inhibition of nitrate uptake at ambient ammonium concentrations above 2 μmol l^{-1} is a recognized phenomenon (McCarthy, 1981). This
interaction is commonly modeled (Wroblewski, 1977) by

$$V = V_m \left(\frac{NO_3}{K_s + NO_3} e^{-\lambda NH_4} + \frac{NH_4}{K_s + NH_4} \right) \tag{1}$$

where $V_m(NO_3) = V_m(NH_4)$
$\qquad K_s(NO_3) = K_s(NH_4)$

(see Appendix I for a definition of the terms in all text equations).

Upper ocean nitrogen availability is under intense investigation. In
one view of nitrogen-depleted mixed layers, the dominant sources of nitrogen
are excreted ammonium from heterotrophic processes (McCarthy and Goldman,
1979; but see Jackson, 1980) and upward fluxing nitrate across the pycno-
cline (Garside, 1982; Cullen et al., 1982). The former may be available to
some phytoplankton as randomly or intermittently scattered point sources
occurring locally at concentrations above 2 μmol l^{-1} but decreasing rapidly
to undetectable levels. The latter may increase exponentially with depth to
the thermocline in the nmol l^{-1} range of concentrations. Syrett (1981) and
McCarthy (1981) suggest that under these conditions phytoplankton may become
nitrogen opportunists that utilize various chemical forms of nitrogen
simultaneously. This activity depends on their nutritional state, their
inherent nutrient uptake capabilities and the regime of exposure to various
nutrient sources. The exposure regime depends on heterotroph density,
distribution and structure, and on vertical mixing processes.

As these various processes and interactions are increasingly investigated, modelers may have to modify the way in which they handle nitrogen
uptake in the upper ocean. As knowledge of the exposure of phytoplankton
to ammonium and nitrate in the upper ocean improves, adjustments could be
made in the time course of changes in the values of the nutrient kinetic
"constants", depending on the spectrum of nitrogen exposure resulting from
the changing vertical position of the organisms in the water column.

Example 2: Dinoflagellate Behavior

Kamykowski (1981a) used a coherent symmetrical multiperiod internal
wave model to determine if a red tide off Southern California could be
caused by the interaction between dinoflagellate diurnal vertical
migration and the water currents associated with the approximate internal
wave field. This was accomplished through an iterative solution of
equations of water and organism motion oriented offshore and with depth.
The following vertical axis equation set for above the thermocline provides
an example:

$$w_1 = \frac{2\pi a_1}{T_1} \cos \left(\frac{2\pi x}{E_1} - \frac{\pi}{2} \right) \tag{2}$$

$$w_2 = \frac{2\pi a_2}{T_2} \cos \left(\frac{2\pi x}{E_2} - \frac{\pi}{2} \right) \tag{3}$$

$$w_3 = \frac{2\pi a_3}{T_3} \cos \left(\frac{2\pi x}{E_3} - \frac{\pi}{2} \right) \tag{4}$$

$$w_p = \frac{z}{z_T} (w_1 + w_2 + w_3) \tag{5}$$

$$w_0 = w_p + w_b \tag{6}$$

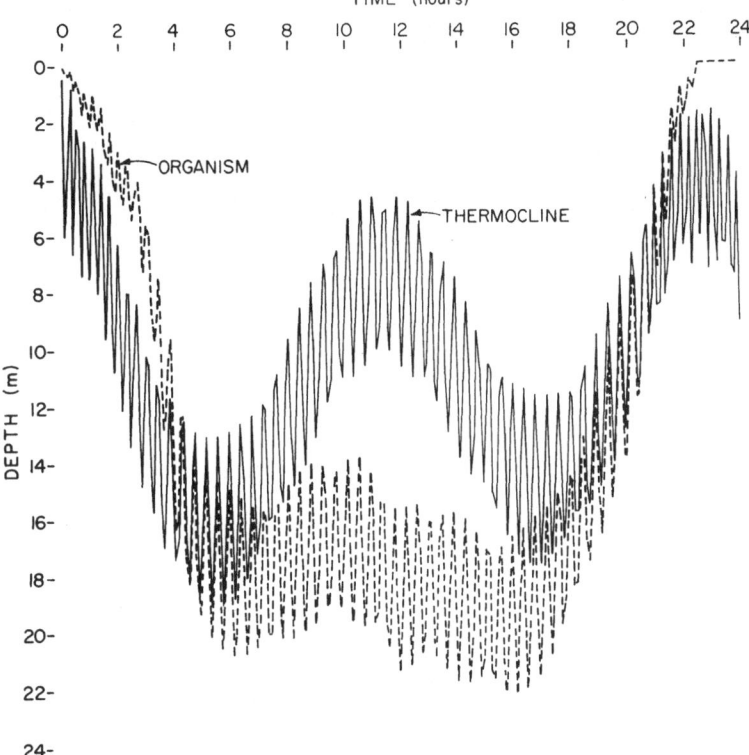

Fig. 1. Representative trajectory (broken line) of a dinoflagellate cell swimming at 1.70 m h^{-1} above the
thermocline and at 1.33 m h^{-1} below the thermocline is shown. The thermocline (solid line)
at a mean depth of 10 m. is oscillating under
the influence of three coherent symmetrical
internal waves with periods of 0.33, 12.40 and
22.13 h. Though physically simplistic, this
figure provides conceptual insight into the
interaction.

The organism swimming velocity is added to the computed vertical flow
velocities associated with the internal waves of different periods. The
biological–physical interaction is represented in Fig. 1. The sequential
organism distributions resulting from the model over several days are
suggestive of red tide formation but remain inconclusive primarily because
the exact behavior of the causative dinoflagellate is unknown. This uncertainty results not only because, as often happens, appropriate biological samples were not collected at the time of the red tide but also because
the pertinent dinoflagellate behavior may have occurred before the red tide.

Though a significant amount of information on dinoflagellate behavior
is available (Forward, 1976), the laboratory data base is inadequate to
provide required insight into the range of possible dinoflagellate behavior. For instance, Holwill (1977) projects that protozoa swim at the
same speed irrespective of size, based on indiscriminate historical data.
A more selective survey of dinoflagellate swimming speeds, however,
suggests a cell length dependence.

The regression line fit for swimming speed *versus* cell length for all

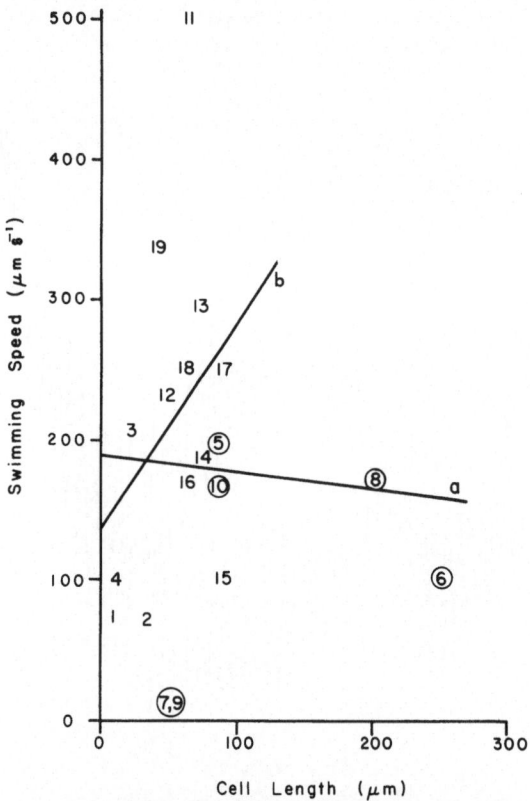

Fig. 2. Linear plot of cell length (μm)
versus swimming speed (μm s^{-1})
for several species of marine
dinoflagellate. Line 'a' represents a least squares fit
(SS = 186.44-0.09CL, r^2 = 0.002,
p ≃ 0.5) to all the data. Line
'b' represents a least squares
fit (SS = 136.02 + 1.45CL, r^2 =
0.10, 0.2>p>0.1) to the same
data without *Ceratium* species
(circled numbers). See Appendix
II for sources of data.

included species (Appendix II) is contrasted in Fig. 2 with that for all
included species except those from the genus *Ceratium*. Species from the
genus *Ceratium* generally possess long projections that may not be hydro-
dynamically significant at low Reynolds numbers and, thus, this genus may
confuse the picture for the swimming speed versus cell size relationship
for dinoflagellates.

Preliminary laboratory data using visual observations of cells support
a swimming speed versus cell length relationship within the dinoflagellates
(D. Kamykowski, unpublished). Increase in swimming speed with cell size is
not necessarily an absolute, however, since dinoflagellate swimming speed
depends on several factors including the intricacies of cell structure, the
flagellar length, and the wavelength, amplitude and frequency of the
flagellar beat (Gray and Hancock, 1955). These comments demonstrate that
even relatively simple swimming speed relationships are poorly documented
within the dinoflagellates.

Orientation must be considered together with swimming speed to define velocity, the true concern of behavior. Preliminary laboratory results based on visual observations of cells suggest that the time course of orientation is species specific (D. Kamykowski, unpublished). Heany and Talling (1980), Tyler and Seliger (1981) and Cullen and Horrigan (1981) provide some of the more innovative laboratory and field discussions based on the sequential observations of biomass demonstrating how changing orientation influences the environmental spectrum to which given dinoflagellates are exposed. These and other studies suggest that many plankton are not simple Lagrangian drifters but add complex components of motion due to buoyancy control or swimming. In the future modelers should be able to specify better the behavioral characteristics of dinoflagellates derived from ongoing laboratory and field studies so that more realistic interactions of these organisms with physical and chemical systems can be formulated.

LARGE SCALE

Example 3: Temperature and Plant Nutrients

Goldman (1977, 1979) has shown that the nutrient uptake and growth of different phytoplankton respond to temperature in several possible ways. In general, temperature modulation of nutrient uptake cannot be treated with a simple functional dependence suggested by the equations

$$\mu = \mu_m \left(\frac{S}{K_m+S}\right) \tag{7}$$

$$\mu_m = Ae^{-E/RT} \tag{8}$$

$$\mu = Ae^{-E/RT} \left(\frac{S}{K_m+S}\right) \tag{9}$$

which combine the Monod and Arrhenius formulations, because the relationship is not this simple for all species. Tilman et al. (1981) have shown, however, that changes in nutrient uptake kinetics along a temperature gradient can affect the outcome of species competition in freshwater algae and thus play an important role in phytoplankton community structure. Therefore, in spite of the reported complexity in the biological response, additional work on the temperature dependence of nutrient uptake seems worthwhile.

One prerequisite for more realistic laboratory experiments on the marine system is the definition of how temperature and plant nutrients are related in the ocean. Zentara and Kamykowski (1977) have demonstrated that temperature and plant nutrients in the eastern Pacific Ocean exhibit a latitudinal pattern in which the temperature at which a given plant nutrient concentration occurs decreases with distance from the equator. Presently an NODC data base, consisting of about 250,000 stations, is being analyzed (D. Kamykowski and S.-J. Zentara, unpublished) for global patterns among temperature and plant nutrients.

An initial attempt to determine the temperatures at which nitrate and silicic acid deplete in the South Atlantic is shown in Fig. 3. These nutrient depletion temperatures decrease with increasing latitude until light limitation intercedes. This pattern defines an important aspect of phytoplankton growth conditions at a given latitude and may therefore provide insight into species biogeography if the environmental responses of selected species are adequately known. Combining these latitudinal regression lines of temperature *versus* plant nutrients with the robust tempera-

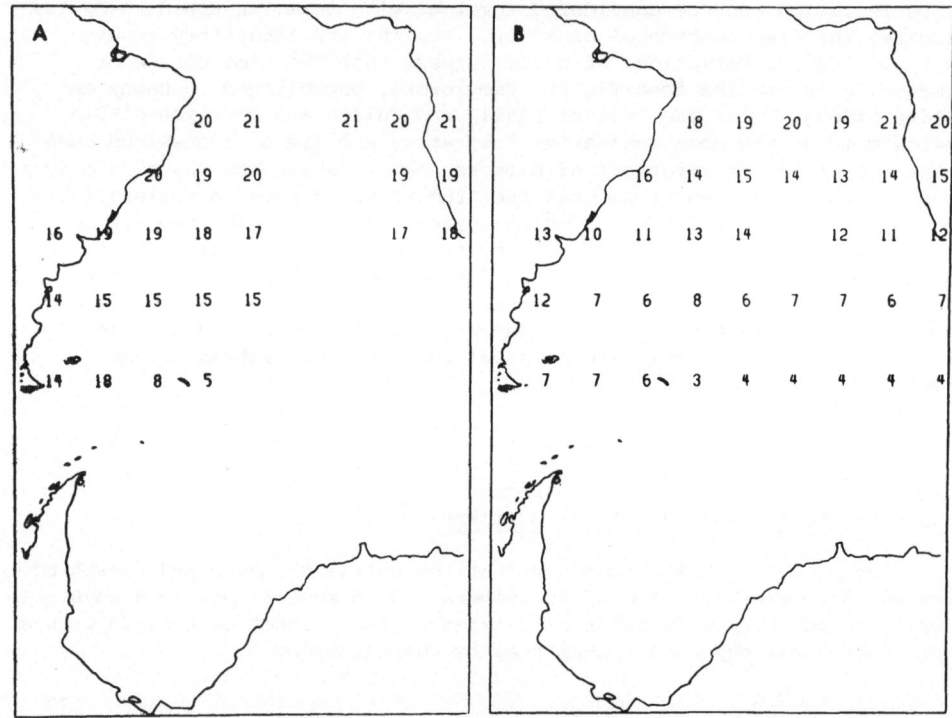

Fig. 3. Plots of the subjective temperatures at which nitrate (A) and
silicic acid (B) deplete in the South Atlantic Ocean, as determined from scatter plots of temperature *versus* nitrate or silicic acid concentrations using the NODC data base for 10° x 10°
blocks of longitude and latitude.

ture-dependent Michaelis-Menten equation of Mack et al. (1981), provides a
fully temperature dependent, functionally robust equation for V, the velocit
of nutrient uptake, that can be applied to different species responses at
selected latitudes:

$$V = \underbrace{\dfrac{\overbrace{\dfrac{V_m(T)}{\alpha_1 Te^{-\alpha_2/T}}}{(1 + \alpha_3 e^{-\alpha_4/T})}}{[\dfrac{\alpha_1}{\gamma_1}(e^{-(\alpha_2-\gamma_2)/T})\underbrace{\left[\dfrac{1+\gamma_3 e^{-\gamma_4/T}}{1+\alpha_3 e^{-\alpha_4/T}}\right]}_{K_s(T)} + (\beta_1+\beta_2 T)]}} \quad \overbrace{(\beta_1+\beta_2 T)}^{S(T)}}_{S(T)} \tag{10}$$

Whether the full complexity of this or some similar equation is actually
required in a given model will depend on the species and on the model
application.

Example 4: Nitrate and Silicic Acid

Tilman (1977) and Tilman et al. (1982) have shown how plant nutrient
ratios affect species competition in freshwater algae and have modeled
these interactions using the Monod form of the resource competition
equations as:

for the *i*th of n different species,

$$dN_i/N_i dt = \min_{1 \le j \le m} [r_i S_j/(K_{ij}+S_j) - D] \qquad (11)$$

for the *j*th of m different resources,

$$dSj/dt = D(_0S_j - S_j) - \sum_{i=1}^{n} N_i r_i S_j/[(K_{ij}+S_j)Y_{ij}] \qquad (12)$$

Zentara and Kamykowski (1977, 1981) have shown that upper ocean nitrate and silicic acid in the Pacific Ocean exhibit a large range of nutrient ratios. Using the depletion of one of the nutrients as a normalizing level, nitrate depletion can be projected with as much as 40 μmol l^{-1} silicic acid remaining in the water while silicic acid depletion can be projected with as much as 25 μmol l^{-1} nitrate remaining in the water. An analysis of an NODC data base, consisting of about 20,000 stations, for geographic patterns in the nitrate and silicic acid relationship is nearing completion (D. Kamykowski and S.-J. Zentara, unpublished). In Fig. 4, areas of the world ocean are identified that exhibit nitrate excess above 5 μmol l^{-1} at silicic acid depletion or silicic acid excess above 10 μmol l^{-1} at nitrate depletion. The theoretical framework of Tilman and associates carries over to the marine

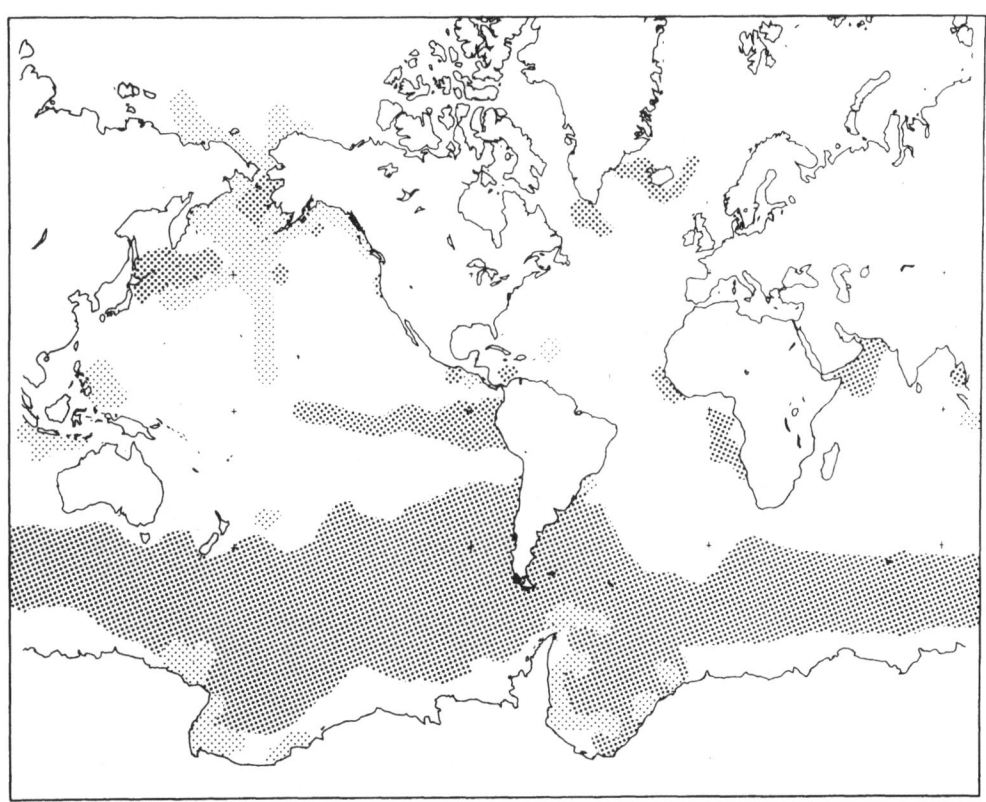

Fig. 4. Subjective contour plot of nitrate excess (>5 μmol l^{-1}; large dots) at projected silicic acid depletion or silicic acid excess (> 10 μmol l^{-1}; small dots) at projected nitrate depletion, as determined in an initial analysis of the NODC data base. Projected nutrient depletion is obtained by extrapolating linear relationships to intercept the nitrate or silicic acid axis.

environment for diatom competition. In certain areas of the world ocean, silicic acid depletion probably limits diatom growth altogether and allows substantial blooms of algae that do not require silicic acid. As with the temperature *versus* plant nutrient relationships, modelers again have the opportunity for new approaches to marine phytoplankton community structure at a given location due to large scale chemical patterns. Opportunities exist for better insight into species level phytoplankton biogeography based not only on physiological tolerances but also on some insight into competitive interactions with other algae.

CONCLUSIONS

The emphasis of this presentation has been on possible future directions for a class of variance orientated ecological models, from an individual perspective. Progress in future models is tied in with technological and conceptual advances that will affect formulation and parameterization The case for details has been purposely overstated to demonstrate frontiers that rarely appear in mass balance models. The required terms for future models will be some distillation of the present conjecture.

Specific conclusions are:

(1) Though mesoscale processes are of special current interest, physical and biological processes on both smaller and larger scales can affect the exact chemical dynamics in a given location.

(2) Continued progress is needed in measuring vertical motion in the upper ocean and integrating the diverse, presently isolated, subcategories of physical processes into more realistic, biologically active models.

(3) Considerable effort is required to incorporate the rapid advances in laboratory studies of plankton physiology and behavior, and in biological interpretations of historical field data, into biological subsystem models to determine the sensitivity of model predictions to these additions.

(4) Based on these sensitivity analyses, advocates of subsystem models may be able to provide laboratory investigations with a more efficient field orientation in a way that could parallel the guiding force of holistic models on field sampling design, as suggested by Platt et al. (1981). These efforts may identify some intermediate level of biological complexity between species—specific response and aggregated trophic level dynamics that will provide a more realistic analogue to natural systems.

(5) For upper ocean chemistry, advances along the lines discussed will provide a broader matrix of environmentally driven biological processes with which to interpret observed chemical patterns

Finally, the caveat of Haury et al. (1978), concerning the effect of the sampling filter on the relationship between "true" field patterns and observed field patterns, bears repeating. For the modeler, this caveat is amplified severalfold. Not only are the observed field patterns the standard against which the model results are judged, but the models are based on simplifications of real environmental variability and on laboratory analogues of the physiology and behavior of natural populations. This caveat is not offered with pessimism but simply to place the discussed perspectives on a realistic base relative to natural phenomena. Within the present context, models serve better as conceptual aids contributing a better understanding of dominant processes rather than as predictive tools of natural phenomena.

ACKNOWLEDGEMENTS

This material is based on research supported by NSF Grants Nos OCE-81-00423 and OCE-8200159 AO1 and by NASA Grant No. NAGW367. The text benefitted from discussion with Dr. T.B. Curtin, Dr. J.S. Wroblewski, Ms. S.J. Zentara and several participants at this NATO Workshop.

REFERENCES

Barber, R.T., and Smith, W.O., Jr., 1981, The role of circulation, sinking and vertical migration in physical sorting of phytoplankton in the upwelling center at 15°C, in: "Coastal Upwelling", F.A. Richards, ed., pp.366–371, American Geophysical Union, Washington, D.C.

Bowman, M.J., and Esaias, W.E., 1978, "Ocean Fronts in Coastal Processes", Springer-Verlag, New York.

Brennen, C., and Winet, H., 1977, Fluid mechanics of propulsion by cilia and flagella, Ann. Rev. Fluid Mech., 9:339.

Broecker, W.S., and Peng, T.H., 1982, "Tracers in the Sea", Eldigio Press, Palisades.

Cullen, J.J., and Horrigan, S.G., 1981, The effects of nitrate on the diurnal vertical migration, carbon to nitrogen ratio, and photosynthetic capacity of a dinoflagellate; Gymnodinium splendens, Mar. Biol., 62:81.

Cullen, J.J., Stewart, E., Renger, E., Reid, F.M.H., Eppley, R.W., and Winant, C.D., 1982, Vertical motion of the thermocline, nitra-cline and chlorophyll maximum layers in relation to currents on the Southern California shelf, Trans. Amer. Geophys. Un., 63:975.

DiToro, D.M., Thomann, R.V., O'Connor, D.J., and Mancini, J.L., 1977, Estuarine phytoplankton biomass models - verification analyses and preliminary applications, in: "The Sea, Volume 6, Marine Modeling", E.D. Goldberg, I.N. McCave, J.J. O'Brien and J.H. Steele, eds., pp.969–1020, Wiley–Interscience, New York.

Evans, G.T., 1978, Biological effects of vertical-horizontal interactions, in: "Spatial Patterns in Plankton Communities", J.H. Steele, ed., pp.157–179, Plenum Press, New York.

Evans, G.T., and Taylor, F.J.R., 1980, Phytoplankton associations in Langmuir cells, Limnol. Oceanogr., 25:840.

Falkowski, P.G., and Wirick, C.D., 1981, A simulation model of the effects of vertical mixing on primary production, Mar. Biol., 65:69.

Forward, R.B., Jr., 1976, Light and diurnal vertical migration: Photobehaviour and photophysiology of plankton, Photochem. Photobiol. Rev., 1:157.

Garside, C., 1982, Nitrate measurements in the oligotrophic photic zone, Trans. Amer. Geophys. Un., 63:995.

Gittleson, S.M., Hotchkiss, S.K., and Valencia, F.G., 1974, Locomotion in the marine dinoflagellate Amphidinium carterae (Hulbert), Trans. Amer. Microscope Soc., 93:101.

Goldberg, E.D., McCave, I.N., O'Brien, J.J., and Steele, J.H., eds, 1977, "The Sea, Volume 6, Marine Modeling", Wiley-Interscience, New York.

Goldman, J.C., 1977, Temperature effects on phytoplankton growth in continuous culture, Limnol. Oceanogr., 22:932.

Goldman, J.C., 1979, Temperature effects on the steady state growth, phosphorus uptake and chemical composition of a marine phytoplankton, Microbial Ecol., 5:153.

Gray, J., and Hancock, G.J., 1955, The movement of sea urchin spermatozoa, J. Exp. Biol., 32:802.

Hand, W.G., Collard, P.A., and Davenport, D., 1965, The effect of temperature and salinity change on the swimming rate of the dinoflagellates, Gonyaulax and Gyrodinium, Biol. Bull., 128:90.

Harris, G.P., 1980, Temporal spatial scales in phytoplankton ecology. Mechanisms, methods and management, Can. J. Fish. Aquatic Sci., 37:877.

Haury, L.R., McGowan, J.A., and Wiebe, P.H., 1978, Plankton and processes in the time-space scales of plankton distributions, in: "Spatial Patterns in Plankton Communities", J.H.Steele, ed., pp.277–328, Plenum Press, New York.

Heaney, S.I., and Talling, J.F., 1980, Dynamic aspects of dinoflagellate distribution pattern in a small productive lake, J. Ecol., 68:75.

Herman, E.M., and Sweeney, B.M., 1976, Cachonina illdefina sp. nov. (Dinophycea): chloroplast tubules and degeneration of the pyrenoid, J. Phycol., 12:198.

Hoffman, E.E., Pietrafesa, L.J., Klinck, J.M., and Atkinson, L.P., 1980, A time-dependent model of nutrient distribution in continental shelf waters, Ecol. Modelling, 10:193.

Holwill, M.E.J., 1977, Some biophysical aspects of ciliary and flagellar motility, Adv. Microbial Physiol., 16:1.

Huthnance, J.M., 1981, Waves and currents near the continental shelf edge, Progr. Oceanogr., 10:193.

Jackson, G.A., 1980, Phytoplankton growth and zooplankton grazing in oligotrophic oceans, Nature, Lond., 284:439.

Jamart, B.M., Winter, D.F., Banse, K., Anderson, G.C., and Lam, R.K., 1977, A theoretical study of phytoplankton growth and nutrient distribution in the Pacific Ocean off the northwestern U.S. coast, Deep-Sea Res., 24:753.

Kamykowski, D., 1974, Possible interactions between phytoplankton and semi-diurnal internal tides, J. Mar. Res., 32:67.

Kamykowski, D., 1976, Possible interactions between plankton and semi-diurnal internal tides. II Deep thermoclines and trophic effects, J. Mar. Res., 34:499.

Kamykowski, D., 1978, Organism patchiness in lakes resulting from the interaction between the internal seiche and plankton diurnal vertical migration, Ecol. Modelling, 4:197.

Kamykowski, D., 1979a, Comparison of the possible effects of internal seiches on the plankton population of selected lakes, in: "State-of-the Art in Ecological Modelling", S.E. Jorjensen, ed., pp. 647–659, International Society of Ecological Modelling, Copenhagen.

Kamykowski, D., 1979b, The growth response of a model Gymnodinium splendens in stationary and wavy water columns, Mar. Biol., 50:289.

Kamykowski, D., 1981a, The simulation of a Southern California red tide using characteristics of a simultaneously measured internal wave field, Ecol. Modelling, 12:253.

Kamykowski, D., 1981b, Laboratory experiments on the diurnal vertical migration of marine dinoflagellates through temperature gradients, Mar. Biol., 62:57.

Kamykowski, D., and Zentara, S.J., 1977, The diurnal vertical migration of motile phytoplankton through temperature gradients, Limnol. Oceanogr., 22:148.

Kofoid, C.A., and Swezy, O., 1921, The free-living unarmored dino-flagellates, Mem. Univ. California, 5:1.

Ledbetter, M., 1979, Langmuir circulations and plankton patchiness, Ecol. Modelling, 7:284.

Longhurst, A.R., ed., 1981, "Analysis of Marine Ecosystems", Academic Press, London.

Mack, T.P., Bajusz, B.A., Nolan, E.S., and Smilowitz, Z., 1981, Development of a temperature-mediated functional response equation, Envir. Entomol., 10:573.

McCarthy, J.J., 1981, The kinetics of nutrient utilization, Can.Bull. Fish. Aquatic Sci., 210:211.

McCarthy, J.J., and Goldman, J.C., 1979, Nitrogeneous nutrition of marine phytoplankton in nutrient-depleted waters, Science, N.Y., 203:670.

Monin, A.S., Kamenkowich, V.M., and Kort, V.G., 1977, "Variability of the Oceans", Wiley-Interscience, New York.

Nihoul, J., 1977, "Modelling of Marine Systems", Elsevier, Amsterdam.

Okubo, A., 1978, Horizontal dispersion and critical scales for phyto-plankton patches, in: "Spatial Patterns in Plankton Commun-ities", J.H. Steele, ed., pp. 21–42, Plenum Press, New York.

Peters, N., 1929, Über Orts-und Geisselbewegeing bei marinen Dinoflagell-
    aten, <u>Archiv. Protistenkd</u>, 67:291.

Platt, T., Mann, K.H., and Ulanowicz, R.E., 1981, "Mathematical Models in
    Biological Oceanography", UNESCO Press, Paris.

Quinby—Hunt, M.S., and Turekian, K.K., 1983, Distribution of elements in
    sea water, <u>Trans. Amer. Geophys. Un.</u>, 64:130.

Radford, P.J., Joint, I.R., and Hibny, A.R., 1981, Simulation models of
    individual production processes, <u>in</u>: "Analysis of Marine
    Ecosystems", A.R. Longhurst, ed., pp. 677—700, Academic Press,
    London.

Riley, G.A., 1976, A model of plankton patchiness, <u>Limnol. Oceanogr.</u>,
    21:873.

Ronkin, R.R., 1959, Motility and power dissipation in flagellated cells,
    especially Chlamydomonas, <u>Biol. Bull.</u>, 116:285.

Syrett, P.J., 1981, Nitrogen metabolism of microalgae, <u>Can. Bull. Fish.
    Aquatic Sci.</u>, 210:182.

Tilman, D., 1977, Resource competition between planktonic algae: an
    experimental and theoretical approach, <u>Ecology</u>, 58:338.

Tilman, D., Mattson, M., and Langer, S., 1981, Competitive and nutrient
    kinetics along a temperature gradient: An experimental test of a
    mechanistic approach to niche theory, <u>Limnol. Oceanogr.</u>,
    26:1020.

Tilman, D., Kilham, S.S., and Kilham, P., 1982, Phytoplankton community
    ecology: The role of limiting nutrients, <u>Ann. Rev. Ecol.
    Systematics</u>, 13:349.

Tyler, M.A., and Seliger, H.H., 1981., Selection for a red tide organism:
    Physiological responses to the physical environment, <u>Limnol.
    Oceanogr.</u>, 26:310.

Vinogradov, M.E., and Menshutkin, V.V., 1977, The modeling of open-sea
    ecosystems, <u>in</u>: "The Sea, Volume 6, Marine Modeling", E.D.
    Goldberg, I.N. McCave, J.J. O'Brien and J.H. Steele, eds, pp.
    891—921, Wiley—Interscience, New York.

Walsh, J.J., 1977, A biological sketchbook for an eastern boundary
    current, <u>in</u>: "The Sea, Volume 6, Marine Modeling", E.D. Goldberg,
    I.N. McCave, J.J. O'Brien and J.H. Steele, eds, pp. 923—968,
    Wiley—Interscience, New York.

Wood, E.J.F., 1968, "Dinoflagellates of the Caribbean Sea and Adjacent
    Seas", University of Miami Press.

Woods, J.D., and Onken, R., 1982, Diurnal variation and primary production
    in the ocean-preliminary results of a Lagrangian ensemble model,
    <u>J.Plankton Res.</u>, 4:735.

Wroblewski, J.S., 1977, A model of phytoplankton plume formation during
    variable Oregon upwelling, <u>J.Mar.Res.</u>, 35:357.

Wroblewski, J.S., 1980, A simulation of the distribution of Acartia clausi
    during Oregon upwelling, August 1973, <u>J.Plankton Res.</u>, 2:43.

Zentara, S.-J., and Kamykowski, D., 1977, Latitudinal relationships among temperature and plant nutrients along the west coast of North and South America, J.Mar.Res., 35:321.

Zentara, S.-J., and Kamykowski, D., 1981, Geographic variations in the relationship between silicic acid and nitrate in the South Pacific Ocean, Deep-Sea Res., 28:455.

APPENDIX I. Definitions of Terms of Equations

Example 1

V -   velocity of nutrient uptake at ambient substrate concentration

V_m -   maximum uptake velocity

NO_3 -   ambient nitrate concentration

K_s -   half saturation constant for Michaelis Menten equation for both nitrate and ammonium.

λ -   coefficient of ammonium effect

NH_4 -   ambient ammonium concentration

Example 2

w_1 -   vertical velocity due to first internal wave

a_1 -   amplitude of first internal wave

T_1 -   period of first internal wave

E_1 -   wavelength of first internal wave

x -   spatial position offshore

w_2, a_2, T_2, E_2 -   same as above for second internal wave

w_3, a_3, T_3, E_3 -   same as above for third internal wave

w_p -   combined vertical velocity for all internal waves

z -   spatial position of depth

z_T -   depth of thermocline

w_0 -   organism velocity due to physical and biological vectors

w_b -   organism swimming velocity

Example 3

μ - organism growth rate at ambient substrate concentration

μ_m - maximum growth rate

K_m - half saturation constant for the Monod equation

S   - ambient substrate concentration

A   - constant

E   - activation energy

R   - universal gas constant

T   - temperature (^oK)

V   - velocity of nutrient uptake of ambient substrate concentration

V_m - maximum uptake velocity

T   - ambient temperature

S   - ambient substrate concentration

K_s - half saturation constant for Michaelis-Menten equation

$\alpha_1, \alpha_2, \alpha_3, \alpha_4, \beta_1, \beta_2, \gamma_1, \gamma_2, \gamma_3, \gamma_4$ - rate constants

Example 4

N_i - number of cells of species i per unit volume

r_i - maximal growth rate of species i

S_j - concentration of resource j external to the cells

K_{ij} - half saturation constant for species i limited by resource j

D   - steady state growth rate

$_oS_j$ - influent concentration of resource j

Y_{ij} - yield of species i limited by resource j

n   - number of species present

m   - number of potentially limiting resources

APPENDIX II. Data for Relationship between Swimming Speed and Body Length

	SPECIES	BODY LENGTH μm	SWIMMING SPEED $\mu m \ s^{-1}$	SOURCE
1.	*Amphidinium carterae*	1	75.1	1,10
2.	*Amphidinium Klebsi*	35	73.9	5,6
3.	*Cachonina illdefina*	23	206	3,9
4.	*Cachonina niei*	13	100	3,10
5.	*Ceratium furca*	85	195	4,6
6.	*Ceratium fusus*	250	100	4,6
7.	*Ceratium horridum*	50	14	4,6
8.	*Ceratium longipes*	200	167	4,6
9.	*Ceratium macroceros*	55	14	4,6
10.	*Ceratium tripos*	82.5	167	4,6
11.	*Dinophysis acuta*	62.5	500	4,7
12	*Gonyaulax polyedra*	48	231	2,7
13.	*Gyrodinium dorsum*	72	294	2,8
14.	*Peridinium claudicans*	75	181	4,6
15.	*Peridinium crassipes*	90	100	4,6
16	*Peridinium ovatum*	62.5	167	4,6
17.	*Peridinium pentagonum*	87.5	250	4,6
18.	*Peridinium subinerme*	62.5	250	4,6
19.	*Prorocentrum micans*	42.5	336	3,6

SPEEDS

1. Gittleson et al. (1974)
2. Hand et al. (1965)
3. Kamykowski (1981b)
4. Peters (1929)
5. Ronkin (1959)

LENGTHS

6. Wood (1968)
7. Brennen & Winet (1977)
8. Kofoid & Swezy (1921)
9. Herman & Sweeney (1976)
10. Kamykowski & Zentara (1977)

WHAT CONTROLS THE VARIABILITY OF CARBON DIOXIDE IN THE SURFACE OCEAN?

A PLEA FOR COMPLETE INFORMATION

Peter G. Brewer[*]

Woods Hole Oceanographic Institution
Woods Hole, Massachusetts 02543 U.S.A.

ABSTRACT

The annual exchanges of carbon dioxide gas between the atmosphere and the surface ocean, and the slow but inexorable oceanic uptake of fossil fuel CO_2, are among the most important problems addressed by ocean chemists today.

There is wide agreement as to experimental procedures in this field, through the measurement of alkalinity, total CO_2, pH and pCO_2 of ocean waters, together with physical properties and ^{14}C. Moreover the uncertainties surrounding ocean physical chemistry and the representation of the various thermodynamic constants necessary to define this system now appear to have been solved. There is however widespread disagreement as to the relative importance of physical and biological controls on the CO_2 chemistry of the surface ocean. Indeed one may divide the scientific papers in this field into two virtually distinct piles: those regarding biological controls as dominant, and those invoking the dominance of physical processes.

What appears to be true is that a complete experiment separating these effects has yet to be done, recognizing that the feedback between ocean mixing and biological productivity prevents, of course, a truly separate set of processes. The great variability of the ocean in space and time suggests the need for a well-coordinated program if progress is to be made in furthering our knowledge of the CO_2 system.

INTRODUCTION

The purpose of this paper is to review the situation regarding current opinions on the controls of the variability of surface ocean CO_2 properties, and to present new data from the TTO (Transient Tracers in the Ocean) experiment in the North Atlantic Ocean.

_____
[*]Paper presented while the author was on leave of absence at the Ocean Sciences Division, National Science Foundation, Washington, DC 20550, U.S.A.

Woods Hole Oceanographic Institution Contribution No. 6170

215

The subject matter is of great interest: the fluctuations in upper ocean CO_2 chemistry potentially provide direct evidence for the photo-synthetic fixation of carbon by primary productivity, the basis for virtually all marine food chains. The heating and cooling of the ocean exerts a strong influence on the gas partial pressure which, because of the relatively slow exchange rate of CO_2 with the atmosphere, causes large scale latitudinal gradients. The rise in atmospheric CO_2 levels, due to fossil fuel burning and land use changes, will probably perturb climate, and ocean uptake of this excess CO_2 provides the principal geochemical constraint.

The fossil fuel CO_2 problem has stimulated a great deal of research in this area. However, even in the absence of a fossil fuel CO_2 signal, there would be compelling reason for study. The glacial to inter-glacial variations in atmospheric CO_2 levels are believed to be controlled by the ocean (Siegenthaler and Wenk, 1984; Sarmiento and Toggweiler, 1984). From the atmospheric CO_2 monitoring network, we now know the total global atmospheric CO_2 content to about 0.5 to 1.0% (Fraser et al., 1983). If the fossil fuel input is known, then fluctuations in this content must result from exchange with the ocean or terrestrial biosphere. If we knew the absolute fluxes from either one of these other reservoirs then our ability to constrain the system would be very greatly enhanced. Absolute measurements of CO_2 fluxes from the complex global terrestrial biosphere are not attainable in the foreseeable future; however, with large scale expeditions to define the rules, and the possible future application of remote sensing of ocean currents, winds and temperatures, a greatly improved knowledge of ocean seasonal and inter-annual CO_2 fluxes could be within our grasp.

We would like eventually to be able to say with some certainty that this must have been a good or bad year for the terrestrial vegetation, and in what hemisphere the changes occurred. Or that unusual oceanic temperatures or primary productivity in say the South Pacific caused thus and such a perturbation in the atmospheric CO_2 budget. The benefits then, of increased knowledge about our world, would be substantial.

The key to this, as mentioned above, is to define the rules. However, as we will see, investigators working in this field have reached quite divergent opinions, and one may divide the scientific papers in this field into two virtually distinct piles: those regarding biological controls as dominant, and those invoking the dominance of physical processes. Our choice of what to measure and where, and when and how, will depend upon our resolution of these conflicts.

LARGE SCALE VARIABILITY

Let us begin with the large scale map (Fig. 1) of Keeling (1968) showing pCO_2, expressed as air-sea disequilibrium, at the ocean surface. A similar map has been prepared by Takahashi (Broecker and Takahashi, 1984) based upon data from the GEOSECS program, and Miyake et al. (1974) have prepared a somewhat more detailed map of surface pCO_2 in the Pacific Ocean. The data shown in Fig. 1 represent largely summertime observations, and no seasonal adjustment has been applied. A complex system of highs and lows is revealed. The most pronounced high is in the equatorial Pacific; a lesser high is observed in the Atlantic, and in the Indian Ocean the equatorial zone appears to be very close to atmospheric equilibrium.

Strong lows in pCO_2 are found in the North Pacific, and particularly in the North Atlantic where summertime values of -120 ppm (or about -33% equilibrium) are observed.

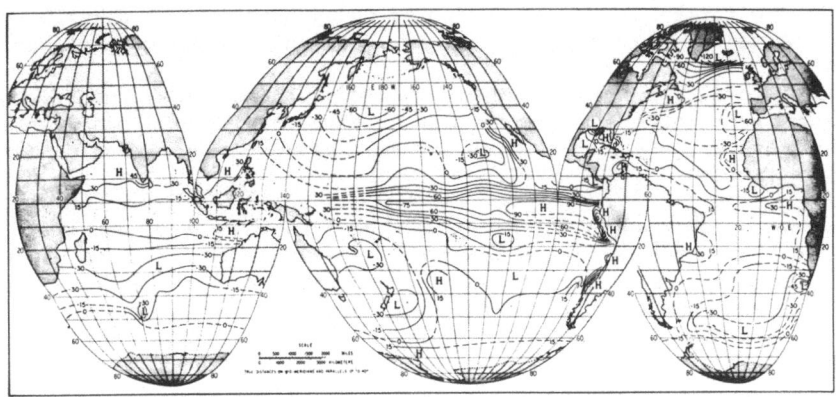

Fig. 1. Map of the distribution of surface ocean pCO_2 expressed as air-sea disequilibrium in ppm (from Keeling, 1968).

The pulling together of this data set enabled investigators to see for the first time (Skirrow, 1975) the large scale systematic variability of the pCO_2 of the surface ocean. Keeling (1968) drew attention to the similarity of this map to the phosphate distribution, suggesting that the circulation patterns resulting in upwelling of phosphate and CO_2 rich water at the equator, connected by the stoichiometry of the Redfield relationship (Redfield, 1934), resulted in this correlation. Low nutrient, low pCO_2 water tends to be found in the sub-tropical oligotrophic gyres.

Lack of biological activity can however hardly result in low pCO_2, and thus additional constraints of heating and cooling must be involved. Moreover phosphate ion cannot be exchanged in large quantities across the sea surface, whereas CO_2 gas exchange with the atmosphere (see Liss and Slinn (1983) for a comprehensive review) must play a large role. All of these features are recognized in a general way in the text of these papers, yet a complete and quantitative description still eludes us.

PERTURBING PROCESSES

In the absence of any perturbation the sea surface pCO_2 would be in equilibrium with the atmosphere. Baes (1982) has neatly summarized the effect of the perturbing processes by means of vectors on a total alkalin-ity-total CO_2 diagram (Fig. 2). On such a diagram lines of derived proper-ties such as pCO_2 or pH may be constructed, and the vectors representing composition changes intersect these lines with characteristic slopes (Deffeyes, 1965). At constant temperature the essential features are:

(1) When gas exchange takes place the alkalinity is unaltered and total CO_2 alone is changed, with predictable response in pCO_2;

(2) Biological activity with the formation or decomposition of soft tissue will change total CO_2, with only small changes in alkalinity. The vectors given by Baes (1982) in Fig. 2 show the effect of photosynthesis on the alkalinity-total CO_2 system for growth on a nitrate—nitrogen nutrient source. It should be noted that photosynthesis on an ammonium-nitrogen nutrient source would have the effect of reducing, rather than increasing, alkalinity (Goldman and Brewer, 1980);

217

(3) The formation or dissolution of $CaCO_3$ changes both alkalinity and
     total CO_2 in the proportion of 2 equivalents per mole of carbon.

The vectors represented here are explicit. Potentially then if the
surface CO_2 properties of the ocean could be reduced to a simple representation in total CO_2-alkalinity space a formal separation of these processes
could be attempted.

Brewer and Dyrssen (1984) have shown that for the near isothermal
surface waters of the Persian Gulf where large chemical gradients occur
these concepts can be useful. However for oceanic surface waters of
strongly varying temperature and salinity the assumptions necessary for a
simple application of the concepts expressed in Fig. 2 break down. Moreover kinetic factors associated with seasonality, or the discrepancies
between oceanic mixing and gas exchange rates, cannot be represented in
this scheme. We should however explore these concepts a little further,
and for illustrative purposes we use the TTO North Atlantic data set.

The CO_2 system results of this large scale experiment have recently
been presented by Brewer et al. (1986). The experiment involved in part
the accurate determination of alkalinity and total CO_2 of ocean waters by
means of potentiometric titration (Dyrssen and Sillén, 1967; Bradshaw et
al., 1981). The precision of individual measurements was approximately
± 3 μeq kg^{-1} ($\pm 0.13\%$) in alkalinity, and ± 4 μmol kg^{-1} ($\pm 0.2\%$) in total CO_2.
Comparison with a more limited set of independent measurements of total CO_2
by a highly accurate gas extraction technique (C.D. Keeling, personal communication) showed excellent agreement. In the deep ocean the difference
between the two techniques was 4.0 ± 2.9 μmol CO_2 kg^{-1} ($0.2 \pm 0.15\%$); for
surface ocean waters the discrepancy increased to 7.2 ± 5.9 ($0.36 \pm 0.3\%$)
μmol kg^{-1}, possibly due to the presence of unrecognised protolytes in seawater participating in the titration scheme. The data obtained therefore
appear to be of exceptional quality.

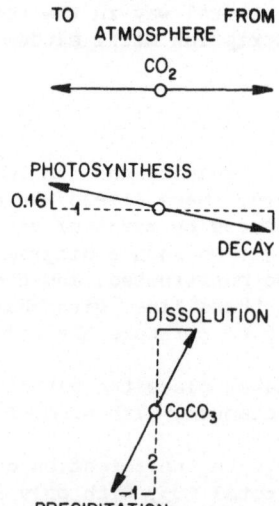

Fig. 2. Vector diagram showing the shifts in alkalinity
        and total CO_2 caused by air-sea exchange, photo-
        synthetic growth or organic decay, and calcium
        carbonate formation or dissolution (from Baes,
        1982).

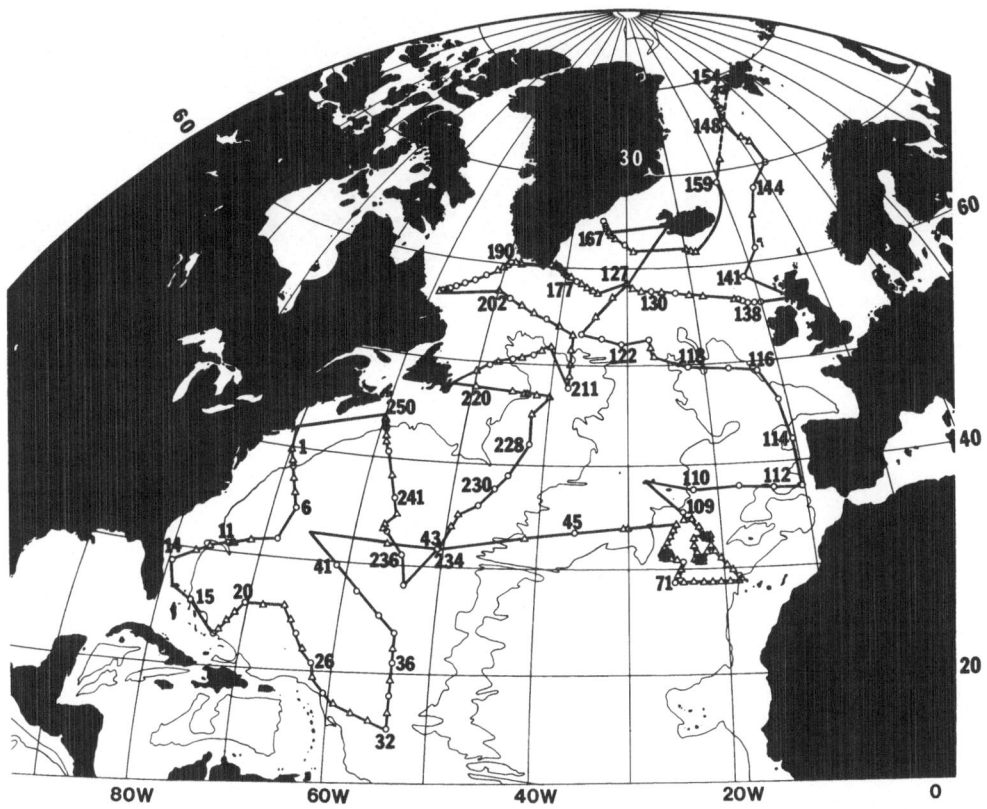

Fig. 3. The cruise track and station locations for the TTO North Atlantic Study, 1981.

The cruise track for the TTO North Atlantic expedition is shown in Fig. 3. The total alkalinity—total CO_2 diagram for these surface waters is shown in Fig. 4, together with atmospheric equilibrium pCO_2 curves at various temperatures. The strong variations in temperature and salinity, and their influence on ocean CO_2 properties negate a simple analysis, and attempts by this author to separate the processes statistically yield inconclusive results.

Faced with the lack of hard conclusions to be drawn from the application of simple concepts, such as those represented by TA-TC diagrams, to real data, investigators have had considerable freedom to invoke the dominance of processes they feel to be important, or that have been observed in their particular experiment. The multitude of processes that could be influential in perturbing the CO_2 system is such that complete constraints are rarely experimentally accessible to the single investigator.

OBSERVATIONS SUPPORTING DOMINANT BIOLOGICAL CONTROLS

There is a widespread belief that biological draw down of CO_2 in the spring and summer, and regeneration in winter, exerts the primary control on ocean surface CO_2 distributions. Indeed many papers strongly emphasize these controls. Perhaps the strongest statement is that of Simpson and Zirino (1980) who measured pH, salinity and chlorophyll fluorescence in surface waters off Peru. They assumed constancy of the salinity–alkalinity relationship, and they calculated pCO_2 from the alkalinity (calculated) —

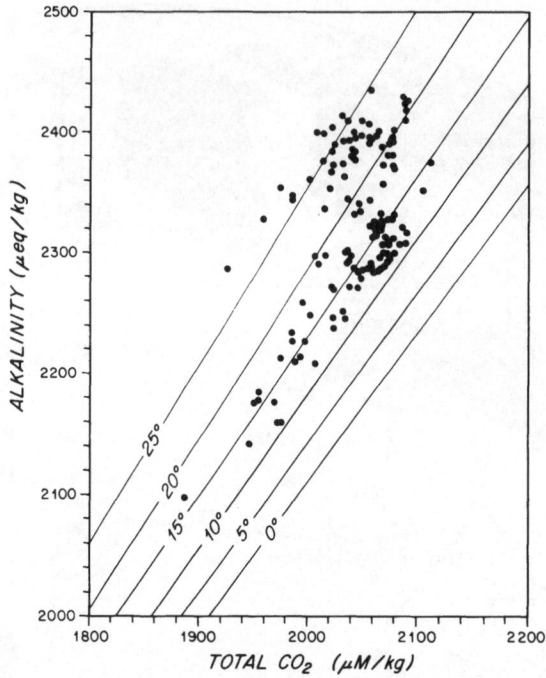

Fig. 4. Total alkalinity — total carbon
dioxide diagram for North
Atlantic surface waters (TTO
data, 1—15m). The lines show
atmospheric equilibrium pCO$_2$ (340
ppm) at various temperatures.

pH (measured) pair. Their data are shown in Fig. 5. The strong correlation between pCO$_2$ and chlorophyll led them to state unequivocally that
biological controls were dominant on the CO$_2$ system. Note however that the
pH scale in Fig. 5 is apparently in error by about 0.3 pH units, since the
value at atmospheric equilibrium today should be close to 8.1. Moreover
since pH is used as the principal measured variable in calculating pCO$_2$,
the x (pH) and y variables are not independent. The chlorophyll data are
however independent. Although biological productivity must clearly have
significant influence on this system, the lack of knowledge of gas exchange
rates, mixing and heat gain or loss makes the argument incomplete.

On a smaller scale Johnson et al. (1979) followed changes in the pH,
alkalinity, and O$_2$ content of surface waters in Stuart Channel, British
Columbia for a 15 day period in July 1976. The flux of O$_2$ across the sea
surface, and its relationship to biological activity, has a rich invest-
igative history (Redfield, 1948). Correlations of these fluxes with
CO$_2$ measurements are however quite rare. Over this short observational
period they found CO$_2$ uptake of 10.8 μmol CO$_2$ l^{-1} day^{-1}, and gaseous
invasion of CO$_2$ from the atmosphere at 0.49 μmol CO$_2$ l^{-1} day^{-1}. The
changes resulting from heat flux or mixing were found to be small. They
concluded that, at this site, "the magnitudes and even the signs of the
changes in O$_2$ and CO$_2$ due to exchange depended primarily on the net amount
of primary production".

Although the small space and time scale of the above experiment precludes its generality, similar results have been obtained on larger

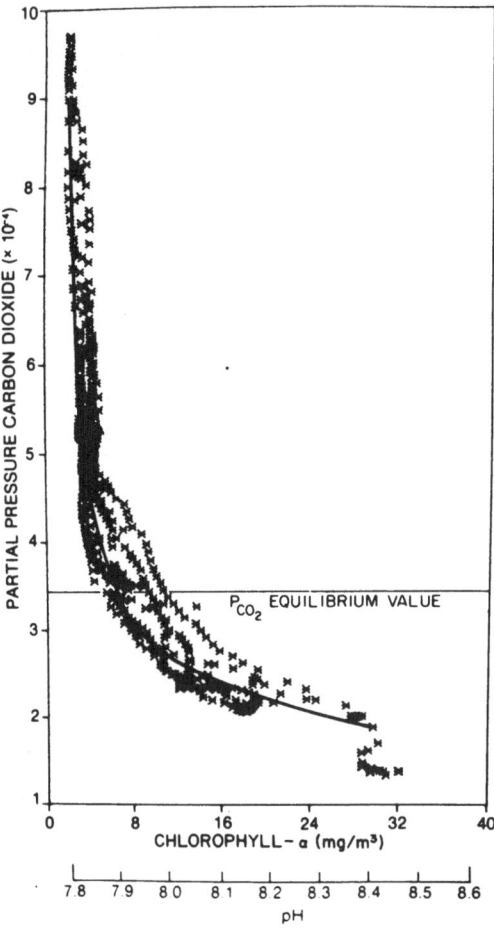

Fig. 5. Correlation between chlorophyll
and pCO₂ for surface waters off
Peru (from Simpson and Zirino,
1980).

scales. Codispoti et al. (1982) reported temporal changes in surface ocean
pCO₂ in the Bering Sea in 1980. Their data (Fig. 6) from repeat occupat-
ions of transects out from the coast showed that during the April-May
period surface pCO₂ fell by about 200 ppm. They estimated that some 30% of
the change in total CO₂ was due to calcium carbonate formation (though
without direct observation of the flux). But the dominant control again
invoked was utilization of CO₂ by phytoplankton during the spring bloom.

Even at the very largest scales the biological control of surface
ocean CO₂ properties is emphasized by some. Newell and Weare (1977) and
Newell et al. (1978) investigated the correlations between changes in
atmospheric CO₂ concentration (after removal of the dominant terrestrial
biosphere signal) and long-term global sea surface temperature variations.
They found the principal correlation to be associated with the variability
of the tropical Pacific Ocean, with atmospheric CO₂ changes following the
water temperature changes by one or two seasons. The correlation was
positive in that warmer Pacific surface temperatures lead to higher
atmospheric CO₂ values. The suggestion was that this was dominated by
upwelling events; more cold, nutrient rich water upwelled leading to more

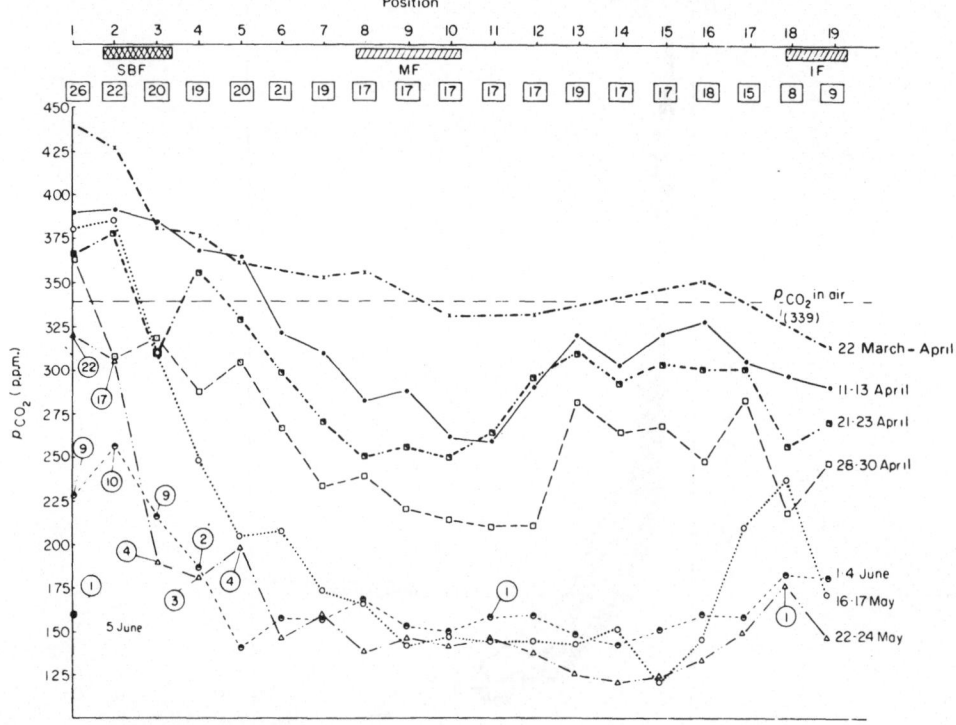

Fig. 6. Change in surface pCO_2 in April—May, 1980 in the Bering Sea (from Codispoti et al., 1982).

photosynthesis and thus more draw down on surface ocean CO_2 and, with a time lag of about one season, atmospheric CO_2 values. The surface ocean biological processes then appear, in this analysis, to exert a significant control on global atmospheric CO_2 fluctuations.

THE CASE FOR THE DOMINANCE OF PHYSICAL PROCESSES

The most categorical statement regarding the dominant control of purely physical processes is perhaps that in the paper by Weiss et al. (1982). These investigators made measurements of pCO_2 and total CO_2 of surface sea water, with remarkable precision and accuracy, during the NORPAX Equatorial Experiment. The experiment utilized repeat transects across the equator from Papeete to Honolulu, and from this time series of measurements the seasonal change in these properties could be observed. Weiss et al. (1982) calculated the fugacity of CO_2 (numerically closely equivalent to the partial pressure) as a function of temperature and salinity and found that the observed variations compared extraordinarily well with predictions based upon temperature and salinity changes alone. Their results are shown in Fig. 7. The inclusion of air-sea exchange fluxes in the model calculations affected the results only slightly.

As a result of these calculations the changes due to alkalinity were predicted to be small, with the specific alkalinity in the surface waters of the north and south Pacific subtropical gyres estimated as varying no more than ±2 μeq kg^{-1}. Biological activity was neglected in these calculations, with the estimation that this contributed no more than 0.3-0.6 μatm variation in the calculated fCO_2; less than 3% of the observed signal.

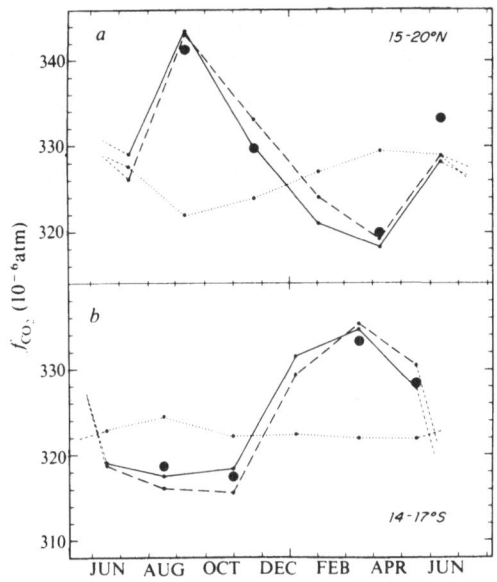

Fig. 7. Seasonal cycle of pCO_2 in
surface waters of the North and
South tropical Pacific gyres
(from Weiss et al., 1982).
Large dots indicate oceanic
analyses; small dots correspond
to atmospheric data. The solid
line connecting the oceanic
points represents a closed
system temperature—salinity fit
to the data; the dotted line
includes air—sea exchange
parameters in the fit.

Takahashi et al. (1980) reviewed the data from the GEOSECS expeditions
on the carbonate chemistry of the surface waters of the world's oceans.
Surface alkalinity in the major oceans covaries strongly with salinity,
with a shift being observed in the Antarctic due to upwelling of deep
waters. The total CO_2 concentration, normalized to a constant salinity and
therefore constant alkalinity, was shown to be closely correlated with
temperature. A recent paper by Brewer et al. (1986), examining the correlations in the TTO North Atlantic data set, supports these conclusions.
Upwelling at the equator produces elevated CO_2 levels which then follow
largely an isochemical cooling curve during poleward flow.

Bacastow (1977) analyzed the effect of the Southern Oscillation on
atmospheric CO_2 levels. The Southern Oscillation Index, an excellent
indicator of El Niño events, and defined as the difference in average
monthly barometric pressures between Easter Island and Darwin, Australia,
is related to the Pacific trade winds. The relationship is positive, high
Southern Oscillation Index indicating strong trade winds.

After removing the seasonal terrestrial vegetation effects, and the
fossil fuel CO_2 signal, from the atmospheric record, the residual anomaly
curves were found to correlate well with the Southern Oscillation Index.
Bacastow (1977) concluded that the correlation resulted from increased
ocean uptake of CO_2, due to physical processes, during periods of high

trade winds, the effect of wind stress overriding the additional CO_2 believed to be brought up with the associated increased upwelling.

Note that although the correlation techniques used here are similar to those of Newell et al. (1978), the conclusions and explanations offered are quite different. Even the sign of the ocean temperature — CO_2 flux correlation is debated. In contrast to the work of Newell cited earlier, Machta et al. (1977) report lower sea surface temperatures associated with higher atmospheric CO_2 levels.

Even over very long time and space scales the dominance of physical processes has been emphasized by some. Pearman et al. (1983) in an elegant paper have used the observed oceanic pCO_2 signal, together with a two-dimensional global atmospheric transport model, to investigate spatial and temporal changes in atmospheric CO_2. Taking the observed meridional distribution of CO_2 in ocean surface waters from Keeling (1968) and Takahashi (1977) they corrected the data for temperature changes so as to produce a meridional distribution for the month of June. The air-sea flux (F) of CO_2 was calculated from the relationship

$$F = 4.8 \times 10^4 (1 + 0.007T)(pCO_2 - C_a)u^* \tag{1}$$

where T is the temperature in °C, C_a is the atmospheric concentration of CO_2 in ppmv, and u^* is the friction velocity in m s^{-1}. The friction velocities were adjusted upwards by a factor of 1.7 to compensate for the observed effect of wind speed on gas exchange rates. Incorporating real data on friction velocities and sea surface temperature, varying with season and latitude, they calculated the changing flux of CO_2 from air to sea over the period of the last century. The rising level of atmospheric CO_2 was shown to have enhanced the oceanic polar uptake by accentuating the disequilibrium in this region, and to have diminished the equatorial outgassing flux by reducing the disequilibrium there.

The model calculations were used to place strict limits on fluxes of CO_2 from the biosphere. The essential features of the calculation are that both seasonality, and the latitudinal dependence, of the pCO_2 signal are driven principally by changes in temperature and mixed layer depth (D), here represented by

$$D = D_0 + (0.44 \, \phi \, \cos(30J°)) \tag{2}$$

where D_0 is the average mixed layer depth (80m), ϕ is the latitude, and J is the month.

Their model fit, and the observations used, are shown in Fig. 8. The paper presents a most clear and convincing account of the latitudinal, seasonal and temporal fluxes of CO_2. The reader will note however that, as clearly stated by Pearman et al. (1983), no attempt was made to simulate the effects of biological activity.

DISCUSSION

It would be a great disservice to the fine scientists whose work is reviewed here to imply that somehow they are not aware of these conflicts. It is also a simple truism to remark that both physical and biological processes are important and must be formally represented if we are to have a working knowledge of the forces controlling seasonal and spatial variability in surface ocean CO_2 properties. The difficulty is to proceed quantitatively beyond this truism.

Broecker and Takahashi (1984) have considered long term (i.e. glacial-interglacial) constraints operating on either a purely "thermodynamic" or a biologically dominated "Redfield" ocean. They reached the natural conclusion that the real ocean lay in between.

Papers devoted to modeling these processes in the ocean, for instance the very detailed studies by Bolin (1983) and Bolin et al. (1983), recognize these complexities in a formal way. This work is most encouraging. Yet the coarse resolution currently attained in these models precludes their application here.

For each of the papers mentioned earlier some caveat may be found, or some conflict with other work exists. Simpson and Zirino (1980) for instance relied upon a simple correlation to emphasize their point, but could not include the effects of mixing.

The different interpretations of the Southern Oscillation Index-CO_2 or temperature correlations by Newell et al. (1978) and Bacastow (1977) may be further complicated by data from the recent El Niño event of 1982-83 (R. Gammon, personal communication). During this time a marked decrease in the atmospheric CO_2 growth rate seems to have occurred. One simple explanation could be that the relaxing of the trade wind fields characterizing such an event permits warm western tropical Pacific surface water to travel eastwards, thereby capping the normal upwelled CO_2-rich water and preventing the normal degassing to the atmosphere. This would support the correlation found by Machta et al. (1977).

In examining the seasonal signal presented by Weiss et al. (1982) we

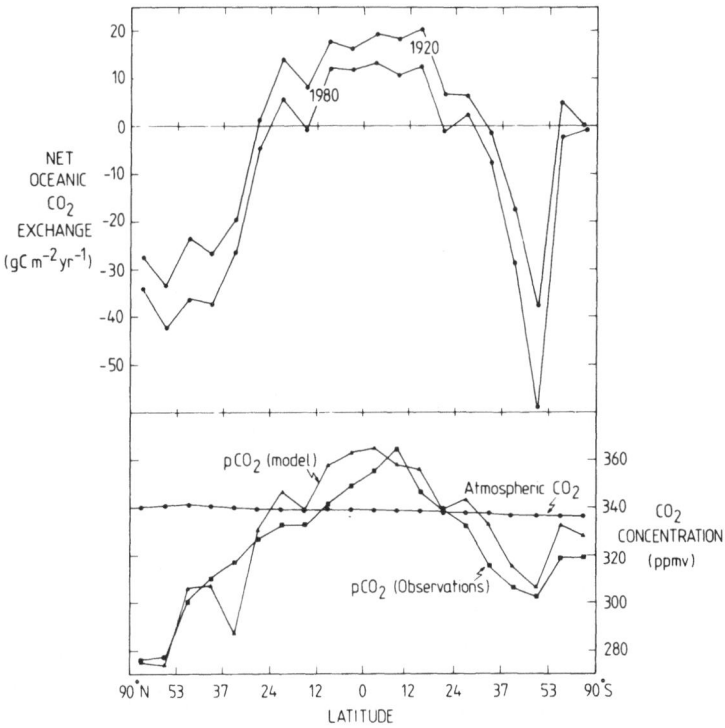

Fig. 8. The North-South latitudinal gradient of surface pCO_2 in the Atlantic Ocean and a model fit (from Pearman et al., 1983).

note that there is no parallel discussion of oxygen cycling. Reference to the original data sets (Williams, 1981) reveals a supersaturation of O_2 gas at the sea surface at these locations in all seasons varying from +8 μmol kg^{-1} to +3 μmol kg^{-1}, with larger excesses of O_2 just below the mixed layer.

The existence of consistently supersaturated O_2 values at the sea surface is discussed in the text by Broecker and Peng (1982). Since the gas exchange rate for O_2 is reasonably well known the rate of supply of O_2, necessary to maintain this excess and probably resulting from photosynthesis, may be calculated. The primary production rates thus inferred are some five times the rates traditionally determined by the ^{14}C method, or about 30 mol m^{-2} yr^{-1}. (A recent paper on seasonal oxygen cycling and primary production in the Sargasso Sea by Jenkins and Goldman (1984) clearly reinforces this conclusion. These authors however had no companion data on CO_2 seasonality!) Even a seasonal oscillation of 5 μmol O_2 kg^{-1} as given above, if correct and if due to photosynthesis, would at a minimum (ignoring the approximately ten-fold slower exchange of CO_2) yield a biologically derived pCO_2 signal of about 8 ppm, or ten times the range estimated by Weiss et al. (1982).

The temperature/mixed layer depth model of Pearman et al. (1983) leads to the prediction of very strong wintertime fluxes of CO_2 into the North Atlantic. However wintertime measurements made there (R.F. Weiss, personal communication) show that, in contrast to the summertime disequilibrium of -30%, wintertime surface pCO_2 values north of Iceland are only about -10% with respect to the atmosphere. Winter ice effects, and summer draw down of CO_2 from primary production, must plainly influence the signal.

RECENT RESULTS FROM THE TTO PROGRAM

There are no perfect answers to these complex problems, but recent results from the North Atlantic may help. The TTO experiment (Fig. 3) yielded a consistent set of data for the North Atlantic Ocean for the period April-October 1981. The partial pressure of CO_2 gas in surface waters along the TTO cruise track was measured directly by automated gas chromatography (Weiss, 1981), but these results are not yet available. The titrator determinations of akalinity and total CO_2 have however been completed (Brewer et al., 1986) and these also yield (calculated) values of pCO_2 which are shown in Fig. 9.

The presentation in Fig. 9 is of a three-dimensional map of the sea surface, seen from the perspective of atmospheric equilibrium (340 ppm) and viewed from the U.S. east coast. The view from Jouy-en-Josas, France would of course be equally attractive. There has been one interesting correction term applied; the sea surface CO_2 data have been corrected for the presence of excess oxygen (negative AOU) by simple subtraction of the Redfield CO_2 equivalent:

$$+ 1 \ \mu\text{mol} \ O_2 \ kg^{-1} = -106/138 \ \mu\text{mol} \ CO_2 \ kg^{-1} \tag{3}$$

The correction term averages about 7 μmol O_2 kg^{-1}. It implies high ocean primary productivity (Broecker and Peng, 1982), but ignores the approximately ten-fold slower CO_2 exchange rate. It is then a crude correction term for biological drawdown of surface CO_2, and has the effect of raising the apparent ocean surface CO_2 levels. The residual signal should then more closely approximate the effects of purely physical processes.

It appears that this is so and we offer the following tentative

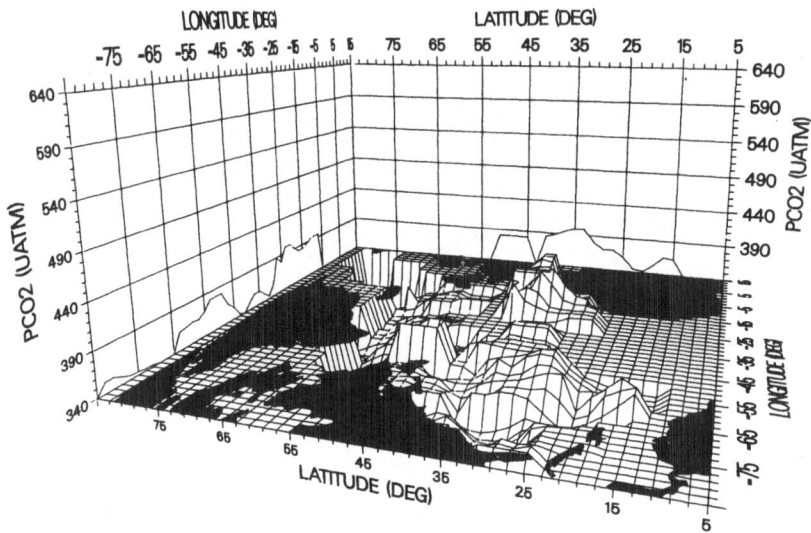

Fig. 9. Three-dimensional plot of surface pCO_2 in the North
Atlantic based upon TTO data (surface, 1—15 m). A
correction term for the surface excess of O_2 has been
applied. The view is from the U.S.A. and from the
perspective of atmospheric equilibrium.

explanation. The signal in Fig. 9 is quite similar to the maps, prepared
by Bunker (1976), of the annual average heat gain of the ocean (Fig. 10).
The qualitative similarity results from heating and cooling of the ocean
faster than exchange of CO_2 with the atmosphere can take place, and thus
anomalies created by these processes persist. Water advected northwards by
the Gulf Stream is swept by cold, dry, continental air over much of the
year, resulting in a large latent heat loss. The negative heat gain here
is greater than in any other area of any ocean, however the very high
velocities mitigate the actual sea surface temperature drop. The cooling
results in lowered pCO_2. Recirculation of surface water around the
temperate anti-cyclonic gyre (Worthington, 1976) results in eventual southward transport and in the vicinity of 35°N 55°W, a region of zero annual
average heat gain, in pCO_2 equilibrium with the atmosphere, is found. To
the south and to the east the gain of heat is matched by pCO_2 levels well
above atmospheric.

Off Newfoundland the cold Labrador Current advects water southwards
along the coast to the Grand Banks. There condensation of water vapor and
solar radiation result in rapid heat gain, and elevated pCO_2. The low to
the east of this results from further cooling of a branch of the North
Atlantic Current. North of 51°N the consistent heat loss, and high
productivity, result in lowered pCO_2 over the entire region. The summertime bias must accentuate the Newfoundland high, and diminish the Gulf
Stream low.

The correlations given here are not perfect, and the attempt to
separate biological and physical perturbing processes is awfully crude.
But the signal revealed does give a tantalizing hint of what might be done.

Ocean heat flux is an objectively measurable quantity. The flux of
carbon from the euphotic zone to the deep ocean can be measured with moored

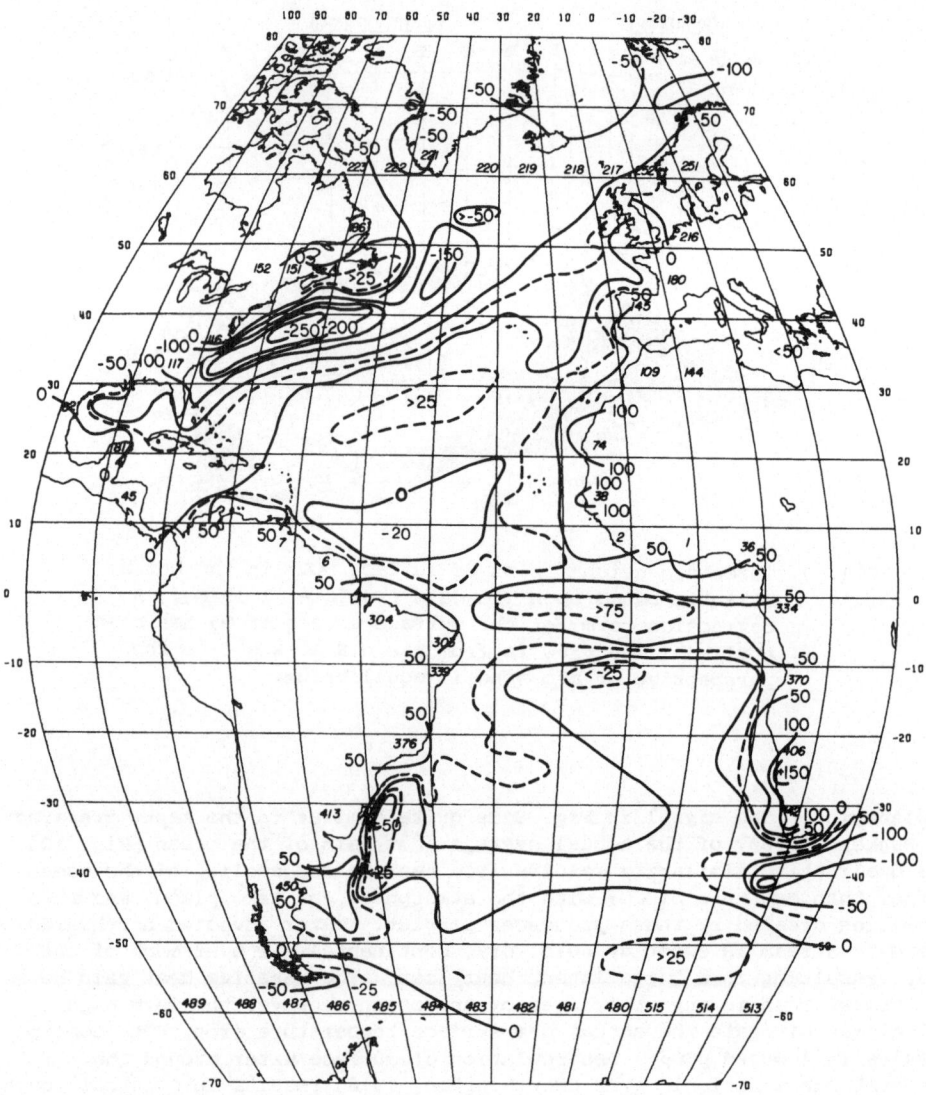

Fig. 10. Map of the annual average heat gain (W m^{-2}) of the North Atlantic (from Bunker, 1976).

sediment traps, and the technology now available permits elegant observations (Honjo et al., 1980) and seasonal studies (Deuser and Ross, 1980; Honjo, 1982). If we can make the connections between physical and biological observables, and ocean CO_2 fluxes, it would be an important scientific advance.

CONCLUSIONS

The divergent views in the literature on the relative importance of physical or biological processes in perturbing surface ocean CO_2 properties are striking. Unless we have a solid grasp of these processes we will not be able adequately to interpret time series measurements, nor to devise sampling schemes, which unambiguously document the changing chemistry of the ocean.

228

A substantial part of the problem seems to be the difficulty in providing enough information from observations at any one site, given the resources usually available to the single investigator. What appears to be needed is a well coordinated attack on the problem combining at a minimum sediment trap, tracer, gas exchange and CO_2 chemistry techniques, together with O_2 measurements of the highest accuracy and precision. Planning such an experiment would take great care, but would surely be well worth the effort.

ACKNOWLEDGEMENTS

The author is indebted to the scientists whose data is discussed here, and to the organizers of the NATO ARI series for a stimulating conference. D. Shafer carefully produced Fig. 9. Helpful comments have been provided by A. L. Bradshaw. This work was supported by grants NSF OCE 81-08160 and 83-16709, and by the U.S. Department of Energy.

REFERENCES

Bacastow, R., 1977, Influence of the southern oscillation on atmospheric carbon dioxide, in: "The Fate of Fossil Fuel CO_2 in the Oceans", N.R. Andersen and A. Malahoff, eds, pp. 33-43, Plenum Press, New York.

Baes, C.F., 1982, Ocean chemistry and biology, in: "Carbon Dioxide Review 1982", W.C. Clark, ed., pp. 187-211, Clarendon Press, Oxford.

Bolin, B., 1983, Changing global biogeochemistry, in: "Oceanography: the Present and Future", P.G. Brewer, ed., pp. 305-326, Springer-Verlag, New York.

Bolin, B., Bjorkstrom, A., Holmen, K., and Moore, B., 1983, The simultaneous use of tracers for ocean circulation studies, Tellus, 35B:206.

Bradshaw, A.L., Brewer, P.G., Shafer, D.K., and Williams, R.T., 1981, Measurements of total carbon dioxide and alkalinity by potentiometric titration in the GEOSECS program, Earth Planet. Sci. Lett., 55:99.

Brewer, P.G., and Dyrssen, D., 1984, Chemical oceanography of the Persian Gulf, Progr. Oceanogr., 14:41.

Brewer, P.G., Bradshaw, A.L., and Williams, R.T., 1984, Measurements of total carbon dioxide and alkalinity in the North Atlantic Ocean in 1981, in: "The Global Carbon Cycle: Analysis of the Natural Cycle and Implications of Anthropogenic Alterations for the Next Century", D. Reichle, ed., pp. 358-381, Springer-Verlag, New York.

Broecker, W.S., and Peng, T.H., 1982, "Tracers in the Sea", Eldigio Press, Palisades, N.Y.

Broecker, W.S., and Takahashi, T., 1984, Is there a tie between atmospheric CO_2 content and ocean circulation?, in: "Climate Processes and Climate Sensitivity", Geophysical Monograph 29, pp. 314-326, American Geophysical Union, Washington, D.C.

Bunker, A.F., 1976, Computations of surface energy flux and annual air-sea interaction cycles of the North Atlantic Ocean, Mon. Weather Rev., 104:1122.

Codispoti, L.A., Friederick, G.E., Iverson, R.L., and Hood, D.W., 1982, Temporal changes in the inorganic carbon system of the south-eastern Bering Sea during Spring 1980, Nature, Lond., 296:242.

Deffeyes, K.S., 1965, Carbonate equilibria: a graphic and algebraic approach, Limnol. Oceanogr., 10:412.

Deuser. W.G., and Ross, E.G., 1980, Seasonal change in the flux of organic carbon to the deep Sargasso Sea, Nature, Lond., 283:364.

Dyrssen, D., and Sillén, L.G., 1967, Alkalinity and total carbonate in sea water, a plea for P-T independent data, Tellus, 19:113.

Fraser, P.J., Pearman, G.I., and Hyson, P., 1983, The global distribution of atmospheric carbon dioxide 2. A review of provisional background observations, 1978-1980, J. Geophys. Res., 88:3591.

Goldman, J.C., and Brewer, P.G., 1980, Effect of nitrogen source and growth rate on phytoplankton-mediated changes in alkalinity, Limnol. Oceanogr., 25:352.

Honjo, S., 1982, Seasonality of biogenic and lithogenic fluxes in the Panama Basin, Science, N.Y., 218:883.

Honjo, S., Connell, J., and Sachs, P.L., 1980, Deep ocean sediment trap; design and function of PARFLUX Mark II, Deep-Sea Res., 27:745.

Jenkins, W.J., and Goldman, J.G., 1985, Seasonal oxygen cycling and primary production in the Sargasso Sea, J. Mar. Res., 43:465.

Johnson, K.S., Pytkowicz, R.M., and Wong, C.S., 1979, Biological production and the exchange of oxygen and carbon dioxide across the sea surface in Stuart Channel, British Columbia, Limnol.Oceanogr., 24:474.

Keeling, C.D., 1968, Carbon dioxide in surface ocean waters, 4. Global distribution, J. Geophys. Res., 73:4543.

Liss, P.S., and Slinn, W.G.N., eds, 1983, "Air-Sea Exchange of Gases and Particles", Reidel, Dordrecht.

Machta, L., Hanson, K. and Keeling, C.D., 1977, Atmospheric carbon dioxide and some interpretations, in: "The Fate of Fossil Fuel CO_2 in the Oceans", N.R. Andersen and A. Malahoff, eds, pp. 131-144, Plenum Press, New York.

Miyake, Y., Sugimura, Y., and Saruhashi, K, 1974, The carbon dioxide content in the surface waters in the Pacific Ocean, Rec. Oceanogr. Works Jpn, 12:45.

Newell, R.E., and Weare, B.C., 1977, A relationship between atmospheric carbon dioxide and Pacific sea surface temperature, Geophys. Res. Lett., 4:1.

Newell, R.E., Navato, A.R., and Hsiung, J., 1978, Long term global sea surface temperature fluctuations and their possible influence on atmospheric CO_2 concentrations, Pageoph., 116:351.

Pearman, G.L., Hyson, P., and Fraser, P.J., 1983, The global distribution of atmospheric carbon dioxide: 1. Aspects of observations and modeling, J. Geophys. Res., 88:3581.

Redfield, A.C., 1934, On the proportions of organic derivatives in sea water and their relation to the composition of plankton, James Johnson Memorial Vol., pp. 177-192, Liverpool University Press.

Redfield, A.C., 1948, The exchange of oxygen across the sea surface, J. Mar. Res., 7:347.

Sarmiento, J.L., and Toggweiler, J.R., 1984, A new model for the role of the oceans in determining atmospheric pCO_2, Nature, Lond., 308:621.

Siegenthaler, U., and Wenk, T., 1984, Rapid atmospheric CO_2 variations and ocean circulation, Nature, Lond., 308:624.

Simpson, J., and Zirino, A., 1980, Biological control of pH in the Peruvian coastal upwelling area, Deep-Sea Res., 27:733.

Skirrow, G., 1975, The dissolved gases -carbon dioxide, in: "Chemical Oceanography," Second Edition, Volume 2, J.P. Riley and G. Skirrow, eds, pp. 1-192, Academic Press, London.

Takahashi, T., 1977, Carbon dioxide chemistry in ocean water, in: "Proceedings of Workshop on the Global Effects of Carbon Dioxide from Fossil Fuels", W.P. Elliott and L. Machta, eds, U.S. Dept. Energy Conf.- 770385-uC-11, Washington, D.C.

Takahashi, T., Broecker, W.S., Werner, S.R., and Bainbridge, A., 1980, Carbonate chemistry of the surface waters of the worlds oceans, in: "Isotope Marine Chemistry", E.D. Goldberg, Y. Horibe, and K. Saruhashi, eds, pp. 291-326, Uchida Rokakuko, Tokyo.

Weiss, R.F., 1981, Determinations of carbon dioxide and methane by dual catalyst flame ionization chromatography and nitrous oxide by electron capture chromatography, J. Chromatogr. Sci., 19:611.

Weiss, R.F., Jahnke, R.A., and Keeling, C.D., 1982, Seasonal effects of temperature and salinity on the partial pressure of CO_2 in sea water, Nature, Lond., 300:511.

Williams, R.T., 1981, Hawaii-Tahiti Shuttle Experiment. Hydrographic Report, Volumes I-IV, Ref. 81-5, Technical Report, Scripps Institution of Oceanography.

Worthington, L.V., 1976, "On the North Atlantic circulation", The John Hopkins Oceanographic Studies, No. 6, John Hopkins University Press, Baltimore.

A PRELIMINARY MODEL OF THE ROLE OF UPPER OCEAN CHEMICAL DYNAMICS IN

DETERMINING OCEANIC OXYGEN AND ATMOSPHERIC CARBON DIOXIDE LEVELS

J.L. Sarmiento and J.R. Toggweiler

Geophysical Fluid Dynamics Program
Princeton University
Princeton, New Jersey 08542   U.S.A.

ABSTRACT

A first version is presented of equations for a three-dimensional model of nutrient and carbon cycling in the oceans. An analytical solution of these equations has been obtained for a one-and-a-half–dimensional "pipe" model. This solution shows that atmospheric CO_2 can be varied by changing the level of preformed nutrients. It is suggested that this mechanism may explain the lower pCO_2 values of the last ice age.

INTRODUCTION

We are in the process of developing a three-dimensional model of nutrient cycling in the oceans. The development of such a model is a two part task requiring the development of ocean circulation models as well as the development of techniques for parameterizing biological and chemical processes. The ocean circulation models that will be used for this study are tracer–calibrated primitive equation models of the type first developed by Bryan (1969). Sarmiento (1983) has described some of the tracer calibration work that is being done with these types of model. This brief account describes our preliminary ideas on how to parameterize the biological and chemical processes. Solutions will be given for a simple one–and-a-half–dimensional model. An analysis of these solutions reveals that atmospheric CO_2 concentrations and oceanic O_2 concentrations are strongly affected by the level of preformed nutrients in deep water. A new mechanism is suggested for the glacial to interglacial increase of atmospheric CO_2 which requires a change in the level of preformed nutrients. This new mechanism does not demand the massive changes in oceanic total carbon alkalinity, and phosphate, required by the Broecker (1982) hypothesis.

This work was completed by mid July 1983. More recent results have since appeared (Sarmiento and Toggweiler, 1984; Toggweiler and Sarmiento, 1985) and related work by other groups has also been published (Siegenthaler and Wenk, 1984; Knox and McElroy, 1984; Wenk and Siegenthaler, 1985; Ennever and McElroy, 1985).

NUTRIENT CYCLING EQUATIONS

A full description of these equations with appropriate attributions will be given in a future paper by Toggweiler and Sarmiento. We owe a great deal to a variety of studies such as those using sediment traps, on trace metals and radionuclides, particularly thorium (e.g., Bacon and Anderson, 1982), nutrient modeling studies such as that of Wyrtki (1962) and Grill (1970), and the investigation of the Narragansett estuary by Kremer and Nixon (1978).

The work on thorium and sediment traps suggests the possibility of treating particles as if they had a biomodal size distribution: small non-settling particles and large fast-settling particles. There is also a need for a means of changing one kind of a particle to another, namely, zooplankton. We thus have three equations governing particulates and animals:

$$\frac{\partial A}{\partial t} = \alpha(g_s P_s + g_L P_L) - RPA(T) \cdot A + ADV/DIFF \tag{1}$$

$$\frac{\partial P_L}{\partial t} = \beta(1 - \alpha)(g_s P_s + g_L P_L) - RPL(T) \cdot P_L$$
$$- g_L P_L - \omega_L \frac{\partial P_L}{\partial z} + ADV/DIFF \tag{2}$$

$$\frac{\partial P_s}{\partial t} = (1 - \beta)(1 - \alpha)(g_s P_s + g_L P_L) - RPS(T) \cdot P_s$$
$$- g_s P_s + ADV/DIFF \tag{3}$$

Animals "graze" small and large particles at rates g_s and g_L, respectively. They convert these into biomass with an efficiency α. Of the remaining material, a fraction β is expelled as large particles, and a fraction $(1 - \beta)$ is expelled as small particles. They metabolize at a rate RPA(T), and the temperature dependence of this rate may be included if desired. Bacteria metabolize small and large particles at rates RPS(T) and RPL(T), respectively. Large particles also fall through the water column at a speed ω_L.

The advection-diffusion terms take the form

$$ADV/DIFF = \vec{\nabla} \cdot (K \vec{\nabla} P) - \vec{V} \cdot \vec{\nabla} P \tag{4}$$

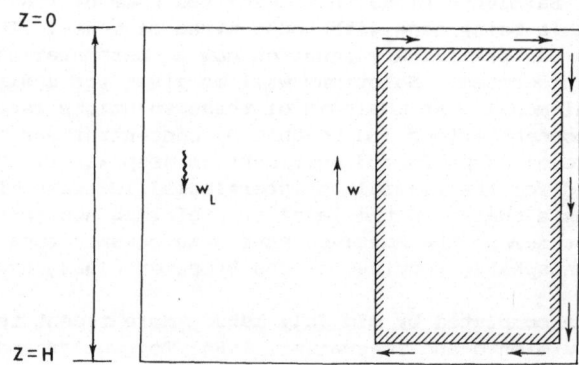

Fig. 1. Schematic diagram of "pipe" model discussed in text.

Solutions to these equations can be readily obtained if one ignores advection/diffusion and assumes steady state:

$$A = \frac{\alpha}{RPA(T)} (g_s P_s + g_L P_L) \qquad (5)$$

$$P_s = \frac{(1 - \alpha)(1 - \beta)g_L P_L}{RPS(T) + g_s[1 - (1 - \alpha)(1 - \beta)]} \qquad (6)$$

$$P_L = P_L^o e^{-az} \qquad (7)$$

where

$$a = \frac{RPL(T) + g_L}{\omega_L} - \frac{\beta(1 - \alpha)}{\omega_L} g_L \frac{RPS(T) + g_s}{\{RPS(T) + g_s[1 - (1 - \alpha)(1 - \beta)]\}}$$

The constant P_L is the large particle concentration at the surface. The distribution of "animals" and small particles is controlled by the shape of the large particle curve, which is exponential. The scaling length for the large particles, $1/a$, is estimated by us to be of the order of 300 m, from sediment trap measurements (e.g., Knauer et al., 1979; Knauer and Martin, 1981). It should be noted, however, that "a" contains many terms that may not be independent of depth. We have included explicitly the possible dependence of respiration rates on temperature. Suess (1980) has discussed evidence from sediment traps that is consistent with a decrease in "a" with increasing depth. We could get the same result by assuming a decrease with depth of respiration or grazing (or both) of large particles. A decrease in either would be consistent with the sign of the temperature dependence (Kremer and Nixon, 1978).

The phosphate conservation equation is:

$$\frac{\partial PO_4}{\partial t} = r_{PO_4:C}(RPA(T) \cdot A + RPS(T) \cdot P_s + RPL(T) \cdot P_L)$$
$$-r_{PO_4:C} \cdot FIXATION + ADV/DIFF \qquad (8)$$

where $r_{PO_4:C}$ is the Redfield ratio of phosphate to carbon in organisms. The fixation term will not be discussed here. We plan to use a formulation similar to that of Kremer and Nixon (1978), which includes the effects of light, nutrient supply, and temperature.

The oxygen equation is:

$$\frac{\partial O_2}{\partial t} = -r_{O_2:C}(RPA(T) \cdot A + RPS(T) \cdot P_s + RPL(T) \cdot P_L)$$
$$+ r_{O_2:C} \cdot FIXATION + GAS EXCHANGE + ADV/DIFF \qquad (9)$$

where $r_{O_2:C}$ is the oxygen to carbon ratio in organisms. The other terms are as explained above.

Solutions to the oxygen and phosphate equations have been obtained for the "pipe" model depicted in Fig. 1. It is assumed that the vertical eddy diffusivity, K, and the vertical advection, ω, are constant everywhere. The flux of phosphate into the surface by mixing is balanced by a loss of phosphate in large particles. The flux of large particles into the bottom is balanced by an upward flux of phosphate. The pipe has the same concentration as the surface.

The solutions are:

$$PO_4 = PO_{4s} + \frac{(\overline{PO_4} - PO_{4s})\, H\, (e^{\frac{\omega}{K}z} - e^{-az})}{\left[\frac{K}{\omega}(e^{\frac{\omega}{K}H} - 1) + \frac{1}{a}(e^{-aH} - 1)\right]} \tag{10}$$

where $\quad (\overline{PO_4}) = [\int_0^H PO_4 dz]/H$

$$O_2 = O_{2s} - r_{O_2:PO_4}(PO_4 - PO_{4s}) \tag{11}$$

The form of the equations is equivalent to what one would obtain by assuming that respiration decreases exponentially with depth with a scale length 1/a. Indeed, the solution to the oxygen equation is identical to the form of the solution obtained by Wyrtki (1962) by assuming that respiration is a simple exponential function of depth.

ANALYSIS OF SOLUTIONS: THE GLACIAL OCEAN

The simplest way to view the foregoing equations is in terms of their vertical averages:

$$\overline{PO_4} = PO_{4s} + (\overline{PO_4} - PO_{4s}) \tag{12}$$

$$\overline{O_2} = O_{2s} - r_{O_2:PO_4}(\overline{PO_4} - PO_{4s}) \tag{13}$$

$$\overline{ALK} = ALK_s + 2\, r_{Ca:PO_4}(\overline{PO_4} - PO_{4s}) \tag{14}$$

$$\overline{\Sigma CO_2} = \Sigma CO_{2s} + r_{\Sigma C:PO_4}(\overline{PO_4} - PO_{4s}) \tag{15}$$

$$\overline{\delta^{13}C} = \delta^{13}C_s + \Delta^{13}C_{PHOTO}\, r_{Corg:PO_4}\left[\frac{\overline{PO_4} - PO_{4s}}{\overline{\Sigma CO_2}}\right] \tag{16}$$

The equations for alkalinity, total carbon, and $\delta^{13}C$ are included for purposes of the discussion below. They are obtained as explained for oxygen and phosphate. The constant "1/a" is larger for $CaCO_3$ than for nutrients, but this drops out in the averaging process. The only alkalinity variations which are considered are those due to removal or addition of $CaCO_3$.

Consider the oxygen equation. This shows that the mean oceanic oxygen levels are determined by the surface oxygen level, the Redfield ratio of oxygen to phosphate, the total phosphate, and the preformed phosphate, PO_{4s}. The first column of Table 1 shows that the mean oxygen level of the oceans and the estimated "preformed" oxygen require quite a high preformed phosphate level. A lower level of preformed phosphate would result in less oxygen in the deep ocean.

A similar sensitivity to the preformed nutrient level is seen with the total carbon and alkalinity. This suggests that one might be able to control the level of atmospheric CO_2 by adjusting the level of preformed nutrients. The second column of Table 1 gives the results of a calculation in which the atmospheric CO_2 was assumed to be 110 ppm lower during the glacial period. This would probably be considered to be an upper limit (e.g., Neftel et al., 1982).

The following equation was used in conjunction with equations (14) and (15) to estimate the change in preformed phosphate required to give the appropriate change in atmospheric CO_2:

$$\frac{\Delta pCO_2}{\left(pCO_2^{IG} - \dfrac{\Delta pCO_2}{2}\right)} = \frac{4\Delta\Sigma CO_{2s} - 2\Delta ALK_s}{\left[2\left(\Sigma CO_{2s}^{IG} - \dfrac{\Delta\Sigma CO_{2s}}{2}\right) - \left(ALK_s^{IG} - \dfrac{\Delta ALK_s}{2}\right)\right]} + \frac{\Delta\Sigma CO_{2s} - \Delta ALK_s}{\left[\left(ALK_s^{IG} + \dfrac{\Delta ALK_s}{2}\right) - \left(\Sigma CO_{2s}^{IG} + \dfrac{\Delta\Sigma CO_{2s}}{2}\right)\right]} \qquad (17)$$

The mean oxygen level and preformed phosphate are predicted to be dramatically lower during the glacial period. The differences between interglacial and glacial periods for $\delta^{13}C_s$ and $\delta^{13}C$ are -0.1 and + 0.7, respectively, if we assume a change in the mean ocean $\delta^{13}C$ of 0.7 due to uptake of carbon by plants and soils, as proposed by Shackleton (1977). The results agree reasonably well with estimates by Broecker (1982) of +0.1 and +0.7 for $\Delta\delta^{13}C_s$ and $\Delta\delta^{13}C$ (interglacial-glacial), respectively, from measurements on foraminifera in cores. Our surface $\delta^{13}C$ can be brought into agreement with the core estimates by increasing the glacial atmospheric CO_2 level by approximately 20 ppm to 220 ppm.

Table 1. Model results for present ocean and an ice age scenario

	Present Ocean (1)	Glacial Ocean (2)	Difference (interglacial -glacial)
pCO_2 (ppm)	310[*]	200[*]	110
O_{2s} (μmol kg^{-1})	~300[*]	~300[*]	0
$\overline{O_2}$ (μmol kg^{-1})	168[*]	50[(a)]	118
PO_{4s} (μmol kg^{-1})	1.1[(a)]	0.3[(e)]	0.8
$\overline{PO_4}$ (μmol kg^{-1})	2.2[*]	2.3[+]	-0.1
$\delta^{13}C_s - \overline{\delta^{13}C_{IG}}$ (‰)	1[(b)]	1.8 (1.1)[‡]	-0.8 (-0.1)
$\delta^{13}C - \delta^{13}C_{IG}$ (‰)	0[*]	0.0[+](-0.7)[‡]	0 (0.7)
ALK_s (μeq kg^{-1})	2308[c]	2344[c]	-36
$\overline{ALK}$ (μeq kg^{-1})	2365[*]	2488[+]	-83
ΣCO_{2s} (μmol kg^{-1})	2106[d]	2067[d]	39
$\overline{\Sigma CO_2}$ (μmol kg^{-1})	2250[*]	2329[+]	-79

$r_{O_2:PO_4} = 120$ $r_{Corg:PO_4} = 105$

$r_{Ca:PO_4} = 26$ $\Delta^{13}C_{PHOTO} = -20‰$

$r_{\Sigma C:PO_4} = 131$

See Table 2 for footnotes

Broecker (1982) proposes that the increase in atmospheric CO_2 from glacial to present is accompanied by a decrease in ΣCO_2 and ALK by 6% due to a loss of carbon and $CaCO_3$ to the sediments. He uses the observed interglacial to glacial changes in $\delta^{13}C_s$ and $\delta^{13}C$ discussed above to estimate the change in PO_4 that must accompany the carbon addition. Table 2 gives various concentrations required by his scenario using our model.

Table 2. Model results for present ocean and an ice age scenario based on Broecker's study

| | Present Ocean (IG) (1) | Glacial Ocean (G) | | | |
| | | Our scenario (2) | | Broecker's scenario (3) | |
			ΔIG-G		ΔIG-G
PCO_2 (ppm)	310*	241##	69	241(f)	69
O_{2s} (μmol kg^{-1})	~300*	~300*	0	~300*	0
$\overline{O_2}$ (μmol kg^{-1})	168*	100(a)	68	62(a)	106
PO_{4s} (μmol kg^{-1})	1.1(a)	0.7(e)	0.4	PO_4-PO_{4s} = 1.9(b)++	
$\overline{PO_4}$ (μmol kg^{-1})	2.2*	2.3+	-0.1		
$\delta^{13}C_s-\overline{\delta^{13}C}_{IG}$ (‰)	1(b)	1.4(b) (0.7)‡	-0.4 (0.3)	0.9**	0.1*
$\overline{\delta^{13}C}-\overline{\delta^{13}C}_{IG}$ (‰)	0*	0+ (-0.7)‡	0 (0.7)	-0.7**	0.7*
τ (yr)		1000		5000	
ALK_s (μeq kg^{-1})	2308c	2365(c)	-57	2480(c)	-172
$\overline{ALK}$ (μeq kg^{-1})	2365*	2448+	-83	2579#	-214
ΣCO_{2s} (μmol kg^{-1})	2106d	2119(d)	-13	2203(d)	-97
$\overline{\Sigma CO_2}$ (μmol kg^{-1})	2250*	2329+	-79	2452#	-202

* Estimated from observations (Broecker, 1982; Levitus, 1982).
+ Column (1) x 1.035 to correct for decrease in ocean volume.
‡ Values in parentheses include a correction of -0.7‰ based on benthic foram observations.
** From Δ_{IG-G} observation and column (1).
# Column (1) x 1.095 to correct for decrease in ocean volume (3.5%) and increase in ΣCO_2 and ALK (6%) (Broecker, 1982).
Taken from column (3).
++ Our model allows us only to define the value shown. Broecker uses PO_4 = 3.2 μmol kg^{-1}.

(a) Equation (13)   (d) Equation (15)
(b) Equation (16)   (e) Equations (14), (15) and (17)
(c) Equation (14)   (f) Equation (17)

It also gives results we obtain with our scenario using the glacial atmospheric CO_2 level predicted by Broecker's scenario with our model. A difference between his scenario and ours that is most critical in determining which is more likely to have occurred is the response time. As the ice core measurements become more precise, the data may allow the two scenarios to be differentiated. The disagreement between our $\delta^{13}C_s$ and Broecker's can be removed by our using a glacial atmospheric CO_2 level of approximately 220 ppm, as explained above. Broecker would like the phosphate level to be 3.2 μmol kg^{-1} in the deep ocean. This is far higher than our model requires, even using Broecker's scenario. The method for estimating phosphate from measurements of cadmium (e.g., Boyle and Kegwin, 1982) should help to identify the phosphate concentrations for the glacial period.

Recent studies have shown that Broecker's scenario of a massive shift of phosphate did not occur, but that there was a massive shift of ΣCO_2 and thus ALK. The shifts of ΣCO_2 and ALK are not included in the present study, but are included in the subsequent studies mentioned in the Introduction.

CONCLUSIONS

A change in the ocean circulation and ventilation patterns during the glacial period may have led to a decreased level of preformed phosphate, and this may explain all or part of the change in atmospheric CO_2 concentrations. This scenario has the advantage of not requiring a massive shift of carbon and nutrients from the shelves to the oceans. It should be possible to decide between the two scenarios by looking at the record of cadmium in foraminifera as an indicator of deep ocean phosphate levels, as well as looking at the time scale of change in CO_2.

It hardly need be pointed out that a shift in preformed nutrient levels could also have a significant impact on the ocean's capacity to take up fossil fuel CO_2, though the time scale of such shifts may be too long to have a significant impact on a decadal time scale. It is extremely important for us to gain a deeper understanding of how the level of preformed nutrients in the oceans is determined and how this can change. An improved understanding of upper ocean chemical dynamics, particularly in deep water formation areas, is crucial.

REFERENCES

Bacon, M.P., and Anderson, R.F., 1982, Distribution of thorium isotopes between dissolved and particulate forms in the deep sea, J. Geophys. Res., 87:2045.

Boyle, E.A., and Kegwin, L.D., 1982, Deep circulation of the North Atlantic over the last 200,000 years: Geochemical evidence, Science, N.Y., 218:784.

Broecker, W.S., 1982, Ocean chemistry during glacial time, Geochim. Cosmochim. Acta, 46:1689.

Bryan, K., 1969, A numerical method for studying the world ocean, J. Comput. Phys., 1:347.

Ennever, F.K., and McElroy, M.B., 1985, Changes in atmospheric CO_2: Factors regulating the glacial to interglacial transition, in:

"The carbon cycle and atmospheric CO_2: Natural variations Archaen to present", Geophys. Monogr. Ser., Volume 32, E.T. Sundquist and W.S. Broecker, eds, pp.154–162, American Geophysical Union, Washington, D.C.

Grill, E.V., 1970, A mathematical model for the marine dissolved silicate cycle, Deep-Sea Res., 17:245.

Knauer, G.A., and Martin, J.H., 1981, Primary production and carbon-nitrogen fluxes in the upper 1500m of the northeast Pacific, Limnol. Oceanogr., 26:181.

Knauer, G.A., Martin, J.H., and Bruland, K.W., 1979, Fluxes of particulate carbon, nitrogen, and phosphorus in the upper water column of the northeast Pacific, Deep-Sea Res., 26:97.

Knox, F., and McElroy, M.B., 1984, Changes in atmospheric CO_2: Influence of the marine biota at high latitude, J. Geophys. Res., 89:4629.

Kremer, J.N., and Nixon, S.W., 1978, "A Coastal Marine Ecosystem", Springer-Verlag, New York.

Levitus, S., 1982, "Climatological Atlas of the World Ocean", NOAA Professional Paper 13, U.S. Department of Commerce, Rockville.

Neftel, A., Oeschger, H., Swander, J., Stauffer, B., and Zumbrunn, R., 1982, Ice core sample measurements give atmosphere CO_2 content during the past 40,000 years, Nature, Lond., 295:220.

Sarmiento, J.L., 1983, A simulation of bomb tritium entry into the Atlantic Ocean, J. Phys. Oceanogr., 13:1924.

Sarmiento, J.L., and Toggweiler, J.R., 1984, A new model for the role of the oceans in determining atmosphere PCO_2, Nature, Lond., 308:621.

Shackleton, N.J., 1977, Tropical rainforest history and the equatorial Pacific carbonate dissolution cycles, in: "The Fate of Fossil Fuel CO_2 in the Ocean", N. Anderson and A. Malahoff, eds, pp. 401–428, Plenum Press, New York.

Siegenthaler, U., and Wenk, T., 1984, Rapid atmospheric CO_2 variations and ocean circulation, Nature, Lond., 308:624.

Suess, E., 1980, Particulate organic carbon flux in the oceans — surface productivity and oxygen utilization, Nature, Lond., 288:260.

Toggweiler, J.R., and Sarmiento, J.L., 1985, Glacial to interglacial changes in atmospheric carbon dioxide: The critical role of ocean surface water in high latitudes. in: "The carbon cycle and atmospheric CO_2: Natural variations Archaen to present", Geophys. Monogr. Ser., Volume 32, ed. E.T. Sundquist and W.S. Broecker, eds, pp.163–184, American Geophysical Union, Washington, C.D

Wenk, T., and Siegenthaler, U., 1985, The high-latitude coean as a control of atmospheric CO_2, in: The carbon cycle and atmospheric CO_2: Natural variations Archaen to present, Geophys. Monogr. Ser., vol. 32, E.T. Sundquist and W.S. Broecker, eds, pp.185–194, American Geophysical Union, Washington, D.C.

Wyrtki, K., 1962, The oxygen minima in relation to ocean circulation, Deep-Sea Res., 9:11.

PARTICIPANTS

P.G.Brewer, Woods Hole Oceanographic Institution, Woods Hole, Massachusetts
    02543, U.S.A. (Steering Committee Member; Working Group III).

H.C.Broecker, Institut für Anorganische und Angewandte Chemie, Universität
    Hamburg, Martin-Luther-King-Platz 6, 2 Hamburg 13, F.R.G. (Working
    Group I).

K.W.Bruland, Institute of Marine Science, University of California, Santa
    Cruz, California 95064, U.S.A. (Working Group II).

J.D.Burton, Department of Oceanography, The University, Southampton
    SO9 5NH, U.K. (Steering Committee Member; Working Group III).

R.Chesselet, Programme Interdisciplinaire de Recherches en Océanographie,
    Centre National de la Recherche Scientifique, 15, Quai Anatole France,
    75700 Paris, France. (Steering Committee Chairman; Working Group
    II).

M.Crepon, Laboratoire d'Océanographie Physique, Muséum d'Histoire
    Naturelle, 43, rue Cuvier, Paris, France. (Working Group III).

R.W.Eppley, Institute of Marine Resources, A—018, Scripps Institution of
    Oceanography, University of California, San Diego, La Jolla,
    California 92093, U.S.A. (Working Group II Chairman).

T.D.Foster, Division of Natural Sciences, University of California, Santa
    Cruz, California 95064, U.S.A. (Working Group III).

G.Holloway, Institute of Ocean Sciences, P.O. Box 6000, 9860 West Saanich
    Road, Sidney, B.C., V8L 4B2, Canada. (Working Group I).

T.M.Joyce, Woods Hole Oceanographic Institution, Woods Hole, Massachusetts
    02543, U.S.A. (Working Group II).

D.Kamykowski, Department of Marine, Earth and Atmospheric Sciences, North
    Carolina State University, Raleigh, North Carolina 27650, U.S.A.
    (Working Group III).

M.Kawase, Geophysical Fluid Dynamics Program, James Forrestal Campus,
    Princeton University, P.O. Box 308, Princeton, New Jersey 08540,
    U.S.A. (Working Group I).

P.S.Liss, School of Environmental Sciences, University of East Anglia,
    Norwich, NR4 7TJ, U.K. (Working Group III Chairman)

K.Mopper, College of Marine Studies, University of Delaware, Lewes,
  Delaware 19958, U.S.A. (Working Group I).
  Present address: University of Miami, Rosentiel School of Marine and
  Atmospheric Science, Division of Marine and Atmospheric Chemistry,
  4600 Rickenbacker Causeway, Miami, Florida 33149–1098, U.S.A.

T.R.Osborn, Department of Oceanography, Naval Postgraduate School,
  Monterey, California 93940, U.S.A. (Working Group I Chairman).
  Present address: Chesapeake Bay Institute, Shady Side, Maryland 20764,
  U.S.A.

W.Roether, Institut für Umweltphysick der Universiät Heidelberg, D–69
  Heidelberg, F.R.G. (Working Group I).

C.G.H.Rooth, School of Marine and Atmospheric Sciences, University of
  Miami, 4600 Rickenbacker Causeway, Miami, Florida 33149, U.S.A.
  (Steering Committee Member; Working Group II).

J.L.Sarmiento, Geophysical Fluid Dynamics Program, James Forrestal Campus,
  Princeton University, P.O. Box 308, Princeton, New Jersey 08540,
  U.S.A. (Working Group II).

W.Seiler, Max–Planck–Institut für Chemie, Postfach 3060, D–6500 Mainz,
  F.R.G. (Working Group III).

J.G.Shepherd, Ministry of Agriculture, Fisheries and Food, Fisheries
  Laboratory, Lowestoft, Suffolk NR33 OHT, U.K. (Working Group III
  Rapporteur).

J.McN.Sieburth, Graduate School of Oceanography, Narragansett Bay Campus,
  University of Rhode Island, Kingston, Rhode Island 02881, U.S.A.
  (Working Group I).

J.J.Simpson, MLR Group A–030, Scripps Institution of Oceanography,
  University of California, San Diego, La Jolla, California 92093,
  U.S.A. (Working Group II).

J.H.Steele, Woods Hole Oceanographic Institution, Woods Hole, Massachusetts
  02543, U.S.A. (Working Group II).

M.Whitfield, Marine Biological Association of the U.K., The Laboratory,
  Citadel Hill, Plymouth, PL1 2PB, U.K. (Working Group I Rapporteur).

O.C.Zafiriou, Woods Hole Oceanographic Institution, Woods Hole,
  Massachusetts 02543, U.S.A.

B.Zeitzschel, Institut für Meereskunde, Universität Kiel, Dusternbrooker
  20, 23 Kiel, F.R.G. (Working Group II Rapporteur).